AGRICULTURAL PROCESSING and FOOD TECHNOLOGY

Objective Type

Er. Suresh Chandra
Assistant Professor
Department of Agricultural Engg. and Food Technology
SV Patel University of Agriculture and Technology, Meerut (UP)

Dr. S.K. Goyal
Post Doctoral Fellow
Department of Agric. Engg. and Food Technology
SV Patel University of Agriculture and Technology, Meerut (UP)

Mrs. Durvesh Kumari
Manager-4 (Quality Assurance)
Parag Dairy, Meerut, (UP

2016

JAIN BROTHERS (NEW DELHI)
'6/873, East Park Road, Karol Bagh, New Delhi-110005,
Phones: (011) 23618426, 45097398
E-mail: jain_bros@hotmail.com
Website: www.theiainbrothers com

Published by :

Shri Sunil Kumar Jain

for

JAIN BROTHERS (New Delhi)

16/873, East Park Road, Karol Bagh, New Delhi-110 005

Phones : 011-23618426, 45097398

Branch Office at :

1. **M/S SUNIL BOOK SERVICE**
   FATIMA NAGAR, KAZIPET-506004
   Ph. (0870) 2459617

2. **M/S JAIN BROTHERS**
   Vidya Vihar, Pilani (Raj.)
   Ph. : (01596) 242144

First Edition : 2009
Second Edition : 2010
Third Edition : 2016

***Price :* ₹ 250.00**

ISBN : 978-81-8360-121-4

Laser Typesetting by :

Sanjay Graphics
9910095684

Dedicated
To
Our Beloved Daughters
Prakriti and Anshika

Preface to the Third Edition

Examination/test has become an extensively accepted and a routine mode and method of selection both for jobs and admissions to college/universities/professional courses now-a-days. The questions in these competitive examinations besides being large number are mostly objective type to be solved within a strictly observable time schedule. It may be added that framing as well as answering of the volley of highly varying (in nature) questions within a time frame is no child's play. Sterling performance in these competitive examinations essentially calls for having a thorough knowledge at finger tips to answer the questions in these hurricanes like tests. The success of a competitor does not only need to secure a predetermined percentage of marks to pass through, but to obtain a percentage enough to find a place in the final merit list.

Agricultural process and food technology like other professional courses is witnessing the generation of newer and newer knowledge to add to the already existing large ocean of knowledge. A would be successful competitor must earn, embody and keep at finger tips these subjects plus recent advances therein. It is no easy jobs and hence all would be successful aspirants look for a comprehensive compilation containing the subjects matter on as many sub divisions as could be possible under one cover. One has to make a herculean effort to find out the correct answers and in the process has to scan through the huge volumes of book which, in effects, requires a great deal of time and effort.

The essential goal of this manuscript is to provide complete and simplified reach out to understanding of the basic fundamentals of ***Agricultural Processing and Food Technology to the students of the Agricultural Engineering, Food Processing, Dairy and Food Engineering, Food Science and Technology, Process and Food Engineering, Food Technology, Dairy Science and Technology, Post Harvest Engineering and Technology, Agricultural Structure and Process Engineering, Horticulture (specialized in Post Harvest Technology) and Home Science (Food and Nutrition).*** Although, both agricultural processing and food technology are a broad subjects which can not confined in a single book, however efforts have made to cover almost syllabus of agricultural process and food technology. This book is also beneficial to those students preparing who are ambitious of higher studies or going to appear in competitive examination such as ICAR/CSIR/UGC fellowships, NET, ARS, SRF, JRF, RES, IAS, IFS, PCS (state level), minor irrigation, NABARD & World bank, etc. and also helpful who are appearing in interviews of RA/SRF/SMS/Assistant Professor and also in national and multinational food industries.

Finally, we are making a plea to those who make use of this book to supply us with information or points of view that differ with those expressed in. We know that there will be errors, for which we alone are responsible, and we will appreciate the opportunity to correct those. This effort should be viewed as a first try at the comprehensive view of agricultural processing and food technology. In subsequent editions or volumes, there will be opportunity not only to correct, but to expand the perspective and thus to achieve more complete our objective of providing the serious students with a consistent and correct answer of how we make better. Feed back from readers of this book will certainly help us to improve its quality in future.

In the revised edition we have incorporated all the corrections brought to our notice by the valuable readers. A new chapter on thermodynamics has been added to enhance the utility of the book. We hope the readers and users of the book shall find the book more useful to them.

In last, the authors express their thanks to their friends, students and family members who encourage for writing such book which is helpful for competitive examinations.

1st June, 2016

Suresh Chandra
S.K. Goyal
Durvesh Kumari

Preface to the Second Edition

Examination/test has become an extensively accepted and a routine mode and method of selection both for jobs and admissions to college/universities/professional courses now-a-days. The questions in these competitive examinations besides being large number are mostly objective type to be solved within a strictly observable time schedule. It may be added that framing as well as answering of the volley of highly varying (in nature) questions within a time frame is no child's play. Sterling performance in these competitive examinations essentially calls for having a thorough knowledge at finger tips to answer the questions in these hurricanes like tests. The success of a competitor does not only need to secure a predetermined percentage of marks to pass through, but to obtain a percentage enough to find a place in the final merit list.

Agricultural process and food technology like other professional courses is witnessing the generation of newer and newer knowledge to add to the already existing large ocean of knowledge. A would be successful competitor must earn, embody and keep at finger tips these subjects plus recent advances therein. It is no easy jobs and hence all would be successful aspirants look for a comprehensive compilation containing the subjects matter on as many sub divisions as could be possible under one cover. One has to make a herculean effort to find out the correct answers and in the process has to scan through the huge volumes of book which, in effects, requires a great deal of time and effort.

The essential goal of this manuscript is to provide complete and simplified reach out to understanding of the basic fundamentals of ***Agricultural Processing and Food Technology to the students of the Agricultural Engineering, Food Processing, Dairy and Food Engineering, Food Science and Technology, Process and Food Engineering, Food Technology, Dairy Science and Technology, Post Harvest Engineering and Technology, Agricultural Structure and Process Engineering, Horticulture (specialized in Post Harvest Technology) and Home Science (Food and Nutrition).*** Although, both agricultural processing and food technology are a broad subjects which can not confined in a single book, however efforts have made to cover almost syllabus of agricultural process and food technology. This book is also beneficial to those students preparing who are ambitious of higher studies or going to appear in competitive examination such as ICAR/CSIR/UGC fellowships, NET, ARS, SRF, JRF, RES, IAS, IFS, PCS (state level), minor irrigation, NABARD &World bank, etc. and also helpful who are appearing in interviews of RA/SRF/SMS/Assistant Professor and also in national and multinational food industries.

Finally, we are making a plea to those who make use of this book to supply us with information or points of view that differ with those expressed in. We know that there will be errors, for which we alone are responsible, and we will appreciate the opportunity to correct those. This effort should be viewed as a first try at the comprehensive view of agricultural processing and food technology. In subsequent editions or volumes, there will be opportunity not only to correct, but to expand the perspective and thus to achieve more complete our objective of providing the serious students with a consistent and correct answer of how we make better. Feed back from readers of this book will certainly help us to improve its quality in future.

In the revised edition we have incorporated all the corrections brought to our notice by the valuable readers. A new chapter on thermodynamics has been added to enhance the utility of the book. We hope the readers and users of the book shall find the book more useful to them.

In last, the authors express their thanks to their friends, students and family members who encourage for writing such book which is helpful for competitive examinations.

15th July, 2010

Suresh Chandra
S.K. Goyal
Durvesh Kumari

Contents

Cereals, Pulses and Oilseed Processing

1. The major carbohydrate of rice is
   (*a*) Glucose (*b*) Fructose
   (*c*) Starch (*d*) Pectin

2. The fibre of rice is the hemicelluloses made up of the
   (*a*) Pentose (*b*) Arabinose
   (*c*) Xylose (*d*) All

3. Glutelin (Oryzenin) is the principal protein of
   (*a*) Rice (*b*) Milk
   (*c*) Wheat (*d*) Pulse

4. The nutritive value of rice protein is superior than
   (*a*) Wheat (*b*) Barley
   (*c*) Oat (*d*) All

5. The rice protein is richer in
   (*a*) Arginine (*b*) Lysine
   (*c*) Threonine (*d*) All

6. Lysine and Threonine is absent in
   (*a*) Rice (*b*) Wheat
   (*c*) Barley (*d*) All

7. The polished rice is poor in
   (*a*) Ca and Fe (*b*) Ca and P
   (*c*) P and Fe (*d*) Fe and Mn

8. The activity of ________________ in fresh rice is probably responsible for its sticky consistency after cooking.
   (*a*) α-Amylase (*b*) Protease
   (*c*) Lipase (*d*) Oxidase

9. The pigments of coloured rice are
   (*a*) Anthocyanin (*b*) Carotenoid
   (*c*) Both (*a*) and (*b*) (*d*) None

10. Parboiling of rice is originated in
    (*a*) Pakistan (*b*) India
    (*c*) Bhutan (*d*) USA

11. High proportion of amylopectin is available in
    (*a*) Flint corn (*b*) Dent corn
    (*c*) Waxy corn (*d*) Pop corn

12. The major lipids of maize oil is
    (*a*) Triglycerides (*b*) Diglycerides
    (*c*) Amylose (*d*) Phospholipid

13. The triglycerides of corn oil have high proportion of ______________ and ______________ acid.
    (*a*) Lysine and Valine (*b*) Alanine and Proline
    (*c*) Oleic and Linoleic (*d*) Tryptophan and Methionine

14. Corn is deficient in the vitamin
    (*a*) Niacin (*b*) Calciferol (B_{12})
    (*c*) Thiamine (B_1) (*d*) Tocopherol (E)

15. The phosphorus in corn is found in the form of
    (*a*) Phytin (*b*) Glutamine
    (*c*) Histidine (*d*) Glycine

16. The keeping quality of breakfast cereals is depended on the
    (*a*) Fat content (*b*) Protein
    (*c*) Starch content (*d*) Vitamin

17. South Indian breakfast food (uppuma) is made from
    (*a*) Flour (*b*) White flour
    (*c*) Semolina (*d*) Grits

18. The cooking time of farina is reduced by adding of
    (*a*) Disodium phosphate (*b*) Sodium benzoate
    (*c*) Salt (*d*) Sugar

19. Bread flour is milled from
    (*a*) Soft wheat (*b*) Hard wheat
    (*c*) Winter wheat (*d*) (*b*) and (*c*) both

20. Pudding, cake, pastries etc. are made from
    (*a*) Self raising flour
    (*b*) Bread flour
    (*c*) Biscuit flour
    (*d*) Cake flour

21. A slight excess of bicarbonate gives rise to an unpleasant odour and a brownish yellow colouration in
    (*a*) Self raising flour
    (*b*) Bread flour
    (*c*) All purpose flour
    (*d*) Pastry flour

22. One of the following is not a leavening agent for baking products
    (*a*) Air
    (*b*) CO_2
    (*c*) Steam
    (*d*) Baking soda

23. One of the following gas is generated during leavening process of baking product by action of chemical leaveners or action of microorganisms on sugars.
    (*a*) O_2
    (*b*) CO_2
    (*c*) NH_3
    (*d*) SO_2

24. The most important chemical agent used in leavening process is
    (*a*) caustic soda
    (*b*) sodium hydrogen carbonate
    (*c*) salt peter
    (*d*) sodium benzoate

25. Baking powder contains an inert filler used in leavening process is
    (*a*) Corn starch
    (*b*) Potato starch
    (*c*) Arrowroot
    (*d*) Egg powder

26. Corn starch acts as buffering agent between
    (*a*) Baking soda and acids
    (*b*) CO_2 and acid
    (*c*) O_2 and CO_2
    (*d*) None

27. Baking powder is a compound of
    (*a*) Sodium bicarbonate and tartrates
    (*b*) Sodium bicarbonate and corn starch
    (*c*) Corn starch and tartrates
    (*d*) None

28. Tartrates in baking powder is implemented in the form of
    (*a*) Tartaric acid
    (*b*) Potassium hydrogen tartrate
    (*c*) (*a*) and (*b*)
    (*d*) None

29. Yeast used form leavening is marketed in the form of
    (*a*) Compressed yeast
    (*b*) Active dry yeast
    (*c*) Both (*a*) and (*b*)
    (*d*) None

30. The activity and shelf life of compressed yeast at 1°C to 3°C remains up to
    (*a*) One week
    (*b*) Two week
    (*c*) Four week
    (*d*) Five week

31. The shelf life of dry yeast is more than two years, if stored at temperature of
(*a*) 5 ºC (*b*) 10 ºC
(*c*) 15 ºC (*d*) 32 ºC

32. The device in which polished rice for further coating of sugar and talcum powder for brightness look is accomplished is known as
(*a*) Trumbol (*b*) Polisher
(*c*) Glazer (*d*) Pearler

33. In countries where rice is staple diet, continuous and exclusive use of polished rice causes *beriberi* due to deficiency of
(*a*) Thiamine (*b*) Riboflavin
(*c*) Pyridoxine (*d*) Cobalamin

34. Flaked rice is made from
(*a*) Raw rice (*b*) Brown rice
(*c*) Parboiled rice (*d*) Bulgur

35. Malting is a basic operation employed in
(*a*) Wheat (*b*) Rice
(*c*) Barley (*d*) Pulse

36. Flour or flour mixtures, when mixed with water form
(*a*) Bread (*b*) Chapatti
(*c*) Dough or batter (*d*) Cake

37. Wheat flour when mixed with the appropriate amount of water forms dough which have a rheological property
(*a*) Elastic (*b*) Plastic
(*c*) Visco-elastic (*d*) None

38. In South India, Idli is a fermented food item made from
(*a*) Wheat and rice (*b*) Black gram and rice
(*c*) Semolina and milk (*d*) White flour and semolina

39. A favourable temperature for fermentation in the making of 'Idli' should be
(*a*) 10-20 ºC (*b*) 20-25 ºC
(*c*) 25-30 ºC (*d*) 30-35 ºC

40. The basic purpose of fermentation in Idli is
(*a*) To increase nutritive value (*b*) Digestibility
(*c*) To decrease the acceptability (*d*) Both (*a*) and (*b*)

41. One of the following is not a south Indian fermented food product
(*a*) Idli (*b*) Dosa
(*c*) Uppuma (*d*) Dokla

42. The appropriate moisture of barley for malting purpose should be between
(*a*) 8-10 % (*b*) 10-14 %
(*c*) 14-17 % (*d*) 20-22 %

43. In milling, cleaned oats are subjected to kiln drying to inactivate lipase. This process is known as
(*a*) Drying (*b*) Sterilization
(*c*) Stabilization (*d*) Dehydration

44. When the proteins in the flour combine with water, a complex between gliadin and glutenin is formed resulting in the development of
(*a*) Gluten (*b*) Casein
(*c*) Oryzenin (*d*) Lactoalbumin

45. During bread making, the elasticity of gluten is controlled by
(*a*) Glutenin (*b*) Gliadin
(*c*) Water (*d*) Yeast

46. Loaf volume of bread during preparation is adjusted by
(*a*) Gliadin (*b*) Glutenin
(*c*) Gluten (*d*) Yeast

47. Flour amylase attacks only starch is responsible for
(*a*) Adverse affect on quality of bread (*b*) Decrease loaf volume
(*c*) Decrease attractiveness in colour (*d*) All of the above

48. The quantities and quality of gluten which a flour can yield is observed by the following instruments
(*a*) Mixograph and farinograph (*b*) Amylograph
(*c*) Extensograph (*d*) Alveogragh

49. The brown colour of crust of bread during baking is due to the maillard reaction between
(*a*) Protein and sugar (*b*) Sugar and vitamin
(*c*) Sugar and salt (*d*) Damage starch and amino acid

50. The activity of yeast fermentation during bread making is controlled by
(*a*) Flour (*b*) Water
(*c*) Sugar (*d*) Salt

51. Propionates are added in bread to inhibit the
(*a*) Mold growth (*b*) Bacterial growth
(*c*) Enzyme reaction (*d*) All

52. Enzyme used in bread making is protease acting on the gluten to alter
(*a*) Dough extensibility (*b*) Loaf volume
(*c*) Colour (*d*) Texture

53. Ammonium sulphate and phosphate in bread making are used as a
(*a*) Yeast nutrient
(*b*) Oxidize agent
(*c*) Antioxidant
(*d*) Emulsifier

54. Maize is deficient in the following amino acids
(*a*) Methionine and lysine
(*b*) Methionine and arginine
(*c*) Lysine and arginine
(*d*) Lysine and tryphtophane

55. In rice polishing
(*a*) A coating is applied on the outer surface of brown rice
(*b*) A layer of bran is removed from brown rice
(*c*) A layer of starch is removed
(*d*) Only husk is removed

56. Shelling out turn of (in percentage) of ground nut is
(*a*) 40
(*b*) 60
(*c*) 75
(*d*) 90

57. Terminal velocity of particle is equal to air velocity for which
(*a*) Particle remains in the bottom
(*b*) Particles is carried away
(*c*) The particle remains in suspended state
(*d*) None of these

58. Scalping is also known as
(*a*) Grading
(*b*) Rough cleaning
(*c*) Separating
(*d*) Dust removal

59. The value of moisture content determined on dry basis is
(*a*) lower than that determined on wet basis
(*b*) higher than moisture content on wet basis
(*c*) equal to moisture content on wet basis
(*d*) none of the above

60. In dry milling process prior to treatment with oil, in pulse the following operation is done
(*a*) grading
(*b*) polishing
(*c*) conditioning
(*d*) pitting

61. The edible portion of paddy is called
(*a*) Rice
(*b*) Bran
(*c*) Endosperm
(*d*) Husk

62. In paddy grain, the weight of hull is about
(*a*) 18-22 kg
(*b*) 18-22 gm
(*c*) 18-22 %
(*d*) 20-22 %

63. Removal of bran layer from the brown rice is done by
(*a*) Polisher
(*b*) Bran aspirator
(*c*) Whitener
(*d*) both (*a*) and (*c*)

64. The basic requirement of parboiling in paddy is done to
(*a*) increase the moisture content
(*b*) minimize breakage during milling
(*c*) minimize the recovery of head rice
(*d*) none of them

65. Rubber roll sheller shelled up the paddy about
(*a*) 85 %
(*b*) 87 %
(*c*) 96 %
(*d*) 95 %

66. Horizontal jet pearler is used in paddy milling for
(*a*) Polishing
(*b*) Whitening
(*c*) Pearling
(*d*) None of them

67. Glazing is done on rice for
(*a*) Polishing
(*b*) Whitening
(*c*) Transparent look
(*d*) Pearling

68. The soaking of cleaned wheat for milling is done at the temperature of
(*a*) 65-75 °F
(*b*) 65-75 °C
(*c*) 75-80 °K
(*d*) 80-85 °B

69. In pulse milling, pulse is treated with linseed oil with range of
(*a*) 1.0-2.0 kg/ton
(*b*) 1.5-2.5 kg/ton
(*c*) 1.5-2.5 kg/quintal
(*d*) 1.5-2.0 kg/ton

70. In CFTRI method of pulse milling, Pre conditioning of pulse is done at
(*a*) 120 °C
(*b*) 125 °F
(*c*) 110 °C
(*d*) 100 °C

71. After the oil treatment, the pulses are kept for about 12 hours to allow proper penetration to take place in the pulse kernel is called
(*a*) Conditioning
(*b*) Oil penetration
(*c*) Tempering
(*d*) Saponification

72. The moisture content of pulse for dehusking should not be more than
(*a*) 8-10 %
(*b*) 10-20 %
(*c*) 20-25 %
(*d*) 20-30 %

73. The efficiency of Gota machine for pulse in single pass is to about
(*a*) 50 %
(*b*) 60 %
(*c*) 80 %
(*d*) 90 %

74. The LSU is a
(*a*) Evaporator
(*b*) Drier
(*c*) Washer
(*d*) Polisher

75. The full name of LSU drier
(*a*) Ludhiana state university
(*b*) Laos state university
(*c*) Lebanon state university
(*d*) Louisiana state university

76. Saturated fatty acid increases
(*a*) Cancer
(*b*) AIDS
(*c*) Heart problems
(*d*) Diabetes

77. The boiling point of *n*-hexane for oil recovery should be
(*a*) 70.5 °C
(*b*) 65.0 °C
(*c*) 65.5 °C
(*d*) 60.5 °C

78. Removal of fibre from cotton seed is known as
(*a*) Flaking
(*b*) Shelling
(*c*) Dehusking
(*d*) Delinting

79. A thin layer grain drying is one in which grain thickness should be
(*a*) 20 cm
(*b*) more than 20 cm
(*c*) less than 50 cm
(*d*) less than 100 cm

80. Break rolls have
(*a*) Smooth surface
(*b*) Corrugated surface
(*c*) Rough surface
(*d*) Hole in surface

81. Reduction rolls have
(*a*) Smooth surface
(*b*) Corrugated surface
(*c*) Rough surface
(*d*) Hole in surface

82. The differential speed ratio of break rolls or reduction rolls is about
(*a*) 3.5:1
(*b*) 2.25:1
(*c*) 2.5:1
(*d*) 2.0:1

83. The purpose of hydrothermal treatment in wheat is for
(*a*) Loosening of hull
(*b*) Loosening of endosperm
(*c*) Loosening of bran layer
(*d*) None of them

84. The tempering of wheat is done for
(*a*) 18-72 h
(*b*) 18-62 h
(*c*) 18-74 h
(*d*) 18-50 h

85. The optimum temperature for conditioning of wheat is subjected at
(*a*) 45 °C
(*b*) 47 °C
(*c*) 40 °C
(*d*) 50 °C

86. Drying is a __________ operation.
(*a*) Thermo-physical
(*b*) Physio-chemical
(*c*) Thermo-physical and physico-chemical
(*d*) None of them

87. Grading refers to
(*a*) Classification of cleaned product
(*b*) Classification of uncleaned product
(*c*) Separation of cleaned product
(*d*) Separation of large impurities

88. Screening is a method of separating grains into two or more fractions according to
(*a*) Size alone
(*b*) Shape alone
(*c*) Colour alone
(*d*) Texture alone

89. Conditioning means
(*a*) Wetting
(*b*) Drying
(*c*) Steaming
(*d*) Wetting and drying

90. Pure starch, pure germ and feed are prepared from
(*a*) Corn dry milling
(*b*) Corn wet milling
(*c*) Above both
(*d*) None

91. Rotary steam tube in corn milling are used for the
(*a*) Drying
(*b*) Washing
(*c*) Cooling
(*d*) Steaming

92. Corns are treated at temperatures of 75°C during drying in
(*a*) dry milling
(*b*) wet milling
(*c*) Above both
(*d*) None

93. Hull is the synonyms of
(*a*) Endosperm
(*b*) Bram
(*c*) Husk
(*d*) Germ

94. Removal of bran from the brown rice is called
(*a*) Grading
(*b*) Polishing
(*c*) Bran aspiration
(*d*) Bran separation

95. During shelling, the speed of faster roll of rubber roll sheller has about
(*a*) 8-10 m/s
(*b*) 10-13 m/s
(*c*) 9-10 m/s
(*d*) 12-15 m/s

96. The full name of MT is
(*a*) Million Ton
(*b*) Micro Ton
(*c*) Metric Ton
(*d*) Mega Ton

97. The kernel of paddy is enclosed by
(*a*) Hull (*b*) Germ
(*c*) Bran (*d*) a and c

98. During polishing, bran from brown rice is removed approximately
(*a*) 4-6 % (*b*) 2-3 %
(*c*) 5-6 % (*d*) 3-4 %

99. In paddy grain, the weight of bran is about
(*a*) 2-3 % (*b*) 5-6 %
(*c*) 4-6 % (*d*) 3-4 %

100. The main products of corn dry milling is
(*a*) Pure starch (*b*) Germ
(*c*) Grits (*d*) Bulgur

101. The percent endosperm is present in corn kernel is about
(*a*) 82 % (*b*) 85 %
(*c*) 89 % (*d*) 87 %

102. The wet and soften corn kernel attain the moisture in wet milling is about
(*a*) 14 % (*b*) 22 %
(*c*) 27 % (*d*) 45 %

103. Moisture content in pegion pea (arhar) is about
(*a*) 10.25 % (*b*) 10.52 %
(*c*) 10.35 % (*d*) 15.54 %

104. The oil treatment in pulse is done to loosen the
(*a*) Bran (*b*) Germ
(*c*) Husk (*d*) Aleurone layer

105. The machine used for making scratch over whole grain of pulse is
(*a*) Gota machine (*b*) Emery roller
(*c*) Concave type machine (*d*) Screw conveyor

106. For dehusking and splitting, the moisture content in pulse should be
(*a*) 5-10 % (*b*) 10-20 %
(*c*) 15-25 % (*d*) 25-30 %

107. The pearler is also called
(*a*) Degermer (*b*) Dehusker
(*c*) Polisher (*d*) Debranner

108. Per capita consumption of edible oil per year in India is
(*a*) 5 kg (*b*) 8 kg
(*c*) 10 kg (*d*) 13 kg

109. Oil percent in soybean is approximately
(*a*) 18-20 % (*b*) 30-35 %
(*c*) 45-50 % (*d*) 20-25 %

110. The term disintegration is related to
(*a*) Paddy processing (*b*) Pulse processing
(*c*) Corn milling (*d*) Oil processing

111. In solvent extraction of oil, the solvent is
(*a*) n-hexane (*b*) iso-hexane
(*c*) n-butane (*d*) hot water

112. Separating of grains or seeds in two or more fraction according to the size alone is called
(*a*) Sorting (*b*) Screening
(*c*) Scalping (*d*) Grading

113. At the time of harvesting, the moisture content in paddy should not more or less than
(*a*) 12-14 % (*b*) 16-18 %
(*c*) 20-22 % (*d*) 24-28 %

114. Reduction of food grains in to various edible end products is called
(*a*) grinding (*b*) milling
(*c*) shelling (*d*) splitting

115. Bulgur is the product of the
(*a*) Rice (*b*) Corn
(*c*) Wheat (*d*) Pulse

116. Talcum powder or glucose are used in the unit operation of
(*a*) Polishing (*b*) Whitening
(*c*) Glazing (*d*) Shining

117. Separation of broken rice and head rice is known as
(*a*) Scalping (*b*) Screening
(*c*) Sorting (*d*) Grading

118. Removal of hull or husk from the cereal grain is called
(*a*) Dehusking (*b*) Splitting
(*c*) Degerming (*d*) Polishing

119. Baell degermer are used for processing of
(*a*) Paddy (*b*) Wheat
(*c*) Oat (*d*) Corn

120. One hundred kg of paddy grain would be yield husk about
(*a*) 30 kg (*b*) 25 kg
(*c*) 22 kg (*d*) 72 kg

121. Composition of endosperm, bran and germ in wheat (in percent) is approximately
(*a*) 82, 14, 4 (*b*) 75, 10, 15
(*c*) 85, 12, 3 (*d*) 80, 5, 15

122. The yield of white flour and by product in wheat processing is
(*a*) 80 % and 20 % (*b*) 70 % and 30 %
(*c*) 85 % and 15 % (*d*) 90 % and 10 %

123. Wheat is harvested at the moisture content of
(*a*) 35 % (*b*) 25 %
(*c*) 20 % (*d*) 15 %

124. Husk separators are used in
(*a*) Pulse processing (*b*) Wheat processing
(*c*) Paddy processing (*d*) Vegetable processing

125. The main purpose of hydrothermal treatment in wheat processing is
(*a*) Remove of moisture content (*b*) Remove of germ
(*c*) Loosening of bran (*d*) Increase the temperature

126. Food processing involving the application of
(*a*) Chemistry and biochemistry (*b*) Biophysics, nutrition and microbiology
(*c*) All branches of engineering (*d*) Above all

127. Grading refers to
(*a*) Cleaning of product
(*b*) Separation of cleaned products according of physical properties
(*c*) Classification of cleaned products according commercial values and usages
(*d*) Both (*b*) and (*c*)

128. Milling refers to the
(*a*) Cleaning and grading (*b*) Dehusking, splitting
(*c*) separating, size reduction and polishing (*d*) Overall operations in mill

129. Dehydration means removal of moisture content to
(*a*) Very high level (*b*) Medium level
(*c*) Vey low level (*d*) Bone dry condition

130. Environmental factors such as ____________ will spoil the food.
(*a*) O_2 and water vapour
(*b*) O_2 and H_2
(*c*) H_2 and N_2
(*d*) N_2 and C

131. The ____________ makes the food products suitable for storage.
(*a*) Scalping
(*b*) Sorting
(*c*) Screening
(*d*) Drying

132. Clean and good quality paddy can give a yield of ________ rice.
(*a*) 70 %
(*b*) 76 %
(*c*) 74 %
(*d*) 87 %

133. The ____________ is done to achieve maximum recovery of head rice and to minimize breakage.
(*a*) Shelling
(*b*) Cleaning
(*c*) Polishing
(*d*) Parboiling

134. Moisture of paddy during milling should be
(*a*) 20-22 %
(*b*) 10-12 %
(*c*) 12-14 %
(*d*) 15-20 %

135. Disc Sheller consists of ____________ coated with emery.
(*a*) Two horizontal iron discs
(*b*) Two vertical iron discs
(*c*) One horizontal and one vertical disc
(*d*) Two vertical stone discs

136. Grading in rice milling usually refers to the separation of
(*a*) Small weed seed
(*b*) Broken from head rice
(*c*) Husk particles
(*d*) Bran particles

137. The weight of hull is ________ of the total weight of paddy grain.
(*a*) 20-22 %
(*b*) 12-14 %
(*c*) 14-16 %
(*d*) 18-22 %

138. Rice bran contains about ________ percent fat
(*a*) 10-12 %
(*b*) 12-15 %
(*c*) 15-17 %
(*d*) 17-19 %

139. Pulses contain following percent of protein by weight
(*a*) 5-10 %
(*b*) 10-15 %
(*c*) 17-25 %
(*d*) 30-35 %

140. Parboiling of paddy is following treatment
(*a*) Cooling
(*b*) Heating
(*c*) Hydrothermal
(*d*) None of these

141. In most of the wheat mills, the yield of flour can be obtained about
(*a*) 25 % (*b*) 50 %
(*c*) 75 % (*d*) 90 %

142. Shelled groundnut contains following percent oil content
(*a*) 10-20 % (*b*) 20-30 %
(*c*) 30-40 % (*d*) 45-50 %

143. Soybean contains _________ % protein and ___________ % oil content.
(*a*) 30 and 10 (*b*) 40 and 20
(*c*) 50 and 30 (*d*) 60 and 30

144. The percentage of germ in wheat grain or the composition of germ in wheat is
(*a*) 5 % (*b*) 1 %
(*c*) 2 % (*d*) 3 %

145. During conditioning of wheat, the temperature of wheat grain should not be
(*a*) Less than 40 °C (*b*) Less than 47 °C
(*c*) More than 47 °C (*d*) None of the above

146. In wheat mills, the rolls that have smooth surface are
(*a*) Reduction roll (*b*) Break roll
(*c*) plan sifter (*d*) Scratch roll

147. The differential speed of the pair of break rolls is
(*a*) 1:2 (*b*) 2:1
(*c*) 2.5:1 (*d*) 1:2.5

148. Which one of the following can be said as the extension of break roll is
(*a*) Reduction roll (*b*) Purifier
(*c*) Plan sifter (*d*) Scratch roll

149. The purpose of tempering of wheat through tempering bin is employed to
(*a*) Raise the moisture (*b*) Reduce the moisture
(*c*) Equalize the moisture in whole grain (*d*) Raise the temperature

150. The number of baking quality of the flour depend on the percentage of
(*a*) Germ (*b*) Bran
(*c*) Gluten (*d*) Grits

151. The number of break rolls which are employed to accomplished the proper breaking of grain is
(*a*) 3 to 4 (*b*) 8 to 10
(*c*) 1 to 3 (*d*) 6 to 8

152. The oils and fats are composed of different mixture of
(*a*) Glycerin
(*b*) Glucose
(*c*) Glycerides
(*d*) Glycerol

153. What are the main constituents of vegetable oils?
(*a*) 16 and 18 carbon acids
(*b*) 11 and 12 carbon acids
(*c*) 16 and 18 carbonic acids
(*d*) 13 and 17 carbon acids

154. The recommended value of oil consumption in India per annum per capita is about
(*a*) 6 MT
(*b*) 19.4 MT
(*c*) 5 kg
(*d*) 18 kg

155. The amount of energy released by oils and fats is more than protein and carbohydrate by
(*a*) 2.25 times
(*b*) 2.5 times
(*c*) 2.00 times
(*d*) remains same

156. The essential fatty acid provided by the oils is
(*a*) Carbonic acid
(*b*) Linoleic acid
(*c*) Lactic acid
(*d*) Gluconic acid

157. Which of the following is not a method of oil processing?
(*a*) Bullock driven ghani
(*b*) Mechanical expression
(*c*) Chemical lye method
(*d*) Solvent oil extraction method

158. The process of mechanically pressing liquid containing solid liquid to leach out of liquid is called
(*a*) Pressing
(*b*) Expression
(*c*) Compression
(*d*) Disintegration

159. The optimum moisture content of corn for degerming in the Baell Degermer should be
(*a*) 20-23 %
(*b*) 25-30 %
(*c*) 21-25 %
(*d*) 10-15 %

160. In which unit operation, germ is detached from the other component of corn kernel
(*a*) Tempering
(*b*) Degerming
(*c*) Rolling
(*d*) Flaking

161. Which one of the following, is not an example of equipment used in expression of oil seed :
(*a*) Screw press
(*b*) Screw conveyor
(*c*) Hydraulic press
(*d*) Pasteurizer

162. The process of separating a liquid from liquid solid medium with the help of solvent is known as:
(*a*) Expression (*b*) Pressing
(*c*) Extraction (*d*) Disintegration

163. The deoiled cake that comes out from the oil extraction unit has a weight of total mass is
(*a*) 25 to 40 % (*b*) 20 to 25 %
(*c*) 15 to 20 % (*d*) 30 to 35 %

164. Stone-separator is a form of
(*a*) Specific gravity separator (*b*) Velvet roll separator
(*c*) Magnetic separator (*d*) Disc separator

165. Disc separator separates material on the basis of
(*a*) Surface area (*b*) Length
(*c*) Density (*d*) Roundness

166. Separation on the basis of roundness is accomplished by
(*a*) Tray separator (*b*) Inclined dapper
(*c*) Spiral separator (*d*) Aspiration separator

167. If the separation on the basis of relative length of grain, disc separator is equivalent to
(*a*) Spiral separator (*b*) Cyclone separator
(*c*) Indented cylinder separator (*d*) Magnetic separator

168. 'R-Pockets' are present to separate desirable grains in
(*a*) Velvet roll separator (*b*) Disc separator
(*c*) Indented cylinder separator (*d*) Spiral separator

169. Dissimilar materials like wheat, barley and mustard from oat are separated by
(*a*) Cyclone separator (*b*) Velvet roll separator
(*c*) Spiral separator (*d*) Disc separator

170. In destoner, separation can be controlled by adjusting
(*a*) Feed rate (*b*) Air flow rate
(*c*) Slope of the deck and its vibration (*d*) All of these

171. In spiral separator, the adjustable component for separation is
(*a*) Slope of deck (*b*) Rate of feeding
(*c*) Air flow rate (*d*) None of these

172. Lack of flexibility is the main limitation of
(*a*) Spiral separator (*b*) Velvet roll separator
(*c*) Disc separator (*d*) Magnetic separator

173. Pneumatic separation is based on
(*a*) Difference in relative length
(*b*) Difference in aerodynamic properties
(*c*) Difference in density
(*d*) All of the above

174. The performance factor ($S = Cf/W$) is related with
(*a*) Spiral separator
(*b*) Magnetic separator
(*c*) Disc separator
(*d*) Cyclone separator

175. Larger crops seeds like peas and beans are separated by
(*a*) Disc separator
(*b*) Cyclone separator
(*c*) Colour separator
(*d*) Velvet roll separator

176. Magnetic separator performs separation on the basis of
(*a*) difference in aerodynamic properties
(*b*) relative length of grains
(*c*) surface texture and stickiness property of the grain
(*d*) all of these

177. Separation on the basis of difference in shape and surface texture is done by
(*a*) destoner and magnetic separator
(*b*) pneumatic and aspirator separator
(*c*) spiral and disc separator
(*d*) inclined draper and velvet roll separator

178. The device which is simple, inexpensive and quite useful for seed cleaning purposes
(*a*) Disc separator
(*b*) Magnetic separator
(*c*) Spiral separator
(*d*) Velvet roll separator

179. The high pressure air blast separates the material in
(*a*) Pneumatic separator
(*b*) Aspirator
(*c*) Both of these
(*d*) None of these

180. The function of cyclone separator is to
(*a*) collect the end products (dust and wastes) during processing of grains
(*b*) separate grains on the basis of relative length
(*c*) separate grains on the aerodynamic properties
(*d*) separate the grains with rough seed coat from smooth surface grain

181. A rice mill has to separate 40 tones of rice and paddy (long grain) per hour using a compartment type paddy separator. Calculate (*i*) the number of compartment required to carry out this separation (*ii*) the number of compartment per deck if number of decks are 4.
(*a*) 250, 1000
(*b*) 1000, 250
(*c*) 100, 25
(*d*) 667, 167

182. How many magnetic drums are used to increase the efficiency in magnetic separator generally,
(*a*) 1 or 2
(*b*) 2 or 3
(*c*) 3 or 4
(*d*) 4 or 5

183. Roll mill separator is also known as
(*a*) Disc separator (*b*) Spiral separator
(*c*) Cyclone separator (*d*) Velvet roll separator

184. Fractioning aspirator is a
(*a*) Dehusker (*b*) Sheller
(*c*) Huller (*d*) Separator

185. On which factor, the particles are acted upon by a cyclone separator
(*a*) Centrifugal force (*b*) Weight of the particle
(*c*) Shape of the particle (*d*) Both (*a*) and (*b*)

186. Tray separator is generally used for separating
(*a*) Wheat (*b*) Paddy
(*c*) Pulse (*d*) All of these

187. Husk aspirator separates immature grains from shelled paddy by which of the following action
(*a*) Gravity flow (*b*) Perforating sieve
(*c*) Winnowing action (*d*) Blowing action

188. The table inclination in case of fixed stroke separator is
(*a*) 2–3º (*b*) 3–4º
(*c*) 4–5º (*d*) 5–6º

189. How many strokes and with what speed of the fixed stroke separator is adjusted in case of long grain rice
(*a*) 90-96, 114 rpm (*b*) 86-92, 120 rpm
(*c*) 75-80, 124 rpm (*d*) 95-100, 110 rpm

190. Which separators separate the grain on the basis of length
(*a*) Disc separator and magnetic separator
(*b*) Indented cylinder separator and disc separator
(*c*) Indented cylinder separator and spiral separator
(*d*) Velvet roll separator and inclined separator

191. The inclination of the deck in the compartment type paddy separator depends on
(*a*) Percentage paddy and brown rice (*b*) Percentage of brown rice
(*c*) Percentage of paddy (*d*) None of these

192. The shape of steel flanks and impact angle in compartment type paddy separator is
(*a*) *Z* and 60º (*b*) *Z* and 30º
(*c*) *V* and 30º (*d*) *V* and 60º

193. The efficiency of hullers doesn't depend upon
(*a*) Husking machine (*b*) Operator skill
(*c*) Uniformity of paddy (*d*) None of these

194. The distance between indented trays of tray separator is
(*a*) 4 cm (*b*) 5 cm
(*c*) 7 cm (*d*) 9 cm

195. Separation on the basis of smooth surface and higher bulk density of brown rice from paddy is done by
(*a*) Disc separator (*b*) Tray separator
(*c*) Spiral separator (*d*) Velvet roll separator

196. The velocity of fixed stroke separators in compartment type paddy separator is
(*a*) 114/100 rpm (*b*) 100/100 rpm
(*c*) 124/100 rpm (*d*) 105/100 rpm

197. The pockets/indentations in disc separator capture
(*a*) Small sized grains (*b*) Large sized grains
(*c*) Bothe of these (*d*) None of these

198. The shape of indent in inclined cylinder separator is
(*a*) Hemispherical (*b*) Circular
(*c*) Spherical (*d*) Triangular

199. Size of all indents particularly in indented cylinder separator should be
(*a*) Variable (*b*) Same
(*c*) Adjustable (*d*) None of these

200. The separator in which centrifugal force is involved for the separation of grain is
(*a*) Indented cylinder separator (*b*) Cyclone separator
(*c*) Spiral separator (*d*) All of these

201. Mustard is separated from wheat mainly by
(*a*) Disc separator (*b*) Magnetic separator
(*c*) Spiral separator (*d*) None of these

202. The power requirement of tray separator is ____________ than the compartment type separator.
(*a*) Equal (*b*) Half
(*c*) Less (*d*) More

203. Performance factor '*S*' in cyclone separator is equal to
(*a*) $V^2/g\ R^2$ (*b*) $V/g\ R^2$
(*c*) $V^2/g\ R$ (*d*) $V/g\ R$

204. The motion of belt in inclined drapper separator is
(*a*) in upward direction (*b*) in downward direction
(*c*) opposite direction (*d*) none of these

205. The roller of velvet roll separator rotate in
(*a*) Opposite direction (*b*) Same direction
(*c*) In arbitrary direction (*d*) None of these

206. Maximum percentage of oil is found in
(*a*) Ground nut (*b*) Mustard
(*c*) Linseed (*d*) Coconut

207. Cracked debranned parboiled wheat is called as
(*a*) Maida (*b*) Bulgur
(*c*) Semolina (*d*) Flour

208. Cleaning of agricultural materials is performed by
(*a*) Washing (*b*) Screening
(*c*) Hand picking (*d*) All of these

209. Classification of cleaned products into various quality fractions on the basis of commercial values is called
(*a*) Sorting (*b*) Grading
(*c*) Cleaning (*d*) Scalping

210. Separation of cleaned products into various quality fractions on the basis of any one physical property is known as
(*a*) Sorting (*b*) Grading
(*c*) Cleaning (*d*) Scalping

211. Removal of few large particles involved in the process is called
(*a*) Screening (*b*) Cleaning
(*c*) Grading (*d*) Scalping

212. The material which retained over a screen surface is called
(*a*) Over size or plus materials (*b*) Undersize or minus material
(*c*) Intermediate materials (*d*) Undesirable material

213. The vibratory screens are agitated by
(*a*) Eccentric unit (*b*) Belt and pulley
(*c*) Chain and sprocket (*d*) None

214. A screen is capable of accurate and complete separation of very fine sizes but its capacity is limited.
(*a*) Shaking screen (*b*) Vibratory screen
(*c*) Rotary screen (*d*) Horizontal screen

215. The inclination of cylinder sorter for dry granular materials is kept upto
(*a*) 125 mm/m
(*b*) 145 mm/m
(*c*) 135 mm/m
(*d*) 155 mm/m

216. The effective screening area in cylinder sorter is calculated by the formula
(*a*) 1/3 length × diameter
(*b*) Length × 1/3 diameter
(*c*) Length × 1/2 diameter
(*d*) 1/2 length × diameter

217. Cylinder sorter is a type of
(*a*) Rotary screen
(*b*) Shaking screen
(*c*) Revolving screen
(*d*) Vibratory screen

218. The effectiveness of which screen is higher than others.
(*a*) Rotary screen
(*b*) Horizontal screen
(*c*) Revolving screen
(*d*) Vibratory screen

219. The technique useful in coarse sifting is
(*a*) Horizontal reciprocating motion
(*b*) Combined vertical and horizontal motion
(*c*) Inclined reciprocating motion
(*d*) Horizontal rotary motion

220. In perforated metal screens, the diameter of round opening having unit in
(*a*) mm
(*b*) Inch
(*c*) Above both
(*d*) None

221. In groundnut grader, the part used to avoid clogging is called
(*a*) Spreader
(*b*) Brushes
(*c*) Both
(*d*) None

222. The two oscillating sieves in groundnut grader are oscillated by
(*a*) Eccentric mechanism
(*b*) Vertically motion
(*c*) Offcentric mechanism
(*d*) Downward/horizontal motion

223. The length of screen surface should be
(*a*) 1-2 times its width
(*b*) 2-3 times its width
(*c*) 3-4 times its width
(*d*) Equal to width

224. The common range in pitch adjustment (angle of inclination) in air screen grain cleaner is
(*a*) 3 to 8°
(*b*) 4 to 12°
(*c*) 12 to 20°
(*d*) 16 to 24°

225. During screening in air screen grain cleaner, the bed thickness can be reduced by
(*a*) Increasing the screen slope at constant feed rate
(*b*) Increasing the feed rate at constant screen slope
(*c*) Decreasing the screen slope at constant feed rate
(*d*) Decreasing the feed rate at constant screen slope

226. Which statement is true?
(*a*) Smallest possible perforation is used for scalping
(*b*) Largest possible perforation is used for grading
(*c*) Both
(*d*) None

227. Drying is a
(*a*) Thermo-physical process
(*b*) Physic-chemical process
(*c*) Both
(*d*) None

228. The specific heat of a product is not dependent on
(*a*) Mass density
(*b*) Specific gravity
(*c*) Weight density
(*d*) All of these

229. To determine the heat transfer rates in solid agricultural product of any shape, thermal diffusivity formula is as
(*a*) $\alpha = k/\rho C_p$
(*b*) $\alpha = C_p/k\rho$
(*c*) $\alpha = k\rho/C_p$
(*d*) $\alpha = \rho/kC_p$

230. The heat content of a material is called as
(*a*) Specific heat
(*b*) Enthalpy
(*c*) Thermal conductivity
(*d*) Entropy

231. Thermal conductivity is determined by the formula
(*a*) $Q = AK(T_1 - T_2)/t$
(*b*) $Q = AK(T_1 - T_2)^2/t$
(*c*) $Q = AK(T_2 - T_1)/t^2$
(*d*) $Q = AK(T_2 - T_1)/t$

232. Thermal conductivity depends on ______________ of materials.
(*a*) Mass density
(*b*) Specific gravity
(*c*) Weight density
(*d*) All of these

233. Thermal conductivity of single grain is
(*a*) 0.1 to 0.2 kcal/m-h-°C
(*b*) 0.1 to 0.2 kcal/m²-h-°C
(*c*) 0.3 to 0.6 kcal/m-h-°C
(*d*) 0.3 to 0.6 kcal/m²-h-°C

234. Advantages of drying in favour of farmers are
(*a*) Drying permits early harvest of crops
(*b*) It helps in proper planning of harvesting season
(*c*) Drying of agricultural products to optimum moisture content results in safe storage of products for a longer period
(*d*) All of above

235. W/m-°C is the unit of
(*a*) Enthalpy
(*b*) Entropy
(*c*) Thermal diffusivity
(*d*) Thermal conductivity

236. At 18 % moisture content, the sample of grain contains 18 kg (100 × 0.18) of water and 82 kg of dry matter. The same amount is dried to 13 % moisture content, and then final weight of water will be
(*a*) 12.25 kg (*b*) 30.25 kg
(*c*) 5.75 kg (*d*) 25.25 kg

237. If moisture content (wet basis) is 14 %. Then, moisture content dry basis will be
(*a*) 14.30 % (*b*) 9.28 %
(*c*) 16.27 % (*d*) 19.22 %

238. Henderson equation for determination of EMC is based on
(*a*) Gibbs adsorption equation (*b*) Chung-pfost equation
(*c*) Kelvin equation (*d*) Harkins –Jura equation

239. The difference between desorption and adsorption curve of drying is known as
(*a*) Hysteresis effect (*b*) Cooling effect
(*c*) Chilling effect (*d*) Refrigeration effect

240. The ratio of actual vapour pressure in air water mixture and the saturated water vapour at same temperature is known as
(*a*) Percentage humidity (*b*) Relative humidity
(*c*) Absolute humidity (*d*) None of these

241. Which statement is correct?
(*a*) Drying is convection process in which moisture from a product is removed.
(*b*) Dehydration is a process in which products are dried to very low moisture level.
(*c*) Both (*a*) and (*b*)
(*d*) None of these

242. When the atmospheric vapour pressure is greater than grain vapour pressure, transfer of moisture is from
(*a*) Grain to atmosphere (*b*) Grain absorbs moisture from atmosphere
(*c*) There is no transfer of moisture
(*d*) Moisture is completely removed from grain.

243. The critical moisture of wheat varies between
(*a*) 50-60 % (db) (*b*) 85-90 % (db)
(*c*) 69-85 % (db) (*d*) 80-90 % (db)

244. The moisture content of agricultural products at which constant rate drying ceases and falling rate starts is called
(*a*) Equilibrium moisture content (*b*) Critical moisture content
(*c*) Moisture content on dry basis (*d*) Hysteresis effect

245. The type of dryer used in big rice mills.
(*a*) Recirculating dryer
(*b*) Deep bed dryer
(*c*) LSU dryer
(*d*) Continuous flow dryer

246. The process of cooling the oil to a low temperature for long time and filtering the solid material from the oil is called
(*a*) Deodourization
(*b*) Decolouration
(*c*) Winterization
(*d*) Bleaching

247. The oil content in raw rice bran is
(*a*) 10-12 %
(*b*) 12-18 %
(*c*) 20-28 %
(*d*) 28-30 %

248. The oil content of parboiled rice bran is
(*a*) 20-28 %
(*b*) 12-18 %
(*c*) 30-35 %
(*d*) 43-48 %

249. The refined crude bran oil for edible purpose is uneconomic if FFA content is above
(*a*) 5 %
(*b*) 20 %
(*c*) 12 %
(*d*) 10 %

250. Bran oil is a rich source of
(*a*) Vitamin D
(*b*) Riboflavin
(*c*) Vitamin E
(*d*) Tocopherol

251. Tocopherol content in crude bran oil is about
(*a*) 2-4 %
(*b*) 4-5 %
(*c*) 1-2 %
(*d*) 6-8 %

252. A significant amount of tocopherol is lost during processing of crude oil. This loss occurs during
(*a*) Winterization
(*b*) Deodourization
(*c*) Decolouration
(*d*) None of these

253. Which one of the following is a liquid fumigant?
(*a*) Hydrogen phosphide
(*b*) Methyl bromide
(*c*) Hydrogen cyanide
(*d*) Ethyl bromide

254. The transfer of moisture from interior of grain faster towards surface during drying due to
(*a*) Osmotic pressure
(*b*) Vapour pressure
(*c*) Surface tension
(*d*) Conduction

255. The amount of moisture removed from grain by drying air depends upon the
(*a*) Length of grain
(*b*) Chemical properties of air
(*c*) Psychometric properties of air
(*d*) Amount of air

256. Drying process means removal of moisture due to simultaneous __________ and __________ transfer.
(*a*) Heat, mass
(*b*) Volume, shape
(*c*) Moisture, vapour
(*d*) Density, specific gravity

257. In which drying process heat is supplied to wet products by conduction
(*a*) Radiation drying
(*b*) Freeze drying
(*c*) Superheated steam drying
(*d*) Contact drying

258. The sensible heat of heated air is transferred to the wet product by convection in
(*a*) Osmotic drying
(*b*) Convective drying
(*c*) Fluidized bed drying
(*d*) Desiccated air drying

259. Sublimation of frozen moisture from wet product is employed in
(*a*) Freeze drying
(*b*) Dehydro-freezing
(*c*) Spray drying
(*d*) Drum drying

260. The wavelength of electromagnetic radiation in radiation drying of foods ranges between
(*a*) 0.76–400 μm
(*b*) 100–200 μm
(*c*) 1–200 μm
(*d*) 400–1000 μm

261. The wave length of 0.76 to 400 μm radiation is known as
(*a*) *x*-rays
(*b*) infrared radiation
(*c*) γ-rays
(*d*) none of these

262. Recommended air flow rate for deep bed drying of grain is
(*a*) 2.94–3.92 m^3/min per ton
(*b*) 2.94–3.43 m^3/min per ton
(*c*) 3.43–3.92 m^3/min per ton
(*d*) 50–95 m^3/min per ton

263. The lower limit of air flow rate of 2.94–3.43 m^3/min per ton in deep bed drying is recommended for the __________ region .
(*a*) Hot
(*b*) Rainy
(*c*) Cool and dry
(*d*) Hot and humid.

264. For hot and dry condition, recommended higher limit of air flow rate in deep bed drying is
(*a*) 2.94–3.92 m^3/min per ton
(*b*) 3.43–3.92 m^3/min per ton
(*c*) 2.94–3.43 m^3/min per ton
(*d*) 50–60 m^3/min per ton

265. For uneven deep bed drying, recommended air flow rate for the grain is
(*a*) 4.92 m^3/min per ton
(*b*) 3.92 m^3/min per ton
(*c*) 3.43 m^3/min per ton
(*d*) 2.94 m^3/min per ton

266. If the moisture content of grain is upto 18%, the layer depth of grain for deep drying should not be more than
(*a*) 3.0 m
(*b*) 2.0 m
(*c*) 4.5 m
(*d*) 1.0 m

267. For grain whose moisture content is 18 %, the recommended maximum depth for deep bed drying is
(*a*) 3.0 m (*b*) 2.5 m
(*c*) 2.0 m (*d*) 1.0 m

268. The net perforated drying area of deep bed dryer should be ____________ of the total flour area.
(*a*) 10 % (*b*) 15 %
(*c*) 25 % (*d*) 40 %

269. Grains are spread 0.6 to 1.2 m deep over perforated flour in
(*a*) Deep bed dryer (*b*) Flat bed dryer
(*c*) Continuous flow dryer (*d*) LSU dryer

270. The capacity of flat bed dryer is usually
(*a*) 2-5 ton (*b*) 1-2 ton
(*c*) 50-60 ton (*d*) 5-50 ton

271. The main advantage of flat bed dryer is
(*a*) Whole batch is dried quickly (*b*) Less chance of mold spoilage
(*c*) Less over drying of grain (*d*) Need of lower air pressure
(*e*) All of the above

272. Which dryer is classified in to mixing and non mixing method?
(*a*) Deep bed dryer (*b*) Flat bed dryer
(*c*) Continuous flow dryer (*d*) LSU dryer
(*e*) Recirculating dryer

273. In which dryer, baffles are employed for proper mixing of grain during drying?
(*a*) Deep bed dryer (*b*) Continuous flow dryer
(*c*) LSU dryer (*d*) Recirculating dryer

274. The temperature of hot air at which grain is dried in mixing type continuous flow dryer
(*a*) 60ºC (*b*) 50ºC
(*c*) 65ºC (*d*) 47ºC

275. Drying air temperature used in non mixing type continuous flow dryer is about
(*a*) 54ºC (*b*) 47ºC
(*c*) 45ºC (*d*) 65ºC

276. A popular drier used in paddy processing industry originated in USA is
(*a*) Recirculating dryer (*b*) LSU dryer
(*c*) Deep bed dryer (*d*) Flat bed dryer

277. For continuous mixing of grain, inverted V-shaped channels are installed in
(*a*) Deep bed dryer (*b*) Recirculating dryer
(*c*) Fluidized bed dryer (*d*) LSU dryer

278. Safe storage of grains is essential
(*a*) To prevent losses and disease
(*b*) To prevent from the attack of birds, insects, pest, rodents and microorganisms
(*c*) To maintain good quality of products
(*d*) All of the above

279. The estimated losses of world food production damaged by insects during storage is about
(*a*) 1–3 % (*b*) 5–10 %
(*c*) 10–15 % (*d*) 20–40 %

280. The estimated losses of total food grains of India due to insect is about
(*a*) 1–3% (*b*) 5–10%
(*c*) 10–15 % (*d*) 20–40 %

281. How much grain is contaminated by rodent of own actual weight?
(*a*) Two times (*b*) Five times
(*c*) Ten times (*d*) Twenty times

282. In India, Morai storage structure is commonly used in the rural area of
(*a*) Southern region (*b*) Eastern region
(*c*) Central region (*d*) Both (*a*) and (*b*)

283. Morai structure is generally suitable for
(*a*) Paddy, maize and sorghum (*b*) Paddy, potato and tobacco
(*c*) Mustard, soybean and wheat (*d*) Maize, bazra and oat

284. The capacities of morai storage structure varies between
(*a*) 3.5–18 tonnes (*b*) 9–35 tonnes
(*c*) 10–15 tonnes (*d*) 1–2 tonnes

285. One of the following structures has inverted truncated cone shape.
(*a*) Bukhari (*b*) Kothari
(*c*) Pusa bin (*d*) Morai

286. A wooden floor surrounded by 22 gauge corrugated metal cylinder upto height of 90 cm is in__________.
(*a*) Pusa bin (*b*) Kothari
(*c*) Morai (*d*) Bukhari

287. The height of grain loaded in morai structure is about
(*a*) 75 cm (*b*) 90 cm
(*c*) 120 cm (*d*) 100 cm

288. Bukhari storage structure generally has capacities between
(*a*) 2–4 tonnes (*b*) 4–5 tonnes
(*c*) 3.5–18 tonnes (*d*) 9–35 tonnes

289. Kothar structure is used to store the paddy, maize, sorghum and wheat, their capacities varies between
(*a*) 2–4 tonnes (*b*) 4–5 tonnes
(*c*) 3.5–18 tonnes (*d*) 9–35 tonnes

290. 'Muda' is a popular grain storage structure of
(*a*) Bihar (*b*) West Bengal
(*c*) Orissa (*d*) UP

291. 'Kanaj' is a type of
(*a*) Grain storage structure (*b*) Drying method
(*c*) Food products (*d*) Grain separator

292. The 'Kanaj storage structure is very common in the rural area of
(*a*) Jammu and Kashmir (*b*) Punjab and Haryana
(*c*) UP and Bihar (*d*) Karnataka and Maharashtra

293. The capacity of 'Kanaj' varies between
(*a*) 1 to 20 tonnes (*b*) 3.5 to 18 tonnes
(*c*) 9 to 35 tonnes (*d*) 35 to 40 tonnes

294. In which states, Kuthla for grain storage is very commonly used in the rural areas.
(*a*) UP and Bihar (*b*) Bihar and Jharkhand
(*c*) MP and Chhattisgarh (*d*) Rajasthan and Gujarat

295. One of the following is not a grain storage structure.
(*a*) Kuthla (*b*) Kanaj
(*c*) Muda (*d*) Bellows

296. In which of grain storage structure, plastic film of 700 gauges is used
(*a*) Kothar (*b*) Morai
(*c*) Pusa bin (*d*) Bukhari

297. The size of an open space on top of Pusa bin for loading should be
(*a*) 60 cm × 60 cm (*b*) 50 cm × 50 cm
(*c*) 50 cm × 50 mm (*d*) No open space

298. One of the following is not improved grain storage structure.
(a) Pusa bin
(b) Bunker storage
(c) Cover and plinth storage structure
(d) Muda

299. The height of stack in CAP grain storage structure should not be more than
(a) 10 bags
(b) 15 bags
(c) 18 bags
(d) 20 bags

300. The capacity of CAP storage structure is
(a) 10 tonnes
(b) 100 tonnes
(c) 1000 tonnes
(d) 2000 tonnes

301. The word 'CAP' used for grain storage is also known as
(a) Cover and plinth
(b) Controlled atmospheric packaging
(c) Both of the above
(d) None of them

302. High moisture and warm climate for stored grain is responsible for
(a) Mold growth
(b) Insect growth
(c) Increase respiration of grain
(d) All of the above

303. At present in India, most quite grain storage structures used are
(a) Mild steel bin and RCC bin
(b) Fibre glass and asbestos sheet bin
(c) Plain or corrugated galvanized sheet bin
(d) Aluminium sheet and mild black sheet

304. The advantages of modern grain storage bins are
(a) Less expansive, less space requirement and easy quality control
(b) Protection from losses due to birds and rodents
(c) Provision of automation and mechanization
(d) All of the above

305. A bin whose relative dimensions are such that the plane of rupture meets the grain surface before it strikes the opposite side is called
(a) Deep bin
(b) Shallow bin
(c) Silo
(d) CAP

306. A bin in which plane of rupture meets the opposite side before it emerges from the grain is called
(a) Deep bin
(b) Shallow bin
(c) CAP
(d) Silo

307. If Φ is the angle of repose of material, then the angle of rupture may be taken as $(90 + \Phi)/2$ with the horizontal as per
(a) Airy's theory
(b) Jenssion theory
(c) Rankine theory
(d) Henderson equation

308. If L and h be breadth and depth of the grain storage structure, then the structure can be taken to shallow if
(*a*) $h < L \tan (90+ \Phi)/2$
(*b*) $h > L \tan (90+ \Phi)/2$
(*c*) $h < L \tan (90-\Phi)/2$
(*d*) $h > L \tan (90-\Phi)/2$

309. If L and h be breadth and depth of the grain storage structure, then the structure is deep, if
(*a*) $h < L \tan (90 + \Phi)/2$ (*b*) $h < L \tan (90 - \Phi)/2$
(*c*) $h > L \tan (90 + \Phi)/2$ (*d*) $h > L \tan (90 - \Phi)/2$

310. A grain bin referred to as shallow bin when the depth of the grain is ____ to the equivalent diameter
(*a*) Less (*b*) Equal
(*c*) Greater (*d*) Less or equal

311. In a deep bin, depth of grain is ____________ than equivalent diameter.
(*a*) Less (*b*) Equal
(*c*) Greater (*d*) Less or equal

312. The equivalent diameter is taken ____________ times the hydraulic radius of the circular bin.
(*a*) Two (*b*) Four
(*c*) Eight (*d*) Sixteen

313. If H_d is depth of grain and 'R' is hydraulic radius, the relationship for deep bin will be
(*a*) $H_d = 4R$ (*b*) $H_d < 4R$
(*c*) $H_d = 8R$ (*d*) $H_d > 4R$

314. If H_d is depth of grain and 'R' is hydraulic radius, the relationship for shallow bin will be
(*a*) $H_d = 4R$ (*b*) $H_d < 4R$
(*c*) $H_d = 8R$ (*d*) $H_d > 4R$

315. As per definition given by Issacson and Boyd, deep bin is calculated by
(*a*) $H_d/D < 0.75\ (1/\mu K)$ (*b*) $H_d/D = 0.75\ (\mu/K)$
(*c*) $H_d/D = 0.75\ (K/\mu)$ (*d*) $H_d/D = 0.75\ (1/\mu K)$

316. A good quality storage structure should be
(*a*) Structurally adequate (*b*) Durable and hygienic
(*c*) Gas sealable for fumigation (*d*) All

317. Vertical grain silo may be
(*a*) Circular (*b*) Hexagonal
(*c*) Square or rectangular (*d*) All

318. One of the following is not a permanent grain storage structure.
(*a*) Shed
(*b*) Vertical silo
(*c*) Squat silo
(*d*) CAP

319. Rankine's coefficient (K) for silo designing is equivalent to
(*a*) (1 – sin Φ)/(1 + sin Φ)
(*b*) (1 + sin Φ)/(1 – sin Φ)
(*c*) (1 – sin Φ)/(1 – cos Φ)
(*d*) (1 – sin Φ)/(1 + cos Φ)

320. The Rankine's formula is used for determination of pressure induced by granular materials against retaining wall in
(*a*) Deep bin
(*b*) Shallow bin
(*c*) Squat bin
(*d*) Shed

321. The process of moving air through stored grain at low flow rates to maintain or improve its quality is called
(*a*) Ventilation
(*b*) Aeration
(*c*) Fumigation
(*d*) Migration

322. Aeration can provide major profit in the storage of grain if
(*a*) It cools the grains and slowdown insect activity
(*b*) Prolongs the effectiveness of pesticides by cooling
(*c*) It can provide an appreciable drying function
(*d*) All of the above

323. Aeration has some minor benefits like
(*a*) It prevents storage odour
(*b*) It reduces moisture accumulation
(*c*) Application of fumigation
(*d*) All of the above

324. The most common air flow rates for aerating paddy range varies from
(*a*) 0.07-0.28 m^3/min/tonne
(*b*) 0.01-0.04 m^3/min/tonne
(*c*) 0.80-1.00 litre/sec/tonne
(*d*) 1.20-1.60 litre/sec/tonne

325. The perforation in duct for aeration should be uniformly spaced with minimum area of
(*a*) 10%
(*b*) 15%
(*c*) 18%
(*d*) 20%

326. Recommended perforation of aeration in grain storage shall be
(*a*) 10%
(*b*) 15%
(*c*) 20%
(*d*) 35%

327. Methyl bromide or phosphine are used as
(*a*) Germicide
(*b*) Pesticide
(*c*) Fungicide
(*d*) Fumigant

328. The protection of food by packaging means safeguard against
(*a*) Physical activity (*b*) Chemical activity
(*c*) Biological activity (*d*) All of them

329. Stored food products have the main source of heats *i.e.*
(*a*) Sensible heat of products (*b*) Heat of respiration
(*c*) Specific heat (*d*) Both (*a*) and (*b*)

330. The heat loads in cold storage system are classified like
(*a*) Chilling load (*b*) Freezing load
(*c*) Cooling load (*d*) All of the above

331. For increasing safe storage period of fruits and vegetables under CAP conditions by
(*a*) Increasing CO_2 level (*b*) Decreasing O_2 level
(*c*) Decreasing the temperature (*d*) All of these

332. Air tight storages are normally used
(*a*) To arrest the insect infestation
(*b*) To protect the wet grain from mold attack
(*c*) To decrease the oxygen content (*d*) Both (*a*) and (*b*)

333. In winter season, grain stored in bin will be spoiled at
(*a*) Top (*b*) Bottom
(*c*) Centre (*d*) Near walls of bin

334. Spoilage of grain in bin during warm season takes place due to moisture accumulation at
(*a*) Top (*b*) Centre
(*c*) Bottom (*d*) Inner surface of walls

335. Grain is spoiled in bin due to accumulation of
(*a*) Temperature (*b*) Moisture
(*c*) Heat (*d*) Relative humidity

336. By controlling insect and moisture levels, the losses in Bunker structure can be reduced upto
(*a*) 0-0.5% (*b*) 1-3%
(*c*) 10-20% (*d*) 20-40%

337. The granular material behave as composite mass having the characteristics of
(*a*) Liquid and solid (*b*) Liquid and gel
(*c*) Solid and gas (*d*) Semi solid and semi liquid

338. The size of food grains is represented by
(*a*) Length (*b*) Width
(*c*) Diameter (*d*) Equivalent diameter

339. The equivalent diameter of a granular particle can be determined by the following formula
(*a*) $(abc)^{1/3}$
(*b*) $(abc)^{2/3}$
(*c*) $(abc)^{3/2}$
(*d*) $(abc)^3$

340. The diameter of a sphere having same ratio of surface to volume as actual particle is called
(*a*) Equivalent diameter
(*b*) Spherecity
(*c*) Roundness
(*d*) Roundness ratio

341. The ratio of largest projected area of the particle when it is in natural rest position (A_p) and area of smallest circumscribing circle (A_c) is called
(*a*) Roundness
(*b*) Spherecity
(*c*) Geometric mean diameter
(*d*) Perimeter

342. The ratio of radius of curvature (r) of the sharpest corner and mean radius of particle (R) is known as
(*a*) Geometric mean diameter
(*b*) Equivalent diameter
(*c*) Roundness ratio
(*d*) Spherecity

343. The ratio of diameter of a sphere of the same volume as that of the particle and the diameter of the smallest circumscribing sphere or generally the largest diameter of particle is defined as
(*a*) Roundness ratio
(*b*) Roundness
(*c*) Spherecity
(*d*) Equivalent mean diameter

344. Drying takes place in
(*a*) Fruits
(*b*) Vegetables
(*c*) Food grains
(*d*) All

345. Which of the following is not a food processing operation?
(*a*) Harvesting
(*b*) Cleaning
(*c*) Screening
(*d*) Sorting

346. Removal of large few foreign materials from food grains comes under
(*a*) Sorting
(*b*) Cleaning
(*c*) Scalping
(*d*) Screening

347. The package does not include
(*a*) Attracting the buyers
(*b*) Identifying the products
(*c*) Information about use of product
(*d*) To improve the flavour

348. Which of the following is a not food constituent?
(*a*) Carbohydrate
(*b*) Fat
(*c*) Minerals
(*d*) Enzyme

349. The removal of colour of rice is called
(*a*) Blanching
(*b*) Bleaching
(*c*) Parboiling
(*d*) Polishing

350. Heat energy supplied to wet material by electromagnetic wave length of 0.76-400 μm in
(*a*) Radiation drying (*b*) Fluidized drying
(*c*) Freeze drying (*d*) Osmotic drying

351. Moisture content of substance is expressed in
(*a*) ml (*b*) litre
(*c*) gm (*d*) percentage

352. The process of water absorption by food material is a
(*a*) Diffusion process (*b*) Osmosis
(*c*) Capillary process (*d*) Migration

353. Which scientist expressed the relation between swelling and moisture content?
(*a*) B. Bandyopadhyaya and NC Raj (*b*) B. Bandyopadhyaya and TK Ghose
(*c*) Ali and Ojha (*d*) KR Bhattacharya and UP Subbarao

354. Vitamin B in milled parboiled rice than milled raw rice is
(*a*) Lower (*b*) Equal
(*c*) Higher (*d*) None

355. Oil content in raw rice bran is
(*a*) 10-15% (*b*) 15-25%
(*c*) 20-25% (*d*) 10-20 %

356. During parboiling, the tank is filled with water heated by steam injection of
(*a*) 95°C (*b*) 85°C
(*c*) 70°C (*d*) 90°C

357. For complete gelatinization of starchy endosperm by open steaming process, the moisture content of wheat is increased upto the level of
(*a*) 85% (db) (*b*) 75% (db)
(*c*) 80% (db) (*d*) 70% (db)

358. The hardness of barley endosperm is
(*a*) 7.0-9.0 kg/mm^2 (*b*) 6.0-9.0 kg/mm^2
(*c*) 8.0-10.0 kg/mm^2 (*d*) 5.0-6.0 kg/mm^2

359. The preheating treatment process in bulgur production, moisture content of wheat is raised upto
(*a*) 10-20 % (db) (*b*) 10-50% (db)
(*c*) 20-30% (db) (*d*) 50-70% (db)

360. Optimum moisture content (db) for milling of parboiled paddy should be
(*a*) 30% (*b*) 40%
(*c*) 50% (*d*) 60%

361. Soaked wheat is conditioned at a temperature between
(*a*) 10–20ºC (*b*) 30–35ºC
(*c*) 40–50ºC (*d*) 50–60ºC

362. The percentage of endosperm in mature corn is about
(*a*) 65% (*b*) 70%
(*c*) 82% (*d*) 85%

363. Optimum moisture content for degerming of corn in Baell Degermer should be
(*a*) 10-15% (*b*) 15-20%
(*c*) 21-25% (*d*) 30-35%

364. Raw pulses may contain impurities
(*a*) 2-5% (*b*) 5-10%
(*c*) 20-25% (*d*) 30-35%

365. Residual oil in the cotton seed deoiled cake is
(*a*) 2-4% (*b*) 4-5%
(*c*) 5-10% (*d*) 10-12%

366. Which machine is used for the cleaning/washing of corn?
(*a*) Whizzer (*b*) Washer
(*c*) Degermer (*d*) Blancher

367. The moisture content at which the drying rate ceases to constant is known as
(*a*) EMC (*b*) CMC
(*c*) MC (db) (*d*) MC (wb)

368. Development of drying equation for heating or cooling is based on
(*a*) Newton's law (*b*) Arrhenius equation
(*c*) Henderson equation (*d*) Kick's law

369. The moisture content at which grain should not be dried in natural air is
(*a*) >20% (*b*) 20%
(*c*) 18% (*d*) < 20%

370. The relationship between heat utilization factor (HUF) and coefficient of performance (COP) of a drier.
(*a*) HUF = 1 – COP (*b*) HUF = 1 + COP
(*c*) HUF = 1/COP (*d*) HUF = COP

371. Natural air drying can not be used if the ambient relative humidity of air exceed
(*a*) 60% (*b*) 65%
(*c*) 70% (*d*) 75%

372. Recommended air flow rate for drying of parboiled paddy is
(*a*) 50-60 m^3/min/tonne (*b*) 60-70 m^3/min/tonne
(*c*) 70-80 m^3/min/tonne (*d*) 80-90 m^3/min/tonne

373. About how much paddy in India is converted in to parboiled of the total production
(*a*) 45% (*b*) 50%
(*c*) 60% (*d*) 65%

374. Deep bed drying theory is given by
(*a*) Becker (*b*) Hukill
(*c*) Huston (*d*) Ives

375. Oil content in parboiled paddy bran is
(*a*) 20-30% (*b*) 25-30%
(*c*) 30-45% (*d*) 40-50%

376. Hardness of rice endosperm is
(*a*) 8.5–10.5 kg/mm^2 (*b*) 11.5–14.5 kg/mm^2
(*c*) 14.5–16.5 kg/mm^2 (*d*) 18.5–19.5 kg/mm^2

377. Coconut shell contains lignin about
(*a*) 20% (*b*) 26%
(*c*) 30% (*d*) 36%

378. At which moistũre content cereals grain should be stored for longer period?
(*a*) 10% (*b*) 12%
(*c*) 13% (*d*) 14%

379. Melting point of the rice bran wax is
(*a*) 20-30°C (*b*) 44-52°C
(*c*) 72-84°C (*d*) 90-96°C

380. Corn syrups are produced by the hydrolysis of
(*a*) corn starch (*b*) corn sweetener
(*c*) corn flour (*d*) corn grits

381. High intensity sweetener are used in confectionary to reduce
(*a*) quantity of sugar (*b*) Calorific value
(*c*) Cost of sweetener (*d*) Above all

382. The term 'food dehydration' refers to
(*a*) artificial drying by thermal means under controlled conditions
(*b*) sun drying under controlled conditions
(*c*) both above
(*d*) none

383. Lower drying temperature and short time are especially important in case of
(*a*) heat sensitive foods
(*b*) highly perishable food
(*c*) staple foods
(*d*) none

384. Thermoplastic foods by heat treatment, become
(*a*) Harder
(*b*) Soften
(*c*) Discolouration
(*d*) Burn

385. How many types of yeasts are used in baking?
(*a*) 2
(*b*) 3
(*c*) 4
(*d*) 5

386. Percentage of carbohydrate in wheat grain is
(*a*) 50%
(*b*) 62%
(*c*) 69%
(*d*) 70%

387. Typical food label shows
(*a*) Principal display
(*b*) Nutrition information
(*c*) Panel and mandatory information
(*d*) Above all

388. The temperature at which denaturation of protein in cereals takes place
(*a*) 30°C
(*b*) 40°C
(*c*) 50°C
(*d*) 70°C

389. Porosity percent in wheat is
(*a*) 40–45
(*b*) 50–55
(*c*) 65–70
(*d*) 48–53

390. Percentage of bran in wheat milling is
(*a*) 5%
(*b*) 10%
(*c*) 12%
(*d*) 15%

391. Percent recovery of white flour in wheat milling is
(*a*) 30%
(*b*) 63%
(*c*) 80%
(*d*) 85%

392. Oil percentage in parboiled rice bran is
(*a*) 20–28%
(*b*) 10–12%
(*c*) 13–17%
(*d*) 20–40%

393. Rice bran oil contains saturated fatty acids about
(*a*) 25–35%
(*b*) 30–42%
(*c*) 35–48%
(*d*) 42–50%

394. The respiratory rate of stored grain depends largely on
(*a*) Moisture content and temperature
(*b*) Relative humidity and temperature
(*c*) Pressure and temperature
(*d*) Relative humidity and moisture content

395. Modern rice milling machine *i.e.* single scalper drum cleaner is developed in
(*a*) Japan (*b*) Australia
(*c*) India (*d*) Germany

396. Fat percentage in corn is
(*a*) 3.0% (*b*) 3.5%
(*c*) 4.0% (*d*) 4.5%

397. Barley has indigestible fibre percentage
(*a*) 6 (*b*) 7
(*c*) 8 (*d*) 9

398. Viscosity of oil is measured in terms of
(*a*) Its resistance to flow (*b*) Its cooling effect
(*c*) Both above (*d*) None

399. Glucose is added to rice as glazing substance with water solution in the ratio of
(*a*) 1:1 (*b*) 1:2
(*c*) 2:1 (*d*) 1:3

400. Spiral separator separates the grain on the basis of
(*a*) size (*b*) length
(*c*) Roundness (*d*) Spherecity

401. Cleaning of materials is performed by
(*a*) Washing (*b*) Screening
(*c*) Hand picking (*d*) All

402. The boiling point of *n*-hexane used as solvent for oil extraction is
(*a*) 60.5°C (*b*) 65.5°C
(*c*) 70.5°C (*d*) 90.5°C

403. For grinding operation, the operating speed of hammer mill should be
(*a*) 700-800 rpm (*b*) 1000-1100 rpm
(*c*) 1400-1500 rpm (*d*) 1500-1600 rpm

404. Hammer mill and Burr mill are used for the grinding of
(*a*) Grain (*b*) Oil seed
(*c*) Fruit powder (*d*) Milk powder

405. Natural drying needs dehumidification if atmospheric humidity exceeds than
(*a*) 30% (*b*) 60%
(*c*) 70% (*d*) 80%

406. In batch type dryer, the air temperature seldom exceeds
(*a*) 35°C (*b*) 40°C
(*c*) 45°C (*d*) 50°C

407. In single boiling method of paddy parboiling, paddy is soaked upto
(*a*) 12–18 hrs (*b*) 12–24 hrs
(*c*) 24–48 hrs (*d*) 24–72 hrs

408. The storage life of non-perishable fruits exceed by
(*a*) 8 weeks (*b*) 10 weeks
(*c*) 12 weeks (*d*) 14 weeks

409. Percentage starch in wheat grain is about
(*a*) 48–55% (*b*) 56–59%
(*c*) 60–68% (*d*) 75–78%

410. In attrition mills, the size reduction of grains is accomplished due to
(*a*) impact and shear (*b*) shear and crushing
(*c*) impact only (*d*) impact and crushing

411. The shelling of paddy by centrifugal dehusker is due to
(*a*) Shear (*b*) Crushing
(*c*) Rubbing (*d*) Impact

412. Brewing and malting is related to
(*a*) Beer (*b*) Juice
(*c*) Milk (*d*) Sausage

413. Generally in thin layer drying, thickness of grain bed with recommended air grain ratio is
(*a*) 10 cm (*b*) 15 cm
(*c*) 20 cm (*d*) 25 cm

414. The moisture contained by a substance in excess of the equilibrium moisture is known as
(*a*) Free moisture (*b*) Bound moisture
(*c*) Unbound moisture (*d*) Critical moisture

415. Fuel consumption per kg of moisture evaporated in convection drying is always ________ than conduction drying.
(*a*) Lower (*b*) Equal
(*c*) Higher (*d*) Unknown

416. Wheat is dried under constant rate drying period when its moisture content exceeds
(*a*) 50% (*b*) 60%
(*c*) 72% (*d*) 75%

417. The drying method particularly suitable for drying of small quantity of seed with respect to conserving strain purity and viability is
(*a*) Sun drying (*b*) Radiation drying
(*c*) Freeze drying (*d*) Sack drying

418. Natural air drying can not be used if the ambient relative humidity exceeds
(*a*) 60% (*b*) 70%
(*c*) 80% (*d*) 85%

419. The parboiled paddy is required to dry the moisture content for desirable milling and storage is
(*a*) 10-12% (*b*) 14-16%
(*c*) 16-18% (*d*) 20-22%

420. Broken in raw rice is about 50%, whereas in parboiled rice the broken is about
(*a*) 10% (*b*) 20%
(*c*) 80% (*d*) 90%

421. Parboiled rice takes __________ time to cook to the same degree of softness than raw rice.
(*a*) Less (*b*) Equal
(*c*) More (*d*) Unknown

422. Addition of moisture and heat to grain for improving their quality and quantity of end product is known as
(*a*) Radiation treatment (*b*) Hydrothermal treatment
(*c*) Blanching (*d*) None

423. The water impregnated with sulphur dioxide for arresting certain fermentation is practiced in the process of
(*a*) Degerming (*b*) Steeping
(*c*) Conditioning (*d*) Thermal treatment

424. The average processing capacity of pulses mills in India varies between
(*a*) 2-5 tonnes/day (*b*) 5-10 tonnes/day
(*c*) 10-20 tonnes/day (*d*) 20-30 tonnes/day

425. In pulses, during pitting process the subsequent apply to facilitate loosing of husk is
(*a*) Water (*b*) Red earth
(*c*) Oil (*d*) $NaHCO_3$

426. The clean pulses contain husk about
(*a*) 10–15% (*b*) 15–20%
(*c*) 20–25% (*d*) 25–30%

427. The highest amount of carbohydrate in the form of starch, cellulose etc. present in
    (*a*) Pericarp
    (*b*) Endosperm
    (*c*) Aleurone layer
    (*d*) Germ

428. Maximum moisture content of wheat for safe storage in leak proof environment for one year is
    (*a*) 8-10%
    (*b*) 12-13%
    (*c*) 16-18%
    (*d*) 18-20%

429. For drying of cereals grains, which is best suitable is
    (*a*) Spray drier
    (*b*) Tray drier
    (*c*) Tunnel drier
    (*d*) Rotary drier

430. The optimum moisture content for harvesting of paddy is
    (*a*) 10-12%
    (*b*) 14-18%
    (*c*) 20-22%
    (*d*) 24-26%

431. Drying medium in desiccated air drying to absorb the RH of air is
    (*a*) Silica gel
    (*b*) Common salt
    (*c*) Baking powder
    (*d*) Electric heater

432. The mature corn kernel is composed of
    (*a*) Endosperm (82%), germ (12%), Pericarp (5%) and tip cap (1%)
    (*b*) Endosperm (80%), germ (10%), Pericarp (7%) and tip cap (3%)
    (*c*) Endosperm (70%), germ (12%), Pericarp (10%) and tip cap (5%)
    (*d*) Endosperm (82%), germ (7%), Pericarp (8%) and tip cap (5%)

433. The optimum moisture content of corn for degerming in Beall Degermer is
    (*a*) 15-20%
    (*b*) 12-14%
    (*c*) 21-25%
    (*d*) 25-30%

434. Steeping of corn in wet milling, water is impregnated with _____ for arresting fermentation.
    (*a*) SO_2
    (*b*) SO_3
    (*c*) NO_2
    (*d*) CO_2

435. The steeping of corn is carried out at 50°C for a period of
    (*a*) 10-18 hrs
    (*b*) 20-25 hrs
    (*c*) 28-48 hrs
    (*d*) 40-50 hrs

436. Corn attains moisture content during steeping in wet milling process at 50°C temperature about
    (*a*) 20%
    (*b*) 40%
    (*c*) 45%
    (*d*) 50%

437. Emery roller known as gota machine are used for dehusking of
(*a*) Conditioned paddy
(*b*) Conditioned wheat
(*c*) Splitted pulse
(*d*) Conditioned pulse

438. According to Simmond's rate of drying of wheat is sharply depended upon the temperature of air varying from
(*a*) 15-65°C
(*b*) 37-96°C
(*c*) 21-77°C
(*d*) 28-80°C

439. In dielectric drying, heat is generated within a solid by placing it in a
(*a*) Fixed light frequency magnetic field
(*b*) Fixed high frequency current
(*c*) Varying frequency magnetic field
(*d*) Low frequency current

440. Modern rice milling machine-single scalper drum cleaner is developed in
(*a*) USA
(*b*) India
(*c*) Japan
(*d*) China

441. Indented tray type paddy separator (Japan) is made up of
(*a*) Iron
(*b*) Steel
(*c*) Aluminium
(*d*) Copper

442. The raw corn for wet milling should have moisture content of
(*a*) 10-12%
(*b*) 12-14%
(*c*) 15-16%
(*d*) 18-20%

443. The husking capacity of a impact type paddy husker (Japan) with a disc of 30 cm in diameter is about
(*a*) 300 kg/h
(*b*) 175 kg/h
(*c*) 280 kg/h
(*d*) 225 kg/h

444. The optimum peripheral speeds of rubber roll husker for paddy and oats respectively are
(*a*) 16 m/s and 20 m/s
(*b*) 14 m/s and 20 m/s
(*c*) 20 m/s and 12 m/s
(*d*) 10 m/s and 16 m/s

445. The pneumatic separation of different compounds is based on the
(*a*) Thermodynamic properties
(*b*) Physical properties
(*c*) Magnetic properties
(*d*) Aerodynamic properties

446. Siddha, a traditional method of paddy parboiling is originated in
(*a*) Bengal
(*b*) Gujrat
(*c*) Japan
(*d*) Pakistan

447. Recommended air flow rate of drying of parboiled paddy is
(*a*) 60-70 m^3/min/tonne
(*b*) 70-80 m^3/min/tonne
(*c*) 16-28 m^3/min/tonne
(*d*) 56-62 m^3/min/tonne

448. Swelling capacity is expressed as the ratio of

(*a*) Final to initial volume of rice
(*b*) Initial to final volume of rice
(*c*) Initial to final weight of rice
(*d*) Final to initial volume or weight of rice

449. Composition of unsaturated fatty acid in rice bran oil is

(*a*) 80-84%
(*b*) 40-50%
(*c*) 20-42%
(*d*) 16-20%

450. Paddy; corn, wheat, buck wheat seeds are specially rich source of

(*a*) Vitamin
(*b*) Fat
(*c*) Oil
(*d*) Carbohydrate

451. The endosperm of cereals grain contains the highest amount of carbohydrates in the form of

(*a*) Pectin
(*b*) Dextrin
(*c*) Starch
(*d*) Sugar

452. The temperature at which denaturation and coagulation of cereals protein take place above

(*a*) 50ºC
(*b*) 60ºC
(*c*) 75ºC
(*d*) 40ºC

453. Useful storage life of dried fruit and dried seed is upto

(*a*) 100 days
(*b*) 200 days
(*c*) 300 days
(*d*) 360 days or more

454. Screening refers to

(*a*) Separation of cleaned products into various quality fractions by any one physical property
(*b*) Removal of few large particles
(*c*) Classification of cleaned products into various quality fractions by commercial values
(*d*) Separating grains into two or more fraction according to size alone.

455. MAQ is defines as

(*a*) Minimum acceptable quality
(*b*) Moderate acceptable quality
(*c*) Modified atmospheric quality
(*d*) Maximum acceptable quality

456. The date beyond which the product is no longer acceptable

(*a*) Pack value
(*b*) Expiration date
(*c*) Manufacturing date
(*d*) Selling date

457. FDA is termed as

(*a*) Fruit and Drug Administration
(*b*) Food and Drug Administration
(*c*) Fruit and dietary allowance
(*d*) Foundation of drug administration

458. The machine used for removal of bran layers from brown rice is known as

(*a*) Polisher
(*b*) Bran aspirator
(*c*) Husk aspirator
(*d*) Paddy sheller

459. The system used for separating husk from the paddy grain is called
(*a*) Husk aspirator
(*b*) Paddy separator
(*c*) Husk separator
(*d*) Rubber roll sheller

460. The equipment applied for separation of husk from the product mixture obtained from sheller.
(*a*) Husk aspirator
(*b*) Paddy separator
(*c*) Husk separator
(*d*) Rubber roll sheller

461. The machines used for separating of broken rice from head rice is known as
(*a*) Rubber roll sheller
(*b*) Polisher
(*c*) Grader
(*d*) Whitener

462. The foreign materials in paddy processing can be grouped in
(*a*) Big impurities like stems, paddy ear heads, stone and iron particles
(*b*) Small impurities like weed seeds, dust, mud particles
(*c*) Equal size impurities like immature grain, stones
(*d*) All of the above

463. Parboiling of paddy is carried out by
(*a*) Soaking
(*b*) Steaming
(*c*) Drying
(*d*) All of them

464. One of the following is not a rice whitener
(*a*) Vertical whitening cone
(*b*) Horizontal abrasive whitener
(*c*) Jet pearler
(*d*) Horizontal polisher

465. The separation of brokens from head rice (grading) is not accomplished by
(*a*) Tray separator
(*b*) Rotating reel grader
(*c*) Rotating indented cylinder
(*d*) Oscillating grading sieves

466. Milling recovery of rice mill is calculated by
(*a*) Weight of head rice × 100 / weight of milled rice
(*b*) Weight of milled rice × 100 / Weight of head rice
(*c*) Weight of brokens × 100 / weight of paddy
(*d*) Weight of total rice × 100 / total weight of paddy

467. Milling efficiency (%) of rice mill is expressed as
(*a*) Coefficient of hulling × coefficient of wholeness of kernel × 100
(*b*) Weight of brown rice × 100 / weight of feed to machine
(*c*) Weight of brown head rice/weight of total brown rice
(*d*) Weight of total rice × 100 / total weight of paddy

468. Coefficient of hulling is formulated as
(*a*) Coefficient of hulling × coefficient of wholeness of kernel × 100
(*b*) Weight of brown rice × 100 / weight of feed to machine
(*c*) Weight of brown head rice/weight of total brown rice
(*d*) Weight of total rice × 100 / total weight of paddy

469. Coefficient of wholeness is expressed as
(*a*) Coefficient of hulling × coefficient of wholeness of kernel × 100
(*b*) Weight of brown rice × 100 / weight of feed to machine
(*c*) Weight of brown head rice/weight of total brown rice
(*d*) Weight of total rice × 100 / total weight of paddy

470. The peripheral speed of vertical whitening cone for whitening of rice should be about
(*a*) 5 m/s (*b*) 7 m/s
(*c*) 10 m/s (*d*) 13 m/s

471. The process used to remove the bran layer from brown rice by the procedure of
(*a*) Shearing and friction (*b*) Impact and cutting
(*c*) Abrasion and friction (*d*) Compression and friction

472. Recovery of dal varies from
(*a*) 40-45% (*b*) 50-60%
(*c*) 60-75% (*d*) 75-80%

473. Pulses are
(*a*) Monocotyledons (*b*) Dicotyledons
(*c*) Multicotyledons (*d*) Polycotyledons

474. The protein content in pulse is about ____________ than cereals.
(*a*) Equal (*b*) Double
(*c*) Triple (*d*) Five times

475. The protein efficiency ratio (PER) of pulses varies
(*a*) 0.1–0.22 (*b*) 0.3–0.41
(*c*) 0.7–2.1 (*d*) 2.9–3.11

476. The main objective of conditioning of pulses is
(*a*) to loose the husk (*b*) to facilitate its separation from the kernel
(*c*) reduce the milling losses (*d*) All of the above

477. Conditioning of grain can be achieved by
(*a*) Water treatment (*b*) Hydrothermal treatment
(*c*) Use of salt and chemicals (*d*) Thermal treatment
(*e*) All of the above

478. The pulse milling means
(*a*) Conditioning of pulse
(*b*) Dehusking and splitting
(*c*) Polishing and grading
(*d*) All of the above

479. The process by which to impart desirable shine and lusture to dal is called
(*a*) Conditioning
(*b*) Dehusking and splitting
(*c*) Grading
(*d*) Polishing

480. One person by hand operated stone chakki can be produced about ________ of dal in eight hours.
(*a*) 10-20 kg
(*b*) 15-25 kg
(*c*) 25-30 kg
(*d*) 30-40 kg

481. An average dal recovery of pigeon pea by wet milling method is reported as
(*a*) 50%
(*b*) 55%
(*c*) 65%
(*d*) 70%

482. The objective of red earth treatment to whole grain dal is
(*a*) To impart yellow colour
(*b*) To provide mild abrasiveness
(*c*) Help in the removal of husk from cotyledons
(*d*) All of the above

483. Which chemical used for preservation of wet paddy is
(*a*) Sodium chloride
(*b*) Calcium propionate
(*c*) Copper or ferrous sulphate
(*d*) Urea
(*e*) All of them

484. When wet paddy is treated with common salt, water is removed from the rice kernel by
(*a*) Diffusion
(*b*) Osmosis
(*c*) Reverse osmosis
(*d*) Osmotic dehydration

485. The method is particularly suitable for drying of small quantity of seed to prevent mixing of varieties and conserve strain purity and variability is known as
(*a*) Dielectric drying
(*b*) Vacuum drying
(*c*) Sack drying
(*d*) Convective drying

486. Natural air drying for in-bin or in-storage drying can not be used if the ambient relative humidity exceeds
(*a*) 60 %
(*b*) 70 %
(*c*) 80 %
(*d*) 90 %

487. The food grains containing moisture higher than ________ percent should not be dried with natural air.
(*a*) 10 %
(*b*) 18 %
(*c*) 20 %
(*d*) 24 %

488. For continuous flow dryer, thickness of grain layer exposed to the heated air is generally restricted to
(*a*) 10 cm (*b*) 15 cm
(*c*) 20 cm (*d*) 22 cm

489. Which one of the flow of drying air is preferred to the grain drying?
(*a*) Cross flow (*b*) Counter flow
(*c*) Co-current flow (*d*) All of the above

490. A hydrothermal treatment followed by drying before milling for the production of milled parboiled grain is called
(*a*) Parboiling (*b*) Soaking
(*c*) Steaming (*d*) None

491. The most important change takes place during parboiling in paddy are
(*a*) Gelatinization of starch
(*b*) Disintegration of protein bodies in the endosperm
(*c*) Destruction of antioxidants
(*d*) All of them

492. Processes in sequence applied in parboiling of paddy are
(*a*) Soaking, steaming and drying (*b*) Steaming, soaking and drying
(*c*) Drying, soaking and steaming (*d*) Drying, steaming and soaking

493. Steaming process during parboiling of grain helps in the
(*a*) Swelling of grain (*b*) Gelatinization of starch
(*c*) Reducing the moisture content (*d*) All

494. The process of water absorption by grain during parboiling is known as
(*a*) Soaking (*b*) Steeping
(*c*) Imbibation (*d*) All

495. Soaking of water in parboiling process to the grain is the result of
(*a*) Molecular absorption (*b*) Capillary absorption
(*c*) Hydration (*d*) All of them

496. The volume of paddy increases due to soaking. But the volume of soaked paddy is always ______ than the sum of the initial volume of paddy and the volume of water absorbed.
(*a*) Less (*b*) Higher
(*c*) Equal (*d*) Unknown

497. The moisture at which the second phase of soaking starts may be called
(*a*) Initial moisture content (*b*) Critical moisture content
(*c*) Equilibrium moisture content (*d*) Final moisture content

498. The parboiled paddy is required to be dried to a moisture content of ________ percent to obtain the desirable milling and storing properties.
(*a*) 8–10 % (*b*) 12–14 %
(*c*) 14–16 % (*d*) 10–12 %

499. Parboiled rice requires ________ to ________ times as much abrasive load as raw rice to obtain the same degree of polish.
(*a*) One to three (*b*) Two to three
(*c*) Sixth to eighth (*d*) Three to four

500. Swelling capacity of grain is expressed as the ratio of
(*a*) Final length to the initial length (*b*) Final mass to the initial mass
(*c*) Final volume to the initial volume (*d*) Both (*b*) and (*c*)

501. Expansion ratio is the ratio of
(*a*) The dimension of cooked and uncooked rice both along the length and width
(*b*) Length to width after cooking
(*c*) Width to length after cooking
(*d*) All of the above

502. Disclouration of rice during parboiling is mainly caused by
(*a*) High temperature (*b*) High moisture content
(*c*) Fermentation (*d*) Non-enzymatic maillard browning

503. The water content of paddy after soaking increases to about ________on dry basis.
(*a*) 10-20 % (*b*) 20-40 %
(*c*) 40-50 % (*d*) 50-55 %

504. Which traditional method of paddy parboiling is not originated in Bengal?
(*a*) Atapa (*b*) Balam
(*c*) Josh (*d*) Siddha

505. Which method of paddy parboiling is originated at Saharanpur (UP)?
(*a*) Sela (*b*) Josh
(*c*) Balam (*d*) Siddha

506. In which method of paddy parboiling, sand is used as drying medium?
(*a*) Atapa (*b*) Sela
(*c*) Siddha (*d*) Josh

507. Which improved method of paddy parboiling is originated in India?
(*a*) Sodium chromate method (*b*) Brine solution method
(*c*) Pressure parboiling method (*d*) RPEC method
(*e*) All of the above

508. Maek process of paddy parboiling was originated at
(*a*) UK (*b*) USA
(*c*) Italy (*d*) Surinam

509. Schule process of paddy parboiling is developed at
(*a*) Italy (*b*) India
(*c*) Germany (*d*) UK

510. Avorio process of parboiling of paddy is patented in
(*a*) Italy (*b*) Indonesia
(*c*) USA (*d*) Germany

511. Debranned, polished and cracked parboiled wheat is known as
(*a*) Sela (*b*) Bulgur
(*c*) Pasta (*d*) Milled wheat

512. Bulgur takes about _________ minutes to cook fully.
(*a*) 10–15 (*b*) 15–20
(*c*) 25–30 (*d*) 30–45

513. Soaking temperature of parboiling of wheat above _______ should not be used so as to minimize the bursting of wheat kernel and leaching losses as well.
(*a*) 40°C (*b*) 45°C
(*c*) 70°C (*d*) 75°C

514. How much percentage of moisture is added during steaming of soaked wheat containing 83 % percent moisture (db)?
(*a*) 1–3% (*b*) 3–5%
(*c*) 5–8% (*d*) 10–13%

515. From the stand point of drying time and quality of parboiled wheat, the drying air temperature should not be more than
(*a*) 45°C (*b*) 65°C
(*c*) 75°C (*d*) 85°C

516. Production of bulgur is done by the method of
(*a*) Batch type (*b*) Preheating treatment process
(*c*) Continuous process (*d*) Chemical lye peeling process
(*e*) All of them

517. To increase the milling quality of the food grains or to improve the quality and quantity of their end products or to facilitate milling operations for desired products, food grains are some times subjected to hydrothermal treatment prior to milling is called
(*a*) Hysteresis (*b*) Reverse osmosis
(*c*) Conditioning (*d*) Product development

518. The aerodynamic properties of a particle depends on the
(*a*) Shape
(*b*) Dimension and weight of the particles
(*c*) The state and position of the particles with respect to the air current
(*d*) All of them

519. The husk of grain is removed by the action of
(*a*) Compression and shear
(*b*) Abrasion and friction
(*c*) Impact and friction
(*d*) All of them

520. Concave type husking machine and rubber roll husker are designed on the basis of
(*a*) Compression and shear
(*b*) Impact and friction
(*c*) Abrasion and friction
(*d*) None of them

521. Centrifugal type paddy sheller works on the principle of
(*a*) Abrasion and friction
(*b*) Impact and friction
(*c*) Compression and shear
(*d*) None of them

522. The minimum clearance between the roll and concave must be ________ than the dimensions of the grain kernel; otherwise the kernel will be crushed.
(*a*) Less
(*b*) Greater
(*c*) Equal
(*d*) None

523. The clearance between rubber roll sheller is ______ than the mean diameter of the paddy.
(*a*) Smaller
(*b*) Greater
(*c*) Equal
(*d*) None

524. Length of husking zone (l_h) for rubber roll sheller can be expressed as follows if d = diameter of rolls (cm), b = size of grain (cm) and C = clearance between rolls (cm).
(*a*) $l_h = (2\pi d/360).\cos^{-1}(d + C/d + b)$
(*b*) $l_h = (2\pi d/360).\cos^{-1}(d + b/d + c)$
(*c*) $l_h = (2\pi d/360).\cos^{-1}(dC/d + b)$
(*d*) $l_h = (2\pi d/360).\cos^{-1}(d + C/db)$

525. The length of the husking zone (l_h) for rubber roll sheller depends on the
(*a*) Diameter of the rolls
(*b*) Thickness of grain to be husked
(*c*) Linear distance between the centres of the rolls
(*d*) All of these

526. In rubber roll sheller, the size of the grain (b) and the clearance (c) are fixed, then the length of husking zone (l_h) increases with ________ in diameter of the roll.
(*a*) Decrease
(*b*) Increase
(*c*) No change
(*d*) Unknown

527. In under runner disc husker, the effective width of the abrasive coating should not be more than ________ of the stone diameter.
(*a*) $1/2^{th}$
(*b*) $1/3^{rd}$
(*c*) $1/4^{th}$
(*d*) $1/6^{th}$

528. The coefficient of hulling ($E_{hulling}$) is expressed as if n_1 = amount of grain before hulling, and n_2 = amount of unhulled grain after hulling.
(*a*) $(n_1 - n_2) / n_1$
(*b*) $(n_2 - n_1) / n_2$
(*c*) $n_1 n_2 / n_2$
(*d*) $n_1 n_2 / n_1$

529. How much power consumption for grinding varies _________ percent of the total power required for all operation.
(*a*) 10-20%
(*b*) 20-35%
(*c*) 35-40%
(*d*) 50-80%

530. Which grain parameter is affected the characteristics of grinding operation?
(*a*) Type and variety of cereal grain
(*b*) Moisture content and mechanical properties
(*c*) Extent of hydrothermal treatment given to the grain
(*d*) All of these

531. Hardness of rice endosperm is as
(*a*) 11.5 to 14.5 kg/mm^2
(*b*) 10.5 to 11.5 kg/mm^2
(*c*) 6.0 to 9.0 kg/mm^2
(*d*) 20.0 to 30.15 kg/mm^2

532. Hardness of barley kernel varies from
(*a*) 11.5 to 14.5 kg/mm^2
(*b*) 10.5 to 11.5 kg/mm^2
(*c*) 6.0 to 9.0 kg/mm^2
(*d*) 20.0 to 30.15 kg/mm^2

533. Degree of grinding is expressed as
(*a*) S_a/S_b
(*b*) S_b/S_a
(*c*) $S_a - S_b$
(*d*) $S_b - S_a$

534. The difference between the percentage of undersized particles (C_1) in the final ground product and the percentage of undersized particles (*I*) in the initial feed entering the grinder is also known as
(*a*) Overall extraction
(*b*) Effectiveness of grinding
(*c*) Coefficient of grinding
(*d*) Grinding efficiency

535. The factor affecting the effectiveness of roller mills is /are
(*a*) Clearance between rolls
(*b*) Geometrical parameter of rolls
(*c*) Kinematic parameter of rolls
(*d*) Capacity of roller mills and power consumption for their operation
(*e*) All of them

536. The hydrothermal treatment can be used for
(*a*) Improving shelling efficiency
(*b*) Nutritional quality and milling quality of paddy
(*c*) Facilitating degermination and dehulling of corn and wheat
(*d*) All of them

537. Which type of changes occurs in grain due to the hydrothermal treatments?
(*a*) Physico-thermal (*b*) Physico-chemical
(*c*) Biochemical (*d*) All

538. Physico-thermal properties of grain is classified into
(*a*) Strength or hardness (*b*) Density and hardness
(*c*) Hysteresis (*d*) Thermal properties
e) All of them

539. Which is not the thermal property of the grain?
(*a*) Specific heat (*b*) Specific gravity
(*c*) Thermal conductivity (*d*) Thermal diffusivity

540. Thermal properties like specific heat, thermal conductivity and thermal diffusivity of grain increase with increase in
(*a*) Temperature (*b*) Moisture content
(*c*) Relative humidity (*d*) Both (*a*) and (*b*)

541. The residual oil percentage in the deoiled cake obtained from mechanical expression is about
(*a*) 1-2 % (*b*) 2-4 %
(*c*) 4-5 % (*d*) 6-8 %

542. The deoiled cake obtained from solvent extraction has the residual oil content of about
(*a*) 1% (*b*) 2%
(*c*) 4% (*d*) 6%

543. Oil content in raw bran varies between
(*a*) 2-4% (*b*) 4-8%
(*c*) 10-12% (*d*) 12-18%

544. Oil content in parboiled rice bran is about
(*a*) 5-10% (*b*) 12-18%
(*c*) 20-28% (*d*) 30-40%

545. After extraction of oil from raw and parboiled bran, the deoiled cake contains about ________ % oil only.
(*a*) 1-3% (*b*) 3-5%
(*c*) 6-8% (*d*) 8-11%

546. Immediately after milling rice, the FFA content in bran is normally bellow
(*a*) 1% (*b*) 3%
(*c*) 8% (*d*) 10%

547. After milling of rice, the rate of increase of FFA in bran may be as high as ________ per hour under favourable conditions.

(*a*) 0.5% (*b*) 1.0%
(*c*) 2.5% (*d*) 4.0%

548. The main problem in processing of rice bran oil for edible purpose and defatted bran for food and feed are

(*a*) Rapid hydrolysis of oil into FFA and glycerol
(*b*) Oxidation of fat (*c*) Flavour reversion
(*d*) High refining losses (*e*) All of the above

549. The main advantages of stabilization of rice bran are

(*a*) It impart hardening effect to the bran for better extrability
(*b*) It increases the particles size and reduces the problem of fines and filtration
(*c*) It increases the bulk density and reduces the hardening problem
(*d*) All of them

550. The factors affecting the rate of formation of FFA in rice bran during storage are

(*a*) Storage temperature and RH (*b*) Moisture content of bran
(*c*) Variety, type and particle size of bran
(*d*) Contamination of bran with microflora or insects
(*e*) All of them

551. In the refining process of bran oil, neutralization process is used for

(*a*) Removing hard wax (*b*) Removal of FFA
(*c*) Removal of colouring matter (*d*) Removal of odours

552. The role of winterization in the rice bran oil refining process is

(*a*) Removal of hard wax, gum, mucilage (*b*) Removal of soft wax
(*c*) Removal of odours matter and unsaponifiable matter
(*d*) Removal of colouring matter

553. A cyclone separator is used for

(*a*) Particles from liquids (*b*) Liquid droplets from gases
(*c*) Fine particles from solids (*d*) All of the above

554. If d_1 and d_2 are average initial and final sizes of material and K is a constant, then E, energy requirement for size reduction of the unit mass is given by

(*a*) $E = K [(d_1 - d_2)/ (d_1.d_2)]$ (*b*) $E = K [(d_1.d_2) / (d_1 - d_2)]$
(*c*) $E = K [(d_2 - d_1)/ (d_1.d_2)]$ (*d*) $E = K [(d_1.d_2) / (d_2 - d_1)]$

555. In a ball mill if '*R*' is the radius of mill, '*g*' is acceleration due to gravity and '*r*' is the radius of ball, then the critical speed is given by

(*a*) $Yc = (1/2\pi)\ \sqrt{g/(R-r)}$ (*b*) $Yc = 2\pi\ \sqrt{g/(R-r)}$

(*c*) $Yc = (1/2p)\ \sqrt{(R-r)/g}$ (*d*) $Yc = 2\pi\ \sqrt{(R-r)/g}$

556. A machine which removes surface moisture from grain after washing by centrifugal force and impact is called

(*a*) Hammer mill (*b*) Ball mill

(*c*) Whizzer (*d*) Degermer

ANSWER KEY (CEREALS, PULSES, AND OILSEED PROCESSING)

1.C	2.D	3.A	4.D	5.A	6.A	7.A	8.A	9.C	10.B
11.C	12.A	13.C	14.A	15.A	16.A	17.C	18.A	19.D	20.A
21.A	22.B	23.B	24.B	25.A	26.A	27.A	28.C	29.C	30.D
31.A	32.A	33.A	34.C	35.C	36.C	37.C	38.B	39.C	40.D
41.D	42.B	43.C	44.A	45.A	46.A	47.D	48.A	49.A	50.D
51.A	52.A	53.A	54.D	55.B	56.C	57.C	58.B	59.B	60.D
61.A	62.C	63.D	64.B	65.D	66.B	67.C	68.B	69.B	70.A
71.C	72.B	73.A	74.B	75.D	76.C	77.C	78.D	79.A	80.B
81.A	82.C	83.C	84.A	85.B	86.C	87.A	88.A	89.D	90.B
91.A	92.B	93.C	94.B	95.B	96.A	97.D	98.A	99.C	100.C
101.A	102.D	103.C	104.C	105.B	106.B	107.C	108.A	109.A	110.D
111.A	112.B	113.C	114.B	115.C	116. C	117.D	118.A	119.D	120.C
121.C	122.C	123.C	124.C	125.C	126.D	127.C	128.D	129.D	130.A
131.D	132.C	133.D	134.C	135.A	136.B	137.D	138.D	139.C	140.C
141.C	142.D	143.B	144.D	145.C	146.A	147.C	148.D	149.C	150.C
151.A	152.C	153.A	154.D	155.A	156.B	157.C	158.B	159.C	160.B
161.D	162.C	163.A	164.A	165.B	166.C	167.C	168.B	169.D	170.D
171.B	172.A	173.B	174.D	175.C	176.C	177.D	178.C	179.A	180.A
181.B	182.B	183.D	184.D	185.D	186.B	187.D	188.B	189.A	190.B
191.C	192.B	193.D	194.B	195.B	196.B	197.A	198.A	199.B	200.D
201.C	202.B	203.C	204.A	205.A	206.D	207.B	208.D	209.B	210.A
211.D	212.A	213.A	214.C	215.A	216.B	217.C	218.B	219.B	220.C
221.B	222.A	223.B	224.B	225.A	226.C	227.C	228.D	229.A	230.B
231.A	232.A	233.C	234.D	235.D	236.C	237.C	238.A	239.A	240.B
241.C	242.C	243.C	244.B	245.C	246.C	247.B	248.A	249.D	250.C
251.A	252.B	253.D	254.A	255.C	256.A	257.D	258.B	259.A	260.A
261.B	262.A	263.C	264.B	265.B	266.A	267.B	268.B	269.B	270.B
271.E	272.C	273.B	274.C	275.A	276.B	277.D	278.D	279.B	280.A
281.C	282.D	283.A	284.A	285.D	286.C	287.B	288.C	289.D	290.A

291.A	292.D	293.A	294.A	295.D	296.C	297.B	298.D	299.C	300.C
301.A	302.D	303.A	304.D	305.B	306.A	307.C	308.A	309.C	310.D
311.C	312.B	313.A	314.B	315.D	316.D	317.D	318.D	319.A	320.B
321.B	322.D	323.D	324.A	325.A	326.B	327.D	328.D	329.D	330.D
331.D	332.D	333.A	334.C	335.B	336.A	337.A	338.D	339.A	340.A
341.A	342.C	343.C	344.C	345.A	346.C	347.D	348.D	349.B	350.A
351.D	352.A	353.A	354.C	355.D	356.B	357.C	358.B	359.B	360.C
361.C	362.C	363.C	364.A	365.B	366.A	367.B	368.A	369.D	370.A
371.C	372.B	373.B	374.B	375.B	376.B	377.D	378.B	379.C	380.A
381.D	382.A	383.A	384.B	385.A	386.C	387.D	388.C	389.B	390.C
391.D	392.A	393.B	394.A	395.A	396.C	397.B	398.A	399.A	400.C
401.D	402.B	403.C	404.A	405.C	406.C	407.C	408.C	409.C	410.B
411.D	412.A	413.C	414.A	415.C	416.C	417.D	418.B	419.B	420.A
421.C	422.B	423.B	424.C	425.C	426.A	427.B	428.B	429.D	430.C
431.A	432.A	433.C	434.A	435.C	436.C	437.D	438.C	439.B	440.C
441.B	442.C	443.A	444.B	445.D	446.A	447.A	448.D	449.A	450.D
451.C	452.A	453.D	454.D	455.A	456.B	457.B	458.A	459.D	460.A
461.C	462.D	463.D	464.D	465.A	466.D	467.A	468.B	469.C	470.D
471.C	472.C	473.B	474.B	475.C	476.D	477.E	478.D	479.D	480.D
481.D	482.D	483.E	484.B	485.C	486.B	487.C	488.C	489.A	490.A
491.D	492.A	493.B	494.D	495.D	496.A	497.B	498.C	499.D	500.A
501.A	502.D	503.D	504.C	505.A	506.B	507.E	508.B	509.C	510.A
511.B	512.C	513.D	514.B	515.C	516.E	517.C	518.D	519.D	520.A
521.B	522.B	523.A	524.A	525.D	526.B	527.D	528.A	529.D	530.D
531.A	532.C	533.A	534.A	535.E	536.D	537.D	538.E	539.B	540.D
541.C	542.A	543.D	544.C	545.A	546.B	547.B	548.E	549.D	550.E
551.B	552.B	553.C	554.A	555.A	556.C				

Fruits and Vegetable Processing

1. Milk and fruit juices are deaerated before they are allowed to flow through pasteurizer. This is done by
   (*a*) Reducing fouling of pasteurizer
   (*b*) Increasing rate of heat transfer
   (*c*) Reducing oxidative deterioration
   (*d*) All of them
2. Jelly is ____________ food product.
   (*a*) Solid
   (*b*) Liquid
   (*c*) Semi-solid
   (*d*) Syrup
3. The end point of cooking of jelly/jam is tested by
   (*a*) Alcohol test
   (*b*) Temperature test
   (*c*) Flake test
   (*d*) both (*b*) and (*c*)
4. How many kinds of marmalade are prepared
   (*a*) One
   (*b*) Two
   (*c*) Three
   (*d*) Four
5. Sweetened juice contains the ______ percent juice.
   (*a*) 85
   (*b*) 20
   (*c*) 25
   (*d*) 45
6. RTS contains the acid content (%) about
   (*a*) 1.5
   (*b*) 0.3
   (*c*) 2.5
   (*d*) 1.0
7. The percentage of fresh juice in fruit squash is added about
   (*a*) 15-20%
   (*b*) 20-25%
   (*c*) 30-35%
   (*d*) 40-45%
8. Iron and copper equipments should not be used in the preparation of
   (*a*) Chutney
   (*b*) Sauce
   (*c*) Ketchup
   (*d*) All

9. Blanching refers to
(*a*) Inactivate the enzyme
(*b*) Cooking the food
(*c*) Sterilization of food
(*d*) Drying of food

10. Invert sugar in Fruit jam should not be more than
(*a*) 40 %
(*b*) 50 %
(*c*) 60 %
(*d*) 70 %

11. Premature setting of jelly is caused by
(*a*) Low TSS
(*b*) High TSS
(*c*) Low Acidity
(*d*) Low pectin

12. Best quality jelly is made from the __________ fruit.
(*a*) Apricot
(*b*) Guava
(*c*) Strawberry
(*d*) Mango

13. A clear sugar solution which has artificial flavour is known as
(*a*) Syrup
(*b*) Cordial
(*c*) Sherbet
(*d*) Squash

14. A good quality chutney should be
(*a*) Palatable
(*b*) Appetizer.
(*c*) Digestive
(*d*) All

15. In sauce/ketch up, tamarind pulp and pomegranate seed powder is used to increase
(*a*) Acidity
(*b*) Digestion
(*c*) Palatability
(*d*) Sweetness

16. Water soluble vitamins are
(*a*) *A*, *D*
(*b*) *E*, *K*
(*c*) *a* and *b*
(*d*) none

17. Oil and fat soluble vitamins are
(*a*) *A*, *D*
(*b*) *E*, *K*
(*c*) *a* and *b*
(*d*) none

18. The process preservation of fruits and vegetable with common salt and vinegar is called
(*a*) Pickling
(*b*) Fermentation
(*c*) Steeping
(*d*) Asepsis

19. Lactic acid forming bacteria is beneficial and grown in
(*a*) Sauce
(*b*) Ketchup
(*c*) Jelly
(*d*) Pickle

20. Mango pickles are kept for proper curing in sun for
(*a*) One week (*b*) Two week
(*c*) Two days (*d*) One day

21. The by product of apple is
(*a*) Rags (*b*) Pomace
(*c*) Peel (*d*) Stones

22. Rags is the by product of
(*a*) Apple (*b*) Citrus fruit
(*c*) Apricot (*d*) Grape

23. Rags (by product of citrus fruits) is used for preparation of
(*a*) Candy (*b*) Vinegar
(*c*) Jam (*d*) Marmalade

24. For the prevention of black neck, sauce should be filled hot in bottle at the temperature of
(*a*) 80 °C (*b*) 85 °C
(*c*) 90 °C (*d*) 75 °C

25. Soy sauce is known by the following name in China
(*a*) Ching-yu (*b*) Kecap
(*c*) Kanjang (*d*) Shoyu

26. Soy sauce is made by fermentation of a combination of
(*a*) Soybean and wheat (*b*) Rice and pulse
(*c*) Milk and sugar (*d*) Soybean and fruits

27. A paste like product made by fermenting cereals, soybeans and salt with molds, yeasts and bacteria is famous in Japan is known as
(*a*) Tempeh (*b*) Miso
(*c*) Kefir (*d*) Koumiss

28. Acidophilus milk is produced by fermentation of milk by
(*a*) *Lactobacillus acidophilus* (*b*) *Saccharomyces rouxii*
(*c*) *Torulopsis* yeasts (*d*) *Lactobacillus caucasicus*

29. Kefir is the main fermented food product of
(*a*) USA (*b*) Russia
(*c*) China (*d*) Japan

30. Which microorganisms is not used in the preparation of cultured butter milk
(*a*) *Streptoccus lactis* (*b*) *Streptoccus cremoris*,
(*c*) *S. lactis subsp. Diacelylactis* (*d*) *Saccharomyces rouxii*

31. Jam contains ______ per cent acid and invert sugar should not be more than ____ per cent.
    (*a*) 0.2-0.3 and 40 (*b*) 0.5-0.6 and 40
    (*c*) 0.1-0.3 and 45 (*d*) 0.5-0.6 and 45

32. Protopectin is converted into pectin by the enzyme
    (*a*) Protopectinase (*b*) Pectinase
    (*c*) Lipase (*d*) Catalase

33. Sugar content in jelly is
    (*a*) 75 % (*b*) 70 %
    (*c*) 72 % (*d*) 65 %

34. Large quantity of acid in jelly may cause
    (*a*) Syneresis (*b*) Crystallization
    (*c*) Gelation (*d*) Premature setting

35. When sugar is boiled with an acid it is hydrolyzed into
    (*a*) Dextrose (*b*) Fructose
    (*c*) Above both (*d*) None

36. Temperature test is not used to judge the end-point of
    (*a*) Jam (*b*) Jelly
    (*c*) Kefir (*d*) Marmalade

37. Sauce is __________ than ketchup.
    (*a*) Thinner (*b*) Thicker
    (*c*) Similar (*d*) None of these

38. Formation of blackening in the neck of sauce bottles is known
    (*a*) Round neck (*b*) Black neck
    (*c*) Gray neck (*d*) All of these

39. How many carbon is constituted in vegetable oils?
    (*a*) 12 & 14 (*b*) 16 & 18
    (*c*) 18 & 20 (*d*) None of above

40. Vitamin A, D, E and K are soluble in
    (*a*) Water (*b*) Fat
    (*c*) Alcohol (*d*) Paraffin

41. Rich source of fatty acid is
    (*a*) Milk (*b*) Cereals
    (*c*) Pulse (*d*) Dry vegetables

42. Which items contain varying quantity of alcohol?
(*a*) Fruit juice (*b*) Fermented beverage
(*c*) Fruit juice powder (*d*) Syrup

43. The grape juice can be preserved by adding of
(*a*) Potassium meta bisulphate (*b*) Sodium benzoate
(*c*) Sulpher dioxide (*d*) Sodium chloride

44. The ratio of sweet orange and khatta orange in jelly marmalade is subjected to
(*a*) 2:1 (*b*) 1:2
(*c*) 1:1 (*d*) 3:1

45. The amount of sugar per kg pulp in the preparation apple jam is subjected to
(*a*) 0.75 kg (*b*) 1.00 kg
(*c*) 1.25 kg (*d*) 1.50 kg

46. In pickles, yeast and molds can be easily grown in the ____________.
(*a*) Basic (*b*) Alkali
(*c*) Acidic (*d*) Salty

47. Lactic acid bacteria are most active at the temperature of
(*a*) 20 ºC (*b*) 23 ºC
(*c*) 30 ºC (*d*) 40 ºC

48. Microbial spoilage in pickles can be prevent at salt concentration of
(*a*) 15 % (*b*) 10 %
(*c*) 5 % (*d*) 3 %

49. Aroma of over ripe fruit of banana is due to
(*a*) Isopenttanol (*b*) Allicin
(*c*) Isothiocyanate (*d*) capsaicin

50. Richest source of vitamin C among the fruits
(*a*) Lemon (*b*) Barbados cherry
(*c*) Aonla (*d*) Guava

51. The ______________ fruit is also known as butter fruit.
(*a*) Banana (*b*) Mango
(*c*) Avacado (*d*) Papaya

52. Zero energy chamber operates on the principle of
(*a*) Evaporative cooling (*b*) Charle's law
(*c*) Second law of thermodynamics (*d*) Boyle's law

53. Garlic contain amino acid
(*a*) Allicin (*b*) Alliin
(*c*) Allinage (*d*) Diallyl disulphide

54. Onion Bulbs are stored well at
(*a*) 20-25 °C (*b*) 15-20 °C
(*c*) 10-15 °C (*d*) 0-10 °C

55. Pungency in onion is due to the presence of
(*a*) Capacin (*b*) Diallyl disulphide
(*c*) Isothiocyanate (*d*) Allyl propyl disulphide

56. Potato contains solanine per 100 gram of fresh weight is considered safe for human consumption
(*a*) 20 mg (*b*) 20 g
(*c*) 12 mg (*d*) 8 g

57. Pungency in reddish is due to volatile compound of
(*a*) Diallyl disulphide (*b*) Isothiocyanate
(*c*) Euric acid (*d*) Acetic acid

58. Lime is good source of vitamin
(*a*) A (*b*) K
(*c*) E (*d*) C

59. Cluster bean contains mucilaginous substance is
(*a*) Glutamine (*b*) Asparagine
(*c*) Mannogalacton (*d*) Galactoronic acid

60. Red colour in tomato is due to pigment
(*a*) Capsaicin (*b*) Capsanthin
(*c*) Isothiocyanate (*d*) Lycopene

61. Balanced diet includes at least gram of vegetables
(*a*) 300 (*b*) 200
(*c*) 100 (*d*) 400

62. Which fruit is richest source of riboflavin?
(*a*) Ber (*b*) Bael
(*c*) Orange (*d*) Mango

63. Flavour in cabbage leaves is due to the glycoside
(*a*) Oleoresin (*b*) Euric acid
(*c*) Sinigrin (*d*) Cucubitacin

64. Coagulating vitamin is
(*a*) A
(*b*) B
(*c*) C
(*d*) K

65. Exhausting temperature of canned fruit and vegetables is
(*a*) 80-82 ºC
(*b*) 90-92 ºC
(*c*) 100-102 ºC
(*d*) 110-112 ºC

66. The ______________ is richest source of iron.
(*a*) Green pea
(*b*) Spinach
(*c*) Parsley
(*d*) Celery

67. Papaya is rich source of Vitamin
(*a*) E
(*b*) B
(*c*) A
(*d*) C

68. Storage temperature to extend shelf life of banana is
(*a*) –10 ºC
(*b*) 2-5 ºC
(*c*) 8-10 ºC
(*d*) 12-15 ºC

69. Fruit for making jelly should be rich in
(*a*) Pectin and acids
(*b*) Acids and proteins
(*c*) Protein and sugars
(*d*) Sugars and acids

70. Controlled atmosphere storage (CAS) of fruits and vegetables is done to
(*a*) Maintain higher concentration of ethylene than the normal atmosphere
(*b*) Maintain higher concentration of CO_2 than the normal atmosphere
(*c*) Maintain higher concentration of N_2 than the normal atmosphere
(*d*) None of these

71. The anti-sterility vitamin is
(*a*) A
(*b*) B
(*c*) E
(*d*) D

72. Fruits on the basis of respiration can be classified as
(*a*) Non climacteric
(*b*) Parthenocarpic
(*c*) Climacteric
(*d*) Climacteric and Non climacteric

73. Sweetest sugar in fruit is
(*a*) Fructose
(*b*) Sucrose
(*c*) Glucose
(*d*) Galactose

74. The yellow pigment in papaya is
(*a*) Xanthophylls
(*b*) Anthocyanin
(*c*) Caricaxanthin
(*d*) Carotene

75. Fruits and vegetables stored in a cold chamber enhance storage life because
(*a*) There is an increase in humidity
(*b*) CO_2 concentration in the environment is increased
(*c*) Exposure to sun light is prevented
(*d*) Rate of respiration is decreased

76. Sodium benzoate is used for preservation of fruit juices in the range of
(*a*) 0.01–0.5 % (*b*) 0.06–0.10 %
(*c*) 1.1–1.5 % (*d*) 1.6–2.0 %

77. Which one of the following fruits is a rich source of fat
(*a*) Cashew nut (*b*) Mango
(*c*) Walnut (*d*) Guava

78. Rich source of protein is
(*a*) Banana (*b*) Almond
(*c*) Mango (*d*) Cashew nut

79. Agriculture and processed products development authority (APEDA) was came in existence in the year
(*a*) 1975 (*b*) 1985
(*c*) 1995 (*d*) 2005

80. The yellow colour in onion is due to the pigment of
(*a*) Lycopene (*b*) Caraotene
(*c*) Quereitin (*d*) Anthocyanin

81. Wax coating treatment enhance the shelf life of fruits because it block the
(*a*) Ripening process (*b*) Respiration
(*c*) Evaporation (*d*) Transpiration

82. Yellow coloured fruits and vegetables are rich source of
(*a*) Vitamin A (*b*) Vitamin B
(*c*) Vitamin C (*d*) Vitamin D

83. Fruit not suitable for jam making is
(*a*) Banana (*b*) Lemon
(*c*) Aonla (*d*) Mango

84. Fruits are blanched before canning at temp.
(*a*) 70 °C (*b*) 90 °C
(*c*) 80 °C (*d*) 100 °C

85. Refractometer is used to determine the
(*a*) Minerals (*b*) Vitamins
(*c*) Acidity (*d*) TSS

86. Vegetables are canned in the
(*a*) Brine (*b*) Syrup
(*c*) Vinegar (*d*) Distilled water

87. Which fruit is a rich source of iron?
(*a*) Datepalm (*b*) Karonda
(*c*) Appie (*d*) None of these

88. Pre^^rvative used for preparation of tomato ketchup is
(*a*) Citric acid (*b*) KMS
(*c*) Sodium benzoate (*d*) Sodium bisulphate

89. Redness in apple is due to
(*a*) Carotene (*b*) Anthocyanin
(*c*) Xanthophylls (*d*) Lycopene

90. CIPHET is situated at
(*a*) Bhopal (*b*) Ludhiana
(*c*) Mysore (*d*) Cochin

91. CIAE is located at
(*a*) Nagpur (*b*) Ludhiana
(*c*) Bhopal (*d*) Budhani

92. Which of the following is a non-reducing carbohydrate?
(*a*) Maltose (*b*) Lactose
(*c*) Glucose (*d*) Sucrose

93. Fat soluble vitamins are
(*a*) A, D, E and K (*b*) A and D
(*c*) A, D and C (*d*) B and C

94. Cryo-preservation is done in liquid nitrogen at a temperature of
(*a*) –196 °C (*b*) 0 °C
(*c*) 200 °C (*d*) 50 °C

95. Recommended temperature for storage of potatoes in cold storage is
(*a*) 0-2 °C (*b*) 3-7 °C
(*c*) 8-10 °C (*d*) 10-15 °C

96. To avoid bruising, the onion should not be stored than
(*a*) 4 m (*b*) 10 m
(*c*) 6 m (*d*) 8 m

97. The acid percentage in jam varies
(*a*) 2.0% (*b*) 0.5-0.6 %
(*c*) 65 % (*d*) 30-50 %

98. Moulds which causes of microbial spoilage in jam when humidity is
(*a*) Less than 90 % (*b*) More than 90%
(*c*) Equal to 90 % (*d*) None of these

99. The jam becomes sticky or gummy because of
(*a*) More citric acid (*b*) Low TSS
(*c*) More pectin content (*d*) High TSS

100. Which of the following is best suitable for jelly making?
(*a*) Apricot (*b*) Apple
(*c*) Strawberry (*d*) Pine apple

101. The effective concentration of sugar in jelly is
(*a*) 65 % (*b*) 80 %
(*c*) 40 % (*d*) 75 %

102. The pectin content for producing good jelly is
(*a*) 1% (*b*) 2 %
(*c*) 20 % (*d*) 10 %

103. The optimum pH for jelly containing 1 % pectin and 65 % TSS should be
(*a*) 3.0 (*b*) 3.4
(*c*) 3.2 (*d*) 3.1

104. The suitable temperature for judging the end point or proper setting of jelly containing 65 %TSS is
(*a*) 105 °C (*b*) 85 °C
(*c*) 82 °C (*d*) 65°C

105. In jelly making, which one of the following does not lead to syneresis:
(*a*) Low concentration of sugar (*b*) Insufficient pectin
(*c*) Less acid (*d*) Fermentation

106. Which one of the following, there is not a problem of jam making?
(*a*) Syneresis (*b*) Crystallization
(*c*) Premature setting (*d*) Microbial spoilage

107. Causes of browning and blackness formation in sauce /ketch and chutney is due to the presence
(*a*) Head less clove
(*b*) Flower head of clove
(*c*) All spice except clove
(*d*) Vinegar only

108. Tannin is responsible for browning and blackness formation in sauce /ketch up and chutney is found in
(*a*) Clove
(*b*) Red chili
(*c*) Cardamom
(*d*) Cinnamon

109. Addition of 100 ppm SO_2 or 100 mg ascorbic acid in sauce /ketch up or chutney to prevent the
(*a*) Black neck ring formation and Browning
(*b*) Blackness
(*c*) Browning
(*d*) Fermentation

110. Fermented beverage undergone in alcoholic fermentation with help of
(*a*) Yeast
(*b*) Mold
(*c*) Bacteria
(*d*) Virus

111. The fruit beverage which contains 20 % juice/pulp, 15 % TSS and 0.3 % acid is called
(*a*) Squash
(*b*) Cordial
(*c*) Syrup
(*d*) Nectar

112. Which one of the following beverage prepared by dissolving 200 g sugar and 1.0 g citric acid in 800 ml water ?
(*a*) Squash
(*b*) Syrup
(*c*) Cordial
(*d*) Sherbet

113. Which fruits are used in the preparation of cordial?
(*a*) Apple and guava
(*b*) Mango and apple
(*c*) Bael and papaya
(*d*) Lime and lemon

114. The problem likes cloudiness and foggy in jelly happens by
(*a*) Use of immature fruits
(*b*) Over cooking
(*c*) Use of non-clarified juice or extract
(*d*) All of the above

115. The kefir is foamy and fizzy in appearance due to its
(*a*) CO_2 content
(*b*) SO_2 content
(*c*) O_2 content
(*d*) O_3 content

116. Aroma in coffee is created due to
(*a*) Fermentation
(*b*) Drying
(*c*) Roasting
(*d*) None

117. In litchi, which portion is edible?
(*a*) Endocarp (*b*) Epicarp
(*c*) Mesocarp (*d*) Rind

118. Minimum TSS required for Jam is
(*a*) 68 % (*b*) 70 %
(*c*) 85 % (*d*) 80 %

119. Vinegar from coconut today is prepared by
(*a*) Yeast (*b*) Bacteria
(*c*) Fermentation (*d*) Fungi

120. The ideal relative humidity (%) for onion storage is
(*a*) 60-65 (*b*) 70-75
(*c*) 80-85 (*d*) 90-95

121. The 'Champaigne' is a type of
(*a*) Juice (*b*) Wine
(*c*) Beverage (*d*) Pod

122. Which are the following used for preparing best Jelly is
(*a*) Jamun (*b*) Guava
(*c*) Mango (*d*) Aonla

123. The Jelly in which fruits pieces are suspended is called as
(*a*) Jam (*b*) Jelly
(*c*) Marmalade (*d*) Candy

124. The sparkling clear liquid free from all suspended solids is called as
(*a*) Concentrate (*b*) Cordial
(*c*) Squash (*d*) Beverage

125. Jelly with high sugar and low acid is called as
(*a*) Weeping jelly (*b*) Crystallized jelly
(*c*) Soft jelly (*d*) Cloudy jelly

126. The preservative recommended by FPO are
(*a*) SO_2 and sodium benzoate (*b*) KMS and KNO_3
(*c*) Acetic acid (*d*) None

127. Chilies are good source of
(*a*) Protein (*b*) Fat
(*c*) Carbohydrates (*d*) Vitamin A & C

128. Guava is a rich source of
(*a*) Carbohydrates
(*b*) Vitamin C
(*c*) Vitamin A
(*d*) β-carotene

129. Dry fruits like cashew and almond are richest source of
(*a*) Vitamin B
(*b*) Fat
(*c*) Vitamin K
(*d*) Carbohydrates

130. Ripe papaya fruits contain
(*a*) Vitamin A
(*b*) Vitamin B
(*c*) Vitamin C
(*d*) Vitamin D

131. Vitamin C is maximum in
(*a*) Guava
(*b*) Aonla
(*c*) Papaya
(*d*) Mango

132. Carrot is a rich source of
(*a*) Vitamin A
(*b*) Vitamin B
(*c*) Vitamin C
(*d*) Vitamin D

133. Pink colour of onion is due to
(*a*) Qucertin
(*b*) Anthocyanin
(*c*) Carotene
(*d*) None

134. Preservative used in tomato ketchup is
(*a*) Sodium benzoate
(*b*) SO_2
(*c*) Both
(*d*) None

135. Low temperature preservation prevent growth of microorganisms due to
(*a*) Reduction in growth of microorganisms
(*b*) Increase in growth of microorganisms
(*c*) Killing of microorganisms
(*d*) All of the above

136. Pectin test is done by
(*a*) Jelmeter
(*b*) Alcohol
(*c*) Acid
(*d*) None

137. Crystal formation in Jelly is due to
(*a*) Excess of sugar
(*b*) Over cooking
(*c*) Both
(*d*) None

138. Treatments of fruits and vegetables by slight temperature for small time is known as
(*a*) Sterilization
(*b*) Blanching
(*c*) Pasteurization
(*d*) None

139. After preservation, nutrients, vitamins and minerals are
(*a*) Less than before preservation
(*b*) More than before preservation
(*c*) Constant
(*d*) None

140. Good storage require
(*a*) High temperature and high humidity
(*b*) Low temperature and high humidity
(*c*) High temperature and low humidity
(*d*) None

141. Good source of carbohydrate, vitamins, fat, minerals and protein is
(*a*) Vegetables
(*b*) Egg
(*c*) Milk
(*d*) None

142. For packing of spinach, cauliflower and sweet potato which type of can is used
(*a*) AR cans
(*b*) SR cans
(*c*) Plain
(*d*) All

143. Papaya contain __________ unit of vitamin A.
(*a*) 2500 IU
(*b*) 2020 IU
(*c*) 3000 IU
(*d*) 4000 IU

144. Standard error formula is
(*a*) σ/n
(*b*) σ^2/n
(*c*) $\sigma/\sqrt{n}$
(*d*) none

145. Pigment in Jamun is due to
(*a*) Anthocyanin
(*b*) Carotene
(*c*) Xanthophylls
(*d*) None

146. Potato is the rich source of
(*a*) Fat
(*b*) Starch
(*c*) Vitamin
(*d*) Minerals

147. Potato contain ______________ percentage of protein.
(*a*) 1.8
(*b*) 2.2
(*c*) 1.4
(*d*) 3.2

148. Which product is made by yeast fermentation?
(*a*) Apple cider
(*b*) Mango pulp
(*c*) Citrus nectar
(*d*) All

149. Guava Jelly is preserved by
(*a*) Pectin
(*b*) Sugar
(*c*) Sodium benzoate
(*d*) None

150. Storage temperature of banana is
(*a*) 2-3 °C (*b*) 10-13 °C
(*c*) 4-5 °C (*d*) 15-20 °C

151. Storage temperature of cucumber is
(*a*) 0 °C (*b*) 4-5 °C
(*c*) 6-7 °C (*d*) 10 °C

152. Ripening fruits exist which gas is
(*a*) NH_3 (*b*) Ethylene
(*c*) Methane (*d*) All

153. Capacity of refrigeration is measured in the term of
(*a*) Gram (*b*) Ton
(*c*) Pascal (*d*) Lux

154. Bacteria are involved in production of
(*a*) Nectar (*b*) Vinegar
(*c*) Jam (*d*) Squash

155. Yeast is used in the production of
(*a*) Nectar (*b*) Cordial
(*c*) Syrup (*d*) Cider

156. Why does blanching in vegetables?
(*a*) microorganism removal (*b*) inactivate enzyme
(*c*) for softening the vegetables (*d*) all

157. Non-climacteric fruit is
(*a*) which ripe after the maturity (*b*) which ripe before the maturity
(*c*) which ripe at time of maturity (*d*) none

158. How much juice is required in making of fruit jelly?
(*a*) 10 % (*b*) 20 %
(*c*) 25 % (*d*) 30 %

159. How much sugar is recommended for making Jam?
(*a*) Above 30 % (*b*) Above 40 %
(*c*) Above 70 % (*d*) Above 60 %

160. Which of the following is responsible for sourness of reddish?
(*a*) Iso thiocynate (*b*) Anthocyanin
(*c*) Auxin (*d*) All

161. Lactic acid bacteria is used in preparation of
    (*a*) Pickle (*b*) Chutney
    (*c*) Marmalade (*d*) Jelly

162. Which of the following fruit contain maximum fat?
    (*a*) Avocado (*b*) Walnut
    (*c*) Apricot (*d*) peanut

163. Which is the rich source of vitamin?
    (*a*) Cabbage (*b*) Tomato
    (*c*) Chili (*d*) All

164. Rich source fruit of Vitamin A is
    (*a*) Ber (*b*) Pomegranate
    (*c*) Indian goose berry (*d*) Mango

165. In which product, TSS is 30° Brix
    (*a*) Cordial (*b*) Squash
    (*c*) Jam (*d*) jelly

166. The pungency of chili is due to
    (*a*) Capsanthin (*b*) Capsacin
    (*c*) Isothiocynate (*d*) none

167. The pungency of onion is due to
    (*a*) Allyl propyl disulphide (*b*) Diallyl disulphide
    (*c*) Both (*d*) None

168. Which of the following is class one preservative?
    (*a*) Silver nitrate (*b*) KMS
    (*c*) Sugar (*d*) Vinegar

169. For measurement of TSS, which instrument is used?
    (*a*) Dendrometer (*b*) Thermometer
    (*c*) Refractometer (*d*) Spectrophotometer

170. Sugar used as preservative have concentration of
    (*a*) 60 % (*b*) 70 %
    (*c*) 40 % (*d*) 50 %

171. Most commonly used oil extraction method from flower is
    (*a*) Distillation (*b*) Maceration
    (*c*) Solvent extraction (*d*) None

172. Time and temperature for vegetable processing in canning depend upon
(*a*) Amount of microbes present in food (*b*) Texture of foods
(*c*) Nutrient (*d*) None

173. Which cucurbit has highest vitamin A?
(*a*) Bitter guard (*b*) Cucumber
(*c*) Pumpkin (*d*) All

174. Protein content in ground nut is
(*a*) 22 % (*b*) 25 %
(*c*) 40 % (*d*) 43 %

175. Most suitable radiation for disinfection of food is
(*a*) UV- rays (*b*) g-rays
(*c*) a-rays (*d*) b-rays

176. TSS of Watermelon is
(*a*) 10-12 °Brix (*b*) 14-16 °Brix
(*c*) 18-22 °Brix (*d*) 20-22 °Brix

177. Pungency of garlic is due to
(*a*) Allyl propyl disulphide (*b*) Diallyl disulphide
(*c*) Isothiocynate (*d*) None

178. According to BIS, tomato has been specialized into
(*a*) 4 grades (*b*) 3 grades
(*c*) 2 grades (*d*) Only 1 grade

179. Fruit which is rich in pectin and acid is
(*a*) Jamun (*b*) Apple
(*c*) Pear (*d*) Pomegranate

180. Permissible limit of SO_2 in jam is
(*a*) 300 ppm (*b*) 40 ppm
(*c*) 70 ppm (*d*) 100 ppm

181. Acetic acid is present in
(*a*) Nectar (*b*) Sweet chutney
(*c*) Vinegar (*d*) Apple cider

182. Vitamin C is commonly known as
(*a*) Citric acid (*b*) Acetic acid
(*c*) Lactic acid (*d*) Ascorbic acid

183. Standard unit used in expressing refrigeration capacity is
(*a*) Calorie (*b*) Ampere
(*c*) Ton (*d*) Joule

184. Ideal storage temperature of papaya is
(*a*) 8–9 ºC (*b*) 9–11 ºC
(*c*) 11–13 ºC (*d*) 5–6 ºC

185. Ideal storage temperature of Brinjal is
(*a*) 10–11 ºC (*b*) 14–15 ºC
(*c*) 8–9 ºC (*d*) 17–18 ºC

186. The pigment present in watermelon is
(*a*) Carotenoid (*b*) Anthocyanin
(*c*) Lycopene (*d*) Both b and c

187. Identify the mango cultivar suitable for canning
(*a*) Alphonso (*b*) Amarapalli
(*c*) Neelum (*d*) Bombay green

188. Potato variety more suitable for chips making
(*a*) Kufri Jyoti (*b*) Kufri Naveen
(*c*) Kufri Badshah (*d*) Kufri Chipsona

189. Softness of Pea is measured by
(*a*) Derometer (*b*) Refractometer
(*c*) Alanometer (*d*) Tenderometer

190. In fruit jelly, pectin and sugar is present in the ratio of
(*a*) 1:65 (*b*) 65:1
(*c*) 65:33 (*d*) 33:65

191. Daily requirement of vitamin D for a healthy adult is
(*a*) 200 IU (*b*) 400 IU
(*c*) 600 IU (*d*) 800 IU

192. Cashew nut contains following % of protein
(*a*) 21 % (*b*) 22.7 %
(*c*) 27 % (*d*) 47 %

193. Yellow colour of onion is due to
(*a*) Qucertin (*b*) Myrcetin
(*c*) Allicin (*d*) Malvelin

194. Wine is preserved at
(*a*) 10 % alcohol (*b*) 14 % alcohol
(*c*) 8 % alcohol (*d*) 5 % alcohol

195. Cryo-preservation of foods is done by liquid N_2 at the temperature of
(*a*) 200 °C (*b*) 0 °C
(*c*) –196 °C (*d*) 4 °C

196. High sugar containing variety of muskmelon may contain TSS of
(*a*) 80 % (*b*) 2 %
(*c*) 17 % (*d*) 4 %

197. Vegetables are called as protective food because they are rich source of
(*a*) Fat (*b*) Vitamins and minerals
(*c*) Proteins (*d*) Sugars

198. In general, fruit and vegetables are not rich source of
(*a*) Vitamin A (*b*) Vitamin B_1
(*c*) Vitamin C (*d*) Minerals

199. Which of following is used as antioxidant?
(*a*) Fat (*b*) Protein
(*c*) Carbohydrate (*d*) Vitamin C

200. Fruit and vegetable are rich source in dietary fibers which protect human body from
(*a*) Colon cancer (*b*) Protein deficiency
(*c*) Fat deficiency (*d*) Sugar deficiency

201. Benzoic acid and SO_2 are added to preserve fruit and vegetable product to check
(*a*) Microorganism's activity (*b*) Ethylene production
(*c*) Reduce rancidity (*d*) None

202. MAP of fruits and vegetables prevent the building up of the following
(*a*) Sugar (*b*) Protein
(*c*) CO_2 and ethylene (*d*) None

203. Vegetable of the 20th century is
(*a*) Cluster bean (*b*) Amaranths
(*c*) Winged bean (*d*) Chikurmanis

204. Which vegetable has high calorific value?
(*a*) Fenugreek (*b*) Amaranths
(*c*) Potato (*d*) Tomato

205. Which vegetables are rich source of niacin
(*a*) Spinach (*b*) Cow pea
(*c*) Radish (*d*) Sweet potato

206. Daily requirement of vegetable per capita/day is
(*a*) 150 g (*b*) 200 g
(*c*) 285 g (*d*) 400 g

207. Which is non-pungent when uncooked?
(*a*) Lettuce (*b*) Radish
(*c*) Onion (*d*) Chili

208. Onion stored at temperature of
(*a*) 4–5 ºC (*b*) 0 ºC
(*c*) 6–8 ºC (*d*) 10 ºC

209. Jam should contain at least nearby percent of sugar
(*a*) 60 % (*b*) 70 %
(*c*) 50 % (*d*) 80 %

210. TSS of jam should be minimum at time of end point
(*a*) 68.5 % (*b*) 70 %
(*c*) 50 % (*d*) 80 %

211. Which is not used as food preservative?
(*a*) KMS (*b*) Sodium benzoate
(*c*) Acetic acid (*d*) Bendate

212. Inactivation of enzyme is done by
(*a*) Canning (*b*) Freezing
(*c*) Blanching (*d*) Dehydration

213. Lye peeling of fruit and vegetables is done by
(*a*) 1 % caustic soda + cool (*b*) 5 % caustic soda + cool
(*c*) 1 % caustic soda + hot (*d*) 5 % caustic soda + hot

214. Curcumin is extracted from
(*a*) Cucumber (*b*) Capsicum
(*c*) Turmeric (*d*) Cashew nut

215. The ideal concentration of sodium benzoate for preserving fruit juice
(*a*) 0.01 - 0.05 % (*b*) 0.06 - 0.10 %
(*c*) 1.1 – 1.5 % (*d*) 1.6 - 2.0 %

216. Chemical formula of potassium meta bisulphate is
(*a*) KMS (*b*) K_2SO_4
(*c*) KSO_2 (*d*) $K_2S_2O_5$

217. Vinegar concentration in finished pickle should not be
(*a*) Less than 10% (*b*) Less than 5%
(*c*) Less than 2 % (*d*) More than 50 %

218. Degree Brix is an unit for measurement of
(*a*) Salt content in juice (*b*) Sucrose content in juice
(*c*) Total solids in juice (*d*) Juice content of fruits

219. The cheapest method of preservation is
(*a*) Canning (*b*) Freezing
(*c*) Drying (*d*) Preserving

220. Palmolein is a/an
(*a*) Anti trasnpirant (*b*) Protein
(*c*) Edible oil (*d*) Essential oil

221. Which is an anti-hemorrhagic vitamin?
(*a*) Vitamin A (*b*) Vitamin K
(*c*) Vitamin K (*d*) Vitamin C

222. 'Essential oil' is extracted from ________________ fruit.
(*a*) Chili (*b*) Mangostein
(*c*) Olive (*d*) Almond

223. An equipment used for removing moisture from fruit and vegetable by evaporation
(*a*) Evaporator (*b*) Transpirator
(*c*) Auto clave (*d*) Dehydrator

224. The most common method of grape preservation is
(*a*) Jam making (*b*) Jelly making
(*c*) Dehydration (*d*) Pickling

225. Flame peeling is useful for
(*a*) Onion and garlic (*b*) Mango and guava
(*c*) Pea and tomato (*d*) Apple and carrot

226. The process of removal of air from cans is known as
(*a*) Appertizing (*b*) Exhausting
(*c*) Brining (*d*) Syruping

227. Which is diluted before serving?
(*a*) Jam
(*b*) Jelly
(*c*) RTS
(*d*) Squash

228. Canned vegetable is dipped in brine solution is made up from
(*a*) Common salt
(*b*) Black salt
(*c*) Sugar
(*d*) Salt and sugar

229. RTS stands for
(*a*) Ready to safe
(*b*) Ready to serve
(*c*) Ready to store
(*d*) Ready to sale

230. Colouring additives are added to foods for
(*a*) Improving shelf life
(*b*) Improving appearance
(*c*) Easily processing
(*d*) Improving taste

231. Father of food preservation is
(*a*) Lavoisier
(*b*) Nicholas Appert
(*c*) Needham
(*d*) Luis Pasteur

232. Food product order was established by GOI in the year
(*a*) 1950
(*b*) 1985
(*c*) 1945
(*d*) 1955

233. First food processing industry was started at
(*a*) Bombay (1935)
(*b*) Lucknow (1934)
(*c*) Mysore (1945)
(*d*) Delhi (1950)

234. World food day is celebrated on
(*a*) 15th July
(*b*) 20th July
(*c*) 16th October
(*d*) 10th July

235. Monoglycerides (GMS) emulsifier is used in
(*a*) Tomato juice
(*b*) Bakery items
(*c*) Ice-cream
(*d*) Fruit juice

236. Field to plate of food process is so called
(*a*) Food processing
(*b*) Food engineering
(*c*) Food technology
(*d*) Food nutrition

237. The process for prevention of decay or spoilage of food is called
(*a*) Preservation
(*b*) Sanitization
(*c*) Filtration
(*d*) Sterilization

238. The KMS of 350 ppm is mostly used in which food products?
(*a*) Juice (*b*) Jelly
(*c*) Jam (*d*) Pickle

239. Phosphoric acid is added in
(*a*) Soft drink (*b*) Jam
(*c*) Jelly (*d*) Juice

240. Ethylbutyrate, a flavouring compound is found in
(*a*) Strawberry and pineapple (*b*) Mango and grape
(*c*) Guava and banana (*d*) None

241. During sealing of can, temperature should not be less than
(*a*) 70ºC (*b*) 74ºC
(*c*) 84ºC (*d*) 89ºC

242. The lye peeling of fruits and vegetables is done at the temperature of
(*a*) 75ºC (*b*) 93ºC
(*c*) 80ºC (*d*) 100ºC

243. Enzymatic browning in apple is due to
(*a*) Oxidase (*b*) Catalase
(*c*) Poly phenol oxidase (*d*) Protease

244. Salt concentration in pickle is measured by
(*a*) Salometer (*b*) pH meter
(*c*) Lactometer (*d*) Refractometer

245. Water core is a serious disorder of
(*a*) Grape (*b*) Mango
(*c*) Apple (*d*) Ber

246. Syrup or brine in can is filled at the temperature of
(*a*) 90-100ºC (*b*) 79-82ºC
(*c*) 50-70ºC (*d*) 45-53ºC

247. Pricking is related to the processing of
(*a*) Papaya (*b*) Apple
(*c*) Mango (*d*) Aonla

248. Cold process is also known as
(*a*) Pre cooling (*b*) Freezing
(*c*) Cold storage (*d*) Ultra filtration

249. Cold sterilization of food is so called as
   (*a*) Freezing (*b*) Irradiation
   (*c*) Waxing (*d*) Chilling

250. Good quality of pectin is found when guava is at
   (*a*) Green stage (*b*) Dark yellow in colour
   (*c*) Over ripe (*d*) Ripe but green stage

251. Most of the enzymes are inactivate at the temperature of
   (*a*) 80°C (*b*) 85°C
   (*c*) 90°C (*d*) 70°C

252. The coconut shells contain cellulose
   (*a*) 40% (*b*) 35%
   (*c*) 34% (*d*) 30%

253. Orange juice contain approximately solid about
   (*a*) 5% (*b*) 12%
   (*c*) 20% (*d*) 25%

254. Fermentation was discovered by
   (*a*) Lavoisier (*b*) Benjamin
   (*c*) Funk (*d*) Louise Pasteur

255. Fruit preservation and canning institute is located at
   (*a*) Hyderabad (*b*) Mumbai
   (*c*) Lucknow (*d*) Allahabad

256. Jelly is pasteurized at temperature of
   (*a*) 90-95°C (*b*) 82-85°C
   (*c*) 70-75°C (*d*) 60-65°C

257. Which vitamin destroyed during freezing due to oxidation?
   (*a*) Vitamin D (*b*) Vitamin C
   (*c*) Vitamin B (*d*) Vitamin A

258. Honey wine is known as
   (*a*) Mead (*b*) Brandy
   (*c*) Bear (*d*) Whisky

259. Which pickle is known as 'National Pickle of India?'
   (*a*) Mango pickle (*b*) Chili pickle
   (*c*) Lime pickle (*d*) garlic pickle

260. Density of liquid is measured by
(*a*) Hydrometer
(*b*) Hygrometer
(*c*) Lactometer
(*d*) Refractometer

261. Food packaging foils consist of pure
(*a*) Iron
(*b*) Aluminium
(*c*) Zinc
(*d*) Lead

262. How much pectin should be present in finished jelly?
(*a*) 2.0%
(*b*) 3.0%
(*c*) 2.5%
(*d*) 1.0%

263. What should be pH of good Jelly?
(*a*) 2.50
(*b*) 3.46
(*c*) 4.73
(*d*) 4.59

264. Basic principle of freeze drying is
(*a*) Water evaporate from ice crystal without ice melting
(*b*) Water evaporate after melting ice crystals
(*c*) Both above
(*d*) None

265. Dry ice sublime at which temperature and pressure
(*a*) 0°C and 2 bar
(*b*) 1°C and 15 bar
(*c*) At room temperature and atmospheric pressure
(*d*) None of these

266. The percentage of sucrose in solution prevents the growth of all microorganisms in foods.
(*a*) 30%
(*b*) 45%
(*c*) 60%
(*d*) 70%

267. The percentage of salt in solution prevents the growth of all microorganisms in foods.
(*a*) 18-25%
(*b*) 25-30%
(*c*) 30-40%
(*d*) 10-15%

268. Atmospheric drying foams is also known as
(*a*) Natural drying
(*b*) Foam mat drying
(*c*) Both above
(*d*) None

269. Heat sensitive foods are mostly concentrated with low temperature by
(*a*) Vacuum evaporator
(*b*) Natural evaporator
(*c*) Flash evaporator
(*d*) Thin film evaporator

270. The freeze drying generally works at pressure of
(*a*) 2 mm to about 0.1 mm Hg (*b*) 3 mm to about 1.0 mm Hg
(*c*) 1 mm to about 0.3 mm Hg (*d*) None

271. At one atmosphere, pure water boils at
(*a*) 30°C (*b*) 82°C
(*c*) 70°C (*d*) 100°C

272. At 250 mm Hg, pure water boils at
(*a*) 72°C (*b*) 82°C
(*c*) 70°C (*d*) 100°C

273. The kiln drier cannot reduce moisture content of food below
(*a*) 10% (*b*) 12%
(*c*) 15% (*d*) 20%

274. Kiln drier is used for drying of
(*a*) Vegetable slices (*b*) Potato slice
(*c*) Cereals (*d*) Apple slices

275. Cabinet, tray and pan drier are best suited for the drying of
(*a*) Fruit and vegetable (*b*) Cereals grains
(*c*) Meat (*d*) Fish

276. In vacuum drying, temperature of foods and rate of water removal is controlled by
(*a*) Degree of vacuum (*b*) Intensity of heat input
(*c*) Both above (*d*) None

277. In vacuum drier, heat transfer from foods takes place by
(*a*) Only conduction (*b*) Only radiation
(*c*) Both above (*d*) Only convection

278. Freeze drying is used for the drying/dehydration of
(*a*) High value liquid foods like coffee, juices etc.
(*b*) Fruit slices
(*c*) Cereals grains (*d*) Live stock products

279. Sugar and salt concentrated solution have
(*a*) High diffusion pressure (*b*) High atmospheric pressure
(*c*) High osmotic pressure (*d*) All above

280. Sodium benzoate is added for preservation to fruit juice at the rate of
(*a*) 0.01-0.5% (*b*) 0.06-0.10%
(*c*) 1.1-1.5% (*d*) 1.6-2.0%

281. Jelly in appearance is
(*a*) Transparent (*b*) Opaque
(*c*) Colloid (*d*) None

282. One of the following preservative is used in squash
(*a*) Formaldehyde (*b*) NaCl
(*c*) KMS (*d*) Sodium benzoate

283. In freeze drying, the liquid phase comes at the pressure above
(*a*) 1.7 mm (*b*) 2.7 mm
(*c*) 4.7 mm (*d*) 3.7 mm

284. In conventional freeze drying, dehydration temperature are kept between
(*a*) 10.6-30.8°C (*b*) 30.2-69.5°C
(*c*) 35.5-75.0°C (*d*) 37.8-93.3°C

285. The heating of infrared drying in comparison to convection drying is higher by
(*a*) 5 times (*b*) 8 times
(*c*) 10 times (*d*) 12 times

286. The toxicant formed by Clostridium botulinum in foods is destroyed in 10 minutes if exposed to moist heat of the temperature of
(*a*) 50°C (*b*) 100°C
(*c*) 200°C (*d*) 300°C

287. Sterilization frequently requires a thermal treatment at least at
(*a*) 100°C (*b*) 121°C
(*c*) 135°C (*d*) 149°C

288. Sterilization increases __________ of foods.
(*a*) Shelf life (*b*) Taste
(*c*) Protection form pathogens (*d*) All

289. Commercially sterilized food has a shelf life more than
(*a*) 3 month (*b*) 6 month
(*c*) 1 year (*d*) 2 years

290. Which combination is required to inactivate the pathogen in particular foods?
(*a*) Time and temperature (*b*) Moisture and temperature
(*c*) RH and temperature (*d*) RH and moisture

291. Acid foods generally having pH of
(*a*) More than 6 (*b*) 4.6 or less
(*c*) 7.5 (*d*) 8.6

292. Lemon juice are generally classified under
(*a*) Low acid food (*b*) Medium acid food
(*c*) Very low acid food (*d*) High acid food

293. The juices are pasteurized at 88°C for a period of
(*a*) 60 seconds (*b*) 900 seconds
(*c*) 30 seconds (*d*) 50 seconds

294. Typically acid fruit juices are first heated in the range of
(*a*) 77-100°C (*b*) 50-80°C
(*c*) 200-225°C (*d*) None

295. Freezing and refrigeration were started in year of
(*a*) 1850 (*b*) 1875
(*c*) 1900 (*d*) 1901

296. Cool storage generally refers to store the foods at temperature of
(*a*) 16°C down to –2°C (*b*) 4°C to – 4°C
(*c*) 0°C (*d*) 10°C

297. Commercial and house hold refrigerators are usually operated at
(*a*) 7.5°C (*b*) 4.5 to 7.0°C
(*c*) 0°C (*d*) –4°C

298. The freezing point of pure water is
(*a*) 0°C (*b*) 4°C
(*c*) – 4°C (*d*) –2°C

299. Storage life of fruits at 0°C is
(*a*) 2-180 days (*b*) 10 days
(*c*) 1 week (*d*) none

300. Air freezing method was introduced around the year
(*a*) 1905 (*b*) 1910
(*c*) 1860 (*d*) 1850

301. Which gas is helpful for freezing of foods at –196°C?
(*a*) Oxygen (*b*) Hydrogen
(*c*) Nitrogen (*d*) Ozone

302. The full form of FDA
(*a*) Food development authority (*b*) Food dehydration agency
(*c*) Food and drugs administration (*d*) None of the above

303. Non carbonated stimulating beverage is
(*a*) Nectar (*b*) Tea and coffee
(*c*) Milk (*d*) Mead

304. Roasting process is employed in making of
(*a*) Coffee (*b*) Milk
(*c*) Jam (*d*) Chips

305. Instant coffee is technically known as
(*a*) Solubilized coffee (*b*) Dry coffee
(*c*) Heated coffee (*d*) Boiled coffee

306. Coffee is made from the
(*a*) Beans (seed) (*b*) Leaves
(*c*) Buds (*d*) Roots

307. Tea is prepared from the
(*a*) Young leaves (*b*) Matures leaves
(*c*) Yellow leaves (*d*) Buds

308. Tea plant is a type of
(*a*) Evergreen (*b*) Herbaceous
(*c*) Annual (*d*) None

309. Tannin compound is found in
(*a*) Tea leaves (*b*) Coffee
(*c*) Fruits (*d*) Milk

310. For the preparation of black tea, air temperature for drying is employed at about
(*a*) 27°C for 2-5 hrs (*b*) 40°C for 2-5 hrs
(*c*) 10°C for 2-5 hrs (*d*) 80°C for 2-5 hrs

311. Caffeine percentage in Indian black tea is
(*a*) 2-3% (*b*) 4-5%
(*c*) 10-11% (*d*) 12-13%

312. Coffee is the major source of
(*a*) Caffeine (*b*) Nicotine
(*c*) Tannin (*d*) Saponin

313. 150 ml coffee beverage contain caffeine about
(*a*) 75-150 mg (*b*) 100-200 mg
(*c*) 75-150 ml (*d*) None of these

314. Full form of FSIS is
(*a*) Food supply and inspection system
(*b*) Food safety and inspection system
(*c*) Food safety and inspection service
(*d*) Food service and inspection scheme

315. One of the following equipment is used for ash determination.
(*a*) Soxlet
(*b*) Macro Jeldhal
(*c*) Micro Jeldhal
(*d*) Muffle furnace

316. Micro Jeldhal apparatus is used for the estimation of
(*a*) Ash
(*b*) Fat
(*c*) Protein
(*d*) Vitamin

317. Function of pulper is to
(*a*) grind the pulp
(*b*) dry the pulp
(*c*) cook the pulp
(*d*) slice the vegetables

318. To estimate the fat content in food which apparatus is employed?
(*a*) Soxlet
(*b*) Macro Jeldhal
(*c*) Microwave oven
(*d*) Muffle furnace

319. To determine the rheological properties of food which apparatus is used?
(*a*) Shaker
(*b*) Mixograph
(*c*) Sedimentation apparatus
(*d*) Texture analyzer

320. Size analysis of flour/powder is done with the help of
(*a*) Pulper
(*b*) Grader
(*c*) Sieve shaker
(*d*) Attrition mill

321. Tray drier is facilitate to
(*a*) cook the food
(*b*) sterilize the food
(*c*) dry the food
(*d*) control the RH

322. The enzyme used as a emulsifying agent is
(*a*) Casein
(*b*) Leucine
(*c*) Rennin
(*d*) Lysine

323. To separate two liquids from their mixture is performed by
(*a*) Double distillation column
(*b*) BOD
(*c*) Spray drier
(*d*) Centrifuge

324. Suitable variety for processing of mango is
(*a*) Chausa
(*b*) Alphanso
(*c*) Amarpalli
(*d*) Malika

325. Grape is rich source of
(*a*) Citric acid
(*b*) Malic acid
(*c*) formic acid
(*d*) Tartaric acid

326. Food adulteration committee was set up in the year
(*a*) 1944
(*b*) 1945
(*c*) 1943
(*d*) 1946

327. Codex Alimetariuos commission was set up in the year
(*a*) 1962
(*b*) 1663
(*c*) 1985
(*d*) 1972

328. Establishment of International Organization for Standards was done in the year
(*a*) 1454
(*b*) 1955
(*c*) 1944
(*d*) 1947

329. Agricultural produce (grading and marketing Act) was started in the year
(*a*) 1935
(*b*) 1937
(*c*) 1947
(*d*) 1955

330. Food product order (FPO) was started in the year
(*a*) 1955
(*b*) 1937
(*c*) 1947
(*d*) 1962

331. Export (quality control and inspection) act was established in the year
(*a*) 1952
(*b*) 1956
(*c*) 1945
(*d*) 1963

332. Solvent extraction oil deoiled meal and edible flour (control) order started in the year
(*a*) 1955
(*b*) 1963
(*c*) 1967
(*d*) 1956

333. Meat product order was set up in the year
(*a*) 1973
(*b*) 1975
(*c*) 1955
(*d*) 1980

334. Quality of Agricultural commodity divided into __________ categories.
(*a*) One
(*b*) Two
(*c*) Three
(*d*) Four

335. The Central Food Advisory Board was established in the year
(*a*) 1955
(*b*) 1937
(*c*) 1956
(*d*) 1938

336. Which chemical is used for controlling sprouting of onion during storage?
(*a*) Malic hydrazide
(*b*) Ethylene
(*c*) GA_3
(*d*) All of these

337. For low sugar content, potato should be stored at
(*a*) 5°C
(*b*) 10°C
(*c*) 15°C
(*d*) 20°C

338. For long term storage potato should be stored at
(*a*) 0-5°C
(*b*) 5-10°C
(*c*) 10-15°C
(*d*) 15-20°C

339. Ratio of sugar to acid determine the
(*a*) pH of fruit
(*b*) sweetness
(*c*) tartness and flavour of fruit
(*d*) all of these

340. Which compound is highly sensitive to pH for changing the colour?
(*a*) Betaloins
(*b*) Carotenoid
(*c*) Anthocyanin
(*d*) Tannins

341. Ascorbic acid (vitamin C) reduces which activities in fruits?
(*a*) Oxidation
(*b*) Dehydration
(*c*) Respiration
(*d*) Decay of fruit by microorganisms

342. The level of vitamin C (%) in apple syrup or peach syrup mixture for acting as antioxidant is
(*a*) 0.05-0.20
(*b*) 0.20-0.50
(*c*) 0.50-1.00
(*d*) 1.00-1.20

343. A fine grind coffee may have a bulk density of
(*a*) 0.29-0.37
(*b*) 0.39-0.47
(*c*) 0.19-0.27
(*d*) 0.49-0.57

344. Instant coffee typically has a moisture content of
(*a*) 1-2 %
(*b*) 2-4 %
(*c*) 4-6 %
(*d*) 6-8 %

345. When roasted whole coffee beans are placed within a closed packaged which gas is released.
(*a*) CO_2
(*b*) O_2
(*c*) NH_3
(*d*) None

346. After picking black tea leaves are placed on trays to wither or dry for about
(*a*) 8 hrs
(*b*) 10 hrs
(*c*) 18 hrs
(*d*) 25 hrs

347. Initial moisture content of black tea is
(*a*) 1-3% (*b*) 3-4%
(*c*) 4-6% (*d*) 7-10%

348. The can must be able to withstand continual internal pressure of upto
(*a*) 1 atm (*b*) 3 atm
(*c*) 5 atm (*d*) 7 atm

349. Malted barley is crushed and mashed by mixing with water upto the temperature of
(*a*) 65°C (*b*) 67°C
(*c*) 68.5°C (*d*) 69.7°C

350. The desserts and appetizer wine contains alcohol about
(*a*) 10-15% (*b*) 15-21%
(*c*) 20-24% (*d*) 5-11%

351. Rate of oxygen uptake by wine may be
(*a*) 0.17 ppm day $^{-1}$ (*b*) 0.18 ppm day $^{-1}$
(*c*) 0.10 ppm day $^{-1}$ (*d*) 0.38 ppm day $^{-1}$

352. Owing to the natural limit of sugar present in grape wine, the alcohol content of such wine is normally about
(*a*) 8 % (*b*) 12%
(*c*) 13.5% (*d*) 18.6 %

353. The initial concentration of SO_2 in wine is
(*a*) 100 ppm (*b*) 200 ppm
(*c*) 350 ppm (*d*) 600 ppm

354. The amount of sodium chloride added upto saturation of cauliflower for making of sauerkraut may vary from
(*a*) 1.0-1.5 % (*b*) 1.5-2.0 %
(*c*) 2.0-2.5 % (*d*) 2.25-2.50 %

355. The lower concentration of salt in sauerkraut permit an
(*a*) Acid fermentation (*b*) Alcoholic fermentation
(*c*) Both (*d*) None

356. The percent of salt increases, the rate of acid production becomes
(*a*) Lower (*b*) Higher
(*c*) Remain same (*d*) None

357. Irradiation can be delay of sprouting for
(*a*) Potato (*b*) Onion
(*c*) Garlic (*d*) All

358. During of preparation of fruits for freezing, undesirable changes may take place such as
(*a*) Darkening
(*b*) Deterioration in flavour
(*c*) Spoilage by molds
(*d*) All of the above

359. The lactic acid fermentation of sugar in apples mostly by heterofermentative lactic acid bacteria such as
(*a*) Lactobacillus pastorianus
(*b*) L. brevis
(*c*) Leuconostoc mesenteroides
(*d*) All

360. The fermentation of organic acid of fruit juices by LAB e.g.
(*a*) Lactobacillus pastorianus
(*b*) L. brevis
(*c*) Leuconostoc mesenteroides
(*d*) All

361. Heat processing kills the bacteria but freezing
(*a*) prevent the growth of bacteria
(*b*) freeze the DNA of bacteria
(*c*) increase the multiplication of bacteria
(*d*) destroy the bacteria

362. Concentration of CO_2 and O_2 for apple in CA storage should be maintained at
(*a*) 1.5 -10 %, 2.5 %
(*b*) 2.5 %, 2.5 %
(*c*) 5-10 %, 3 %
(*d*) 5.0 %, 0%

363. One of the following is predominant mold organisms in frozen fruits is
(*a*) Aspergillus and Penicillium
(*b*) Mucor aad Rhizopus
(*c*) Botrytis, Fusarium and Alternaria
(*d*) All of the above

364. Yeasts are mostly likely to grow in food during
(*a*) Fast thawing
(*b*) Slow thawing
(*c*) Fast freezing
(*d*) Slow drying

365. The number of microorganisms in frozen fruits juices depend on the
(*a*) Condition of fruits and washing process
(*b*) Method of juice filtration
(*c*) Opportunity of contamination and growth
(*d*) All of the above

366. The deterioration of raw vegetable and fruits may results from
(*a*) Physical factor
(*b*) Enzyme activity
(*c*) Microbial action
(*d*) All of the above

367. Low moisture content syrup containing a small quantity of invert syrup has been rapidly cooled so that the sugar crystals are small in size is also called
(*a*) Cane sugar
(*b*) Fondants
(*c*) Syrup
(*d*) Sherbet

368. A mixture of dextrose and levulose made by inverting sucrose with acid or enzymes is called
(*a*) Reducing sugar
(*b*) Non-reducing sugar
(*c*) Invert sugar
(*d*) All of them

369. The process of blending together fat and fat solution of ingredients to produce a stable mixture which will not separate on standing is called

(*a*) Fermentation (*b*) Emulsification

(*c*) Fortification (*d*) Condensation

370. The chemical changes of an organic compound due to the action of living organisms (yeast or bacteria) usually producing a leavening gas is called

(*a*) Dough (*b*) Fortification

(*c*) Fermentation (*d*) Stabilization

371. A combination of fruit juice and sugar, stiffened by the action of the pectin of the fruit, as a result of heating is called

(*a*) Jam (*b*) Marmalade

(*c*) Cordial (*d*) Jelly

372. Food industry as a separate department of Government of India was established in the year

(*a*) 1960 (*b*) 1965

(*c*) 1980 (*d*) 1995

373. Titrable acidity (TA) can be determined by titrating a known volume of fruit juice with

(*a*) 0.1N NaOH (*b*) 0.01N NaOH

(*c*) 1N NaOH (*d*) 10N NaOH

ANSWER KEY (FRUITS AND VEGETABLE PROCESSING)

1.D	2. C	3.D	4.B	5.A	6.B	7.D	8.D	9.A	10.A
11. A	12.B	13.C	14.D	15.A	16.D	17.C	18.A	19.D	20.A
21.B	22.B	23.D	24.B	25.A	26.A	27.B	28.A	29.B	30.D
31.B	32.A	33.D	34.A	35.C	36.C	37.B	38.B	39.B	40.B
41.A	42.B	43.B	44.A	45.A	46.C	47.C	48.A	49.A	50.B
51.C	52.A	53.A	54.D	55.D	56.A	57.B	58.D	59.C	60.D
61.A	62.B	63.C	64.D	65.A	66.B	67.C	68.D	69.A	70.B
71.C	72.D	73.A	74.C	75.D	76.B	77.C	78.D	79.B	80.C
81.D	82.A	83.B	84.C	85.D	86.A	87.B	88.C	89.B	90.B
91.C	92.D	93.A	94.A	95.B	96.A	97.B	98.A	99.D	100.B
101.A	102.A	103.C	104.A	105.C	106.A	107.B	108.A	109.A	110.A
111.D	112.B	113.D	114.D	115. A	116. C	117.C	118.A	119.C	120.B
121.B	122.B	123.C	124.B	125.B	126.A	127.D	128.B	129.B	130.A
131.B	132.A	133.A	134.A	135.A	136.A	137.C	138.B	139.A	140.B
141.A	142.B	143.B	144.C	145.A	146.B	147.C	148.A	149.B	150.B
151.D	152.B	153.B	154.B	155.D	156.D	157.C	158.C	159.D	160.A
161.A	162.B	163.C	164.D	165.A	166.B	167.A	168.C	169.C	170.A
171C	172.B	173.C	174.A	175.B	176.A	177.B	178.A	179.A	180.B
181.C	182.D	183.C	184.B	185.B	186.D	187.A	188.D	189.D	190.A
191.A	192.A	193.A	194.B	195.C	196.C	197.B	198.B	199.D	200.A
201.A	202.C	203.C	204.C	205.C	206.C	207.A	208.B	209.B	210.A
211.D	212.C	213.B	214.C	215.B	216.D	217.C	218.C	219.C	220.C
221.C	222.A	223.A	224.C	225.A	226.B	227.D	228.A	229.B	230.B
231.B	232.D	233.A	234.C	235.C	236.A	237.A	238.A	239.A	240.A
241.B	242,B	243.C	244.A	245.C	246.B	247.D	248.D	249.B	250.D
251.D	252.C	253.B	254.A	255.C	256.B	257.B	258.A	259.C	260.A
261.B	262.D	263.B	264.A	265.C	266.D	267.A	268.B	269.A	270.A
271.D	272.A	273.A	274.D	275.A	276.C	277.C	278.A	279.C	280.B
281.A	282.C	283.C	284.D	285.C	286.B	287.B	288.D	289.D	290.A
291.B	292.D	293.B	294.A	295.B	296.A	297.B	298.A	299.A	300.C
301.C	302.C	303.B	304.A	305.A	306.A	307.A	308.A	309.A	310.A
311.A	312.A	313.A	314.B	315.D	316.C	317.A	318.A	319.D	320.C
321.C	322.C	323.D	324.B	325.D	326.C	327.A	328.D	329.B	330.A
331.D	332.C	333.A	334.B	335.B	336.A	337.B	338.C	339.C	340.C
341.A	342.A	343.B	344.B	345.A	346.C	347.B	348.C	349.B	350.B
351.A	352.B	353.A	354.D	355.A	356.A	357.D	358.D	359.D	360.A
361.A	362.A	363.D	364.B	365.D	366.D	367.B	368.C	369.B	370.C
371.D	372.D	373.A							

Milk and Milk Product Processing

1. HTST milk pasteurization destroys
   (*a*) all bacteria
   (*b*) all disease producing pathogens
   (*c*) only selected disease producing bacteria
   (*d*) no bacteria
2. Milk sugar is
   (*a*) Lactose
   (*b*) Glucose
   (*c*) Sucrose
   (*d*) Maltose
3. The yellow colour of cow milk is due to the presence of
   (*a*) Vitamin B
   (*b*) Carotene
   (*c*) Anthocyanin
   (*d*) None of these
4. Cow milk's protein is
   (*a*) Albumin
   (*b*) Zein
   (*c*) Casein
   (*d*) Lactoalbumin
5. Temperature of LTLT pasteurization is
   (*a*) 42-49 ºC
   (*b*) 51-65 ºC
   (*c*) 62-65 ºC
   (*d*) 61-63 ºC
6. Element deficiency in milk is
   (*a*) Fe
   (*b*) Mg
   (*c*) P
   (*d*) Ca
7. National Dairy Research Institute is located at
   (*a*) Delhi
   (*b*) Izzatnagar
   (*c*) Jhansi
   (*d*) Karnal
8. Casein is _______________ % of total milk protein
   (*a*) 80-83
   (*b*) 70-71
   (*c*) 60-65
   (*d*) 42-45

9. Pasteurization kills
   (*a*) Bacteria +virus
   (*b*) Bacteria spores alive
   (*c*) Bacteria along spores
   (*d*) None of these

10. Most successful example of cooperative in milk processing is
   (*a*) Raksha
   (*b*) Proagro
   (*c*) Syngenta
   (*d*) Anand

11. Operation flood is related to
   (*a*) Milk
   (*b*) Rice
   (*c*) Potato
   (*d*) None of these

12. White revolution is related to
   (*a*) Cotton
   (*b*) Milk
   (*c*) Sheep
   (*d*) Rice

13. The most suitable temperature for separation of cream by centrifugal cream separator is
   (*a*) 40 °C
   (*b*) 30 °C
   (*c*) 20 °C
   (*d*) 10 °C

14. The moisture content of paneer is not less than
   (*a*) 65 %
   (*b*) 70 %
   (*c*) 60 %
   (*d*) 50 %

15. Better quality Khoa is prepared from
   (*a*) Sheep milk
   (*b*) Cow milk
   (*c*) Buffalo milk
   (*d*) None of these

16. Specific gravity of milk is
   (*a*) 1.03
   (*b*) 1.05
   (*c*) 0.85
   (*d*) 0.93

17. The fat content of double toned milk is
   (*a*) 1.5 %
   (*b*) 2.0 %
   (*c*) 2.5 %
   (*d*) 3.0 %

18. Better quality chhana is prepared from
   (*a*) Camel milk
   (*b*) Cow milk
   (*c*) Sheep milk
   (*d*) Buffalo milk

19. Paneer is a
   (*a*) Concentrated milk product
   (*b*) Dried milk product
   (*c*) Coagulated milk product
   (*d*) Fermented product

20. Ash content of human milk is
(*a*) 0.8 % (*b*) 0.7 %
(*c*) 0.5 % (*d*) 0.2 %

21. Lactic acid content in fresh milk is
(*a*) 0.10 % (*b*) 0.20 %
(*c*) 0.30 % (*d*) 0.40 %

22. Flavouring component for butter is
(*a*) Acetyl methyl carbinol (*b*) Diacetyl
(*c*) Lactic acid (*d*) Protein

23. UHT milk is heated to a temperature of
(*a*) 100 ºC (*b*) 130 ºC
(*c*) 145 ºC (*d*) 165 ºC

24. Enzyme coagulated milk product is
(*a*) Dahi (*b*) Paneer
(*c*) Chhana (*d*) Cheese

25. Curd tension of dahi
(*a*) is lower than milk (*b*) remains constant
(*c*) is higher than milk (*d*) None of these

26. The time temperature combination in plate pasteurization is
(*a*) 60 ºC for 30 minutes (*b*) 72 ºC for 15 seconds
(*c*) 72 ºC for 15 minutes (*d*) 121 ºC for 15 seconds

27. Protein in milk exits in the form of
(*a*) Solution (*b*) Emulsion
(*c*) Colloidal (*d*) None of these

28. The fat content of plain ice cream is
(*a*) 6 % (*b*) 8 %
(*c*) 12 % (*d*) 10 %

29. Lactose in milk exits in the form of
(*a*) Solution (*b*) Emulsion
(*c*) Colloidal (*d*) Nine of these

30. Fat in separated milk is not more than
(*a*) 0.1 % (*b*) 0.5 %
(*c*) 1.0% (*d*) 1.5 %

31. Khoa contains milk fat not less than
(*a*) 10 % (*b*) 20 %
(*c*) 30 % (*d*) 40 %

32. The SNF content of toned milk is
(*a*) 6.0 % (*b*) 7.0 %
(*c*) 8.5 % (*d*) 9.0 %

33. Methylene blue reduction test is conducted to know the efficiency of
(*a*) Sterilized milk (*b*) Toned milk
(*c*) Double toned milk (*d*) Pasteurized milk

34. The optimum temperature for making dahi is
(*a*) 22 °C (*b*) 18 °C
(*c*) 15 °C (*d*) 10 °C

35. Gerber method is used for testing the following
(*a*) Lactose (*b*) Fat
(*c*) Vitamins (*d*) Protein

36. A good quality dahi should have acidity less than
(*a*) 0.3 % (*b*) 0.5 %
(*c*) 0.8 % (*d*) 1.0%

37. Buffalo milk is rich source of following minerals
(*a*) Calcium and iron (*b*) Iron and copper
(*c*) Copper and phosphorus (*d*) Calcium and phosphorus

38. Buffer value of milk is due to
(*a*) Protein (*b*) Lipid
(*c*) Fat (*d*) Carbohydrate

39. Polenske value of milk fat is due to
(*a*) Fats (*b*) Water insoluble fatty acids
(*c*) Protein nitrogen (*d*) Water soluble fatty acids

40. Lactose is a disaccharide made of
(*a*) Glucose and fructose (*b*) Maltose and fructose
(*c*) Glucose and galactose (*d*) Maltose and glucose

41. Average pH of normal milk is
(*a*) 3.6 (*b*) 4.6
(*c*) 5.6 (*d*) 6.6

42. Most variable of all constituents present in milk is
(*a*) Fat
(*b*) Protein
(*c*) Minerals
(*d*) Carbohydrate

43. Which vitamin remains most resistant on heat treatment of milk
(*a*) Vitamin C
(*b*) Vitamin A
(*c*) Vitamin B_1
(*d*) Vitamin B_{12}

44. The normal size of fat globules in raw milk is
(*a*) 0.5-1.0 μm
(*b*) 1.0-1.5 μm
(*c*) 2.0-6.0 μm
(*d*) 6.0-8.0 μm

45. The pasteurization process was developed by
(*a*) Louis Pasteur
(*b*) Alexander Fleming
(*c*) Newton
(*d*) Robert Hook

46. Good milk dryer expose the milk to __________ temperature for a very __________ time.
(*a*) High, high
(*b*) High, short
(*c*) Low, short
(*d*) None of these

47. The separation of milk powder from air may be accomplished by
(*a*) Cyclone separator
(*b*) Indented cylinder
(*c*) Compartment separator
(*d*) None of these

48. The cheese and curd of milk is used for __________ the curd to proper size.
(*a*) recusing
(*b*) increasing
(*c*) freezing
(*d*) none of these

49. Butter is an emulsion of
(*a*) Water in oil
(*b*) Oil in oil
(*c*) Water in water
(*d*) Oil in water

50. The lactometer will sink __________ in milk without cream when compared to whole milk.
(*a*) more
(*b*) less
(*c*) no change
(*d*) none of these

51. According to PFA, table butter should not contain fat less than
(*a*) 50 %
(*b*) 60 %
(*c*) 70 %
(*d*) 80 %

52. Table butter contains minimum __________ % fat and maximum __________ % moisture.
(*a*) 80, 16
(*b*) 70, 10
(*c*) 60, 15
(*d*) 50, 20

53. Milk – milk serum =
(*a*) Skim milk (*b*) Milk fat
(*c*) Coagulation (*d*) Casein

54. Casein is present in milk as which of the following form
(*a*) Emulsion (*b*) Colloidal suspension
(*c*) True solvent (*d*) Homogenized

55. Fat is present in milk as which of the following form
(*a*) Emulsion (*b*) Colloidal suspension
(*c*) True solvent (*d*) Homogenized

56. Which of the following is the protein of milk?
(*a*) 3.1 (*b*) 3.2
(*c*) 3.3 (*d*) 3.4

57. Which of the following is known as spoon eating milk?
(*a*) Recombined milk (*b*) Tattoo milk
(*c*) Standard milk (*d*) Toned milk

58. Milk viscosity is due to which constituent of milk
(*a*) Albumin (*b*) Globulin
(*c*) Casein (*d*) Phospholipids

59. Which of the following is milk protein?
(*a*) α- casein (*b*) β-casein
(*c*) γ-casein (*d*) All of these

60. Which of the following have maximum casein in its milk?
(*a*) Buffalo (*b*) Cow
(*c*) Human (*d*) None

61. Fishy odour of milk is due to
(*a*) γ- casein (*b*) kephalin
(*c*) α- casein (*d*) β-casein

62. Which of the following milk does not have 8.5 % SNF?
(*a*) Toned (*b*) Double toned
(*c*) Combined (*d*) Standardized

63. Standard fat percentage in toned and double toned milks are
(*a*) 3.5 and 1.5 (*b*) 1.5 and 3.0
(*c*) 3.0 and 1.5 (*d*) 1.5 and 3.5

64. Milk boils _______ than water.
(*a*) Slowly (*b*) Equally
(*c*) Quickly (*d*) Not compared

65. Effect of addition of water in milk on the specific gravity is that specific gravity
(*a*) increases (*b*) decreases
(*c*) irregular (*d*) no effect

66. Effect of addition of fat in milk on specific gravity
(*a*) increase (*b*) decrease
(*c*) irregular (*d*) no effect

67. Phosphatase test for standardized milk
(*a*) Positive (*b*) Negative
(*c*) Neutral (*d*) Depend upon temperature

68. Maximum pesticide tolerance (mg/kg of milk fat) is for
(*a*) Lindane (*b*) Chlorane
(*c*) DDT (*d*) Finitrothin

69. Minimum pesticide tolerance (mg/kg of milk fat) is for
(*a*) Chlorane (*b*) Finitrothin
(*c*) Both (*a*) and (*b*) (*d*) None of these

70. Ratio of milk powder and water in recombined milk is
(*a*) 1:3 (*b*) 1:2
(*c*) 1:5 (*d*) 1:7

71. Soft curd milk is low in water of the following
(*a*) Calcium (*b*) Casein
(*c*) Both (*a*) and (*b*) (*d*) None of these

72. In pasteurization and sterilization which vitamin get lost
(*a*) A and D (*b*) B_1 and C
(*c*) B_2 and B_{12} (*d*) E and K

73. Yellow colour of cow milk is due to the presence of
(*a*) Fat (*b*) Casein
(*c*) Lactochrome (*d*) Carotene

74. Greenish colour of skim milk is due to
(*a*) Lacto-chrome (*b*) Fat
(*c*) Carotene (*d*) Casein

75. Surface tension of milk (dyne/cm) is which of the following
    (*a*) 60 (*b*) 55
    (*c*) 50 (*d*) 45

76. Milk acidity is
    (*a*) 0.10 % (*b*) 0.11 %
    (*c*) 0.12 % (*d*) 0.13 %

77. Albumin and globulin occur in milk in the state of
    (*a*) Colloidal suspension (*b*) Colloidal solution
    (*c*) True solution (*d*) None of these

78. Flavour is composed of
    (*a*) Smell (*b*) Odour
    (*c*) Smell and taste (*d*) None of these

79. The curd tension of human milk is
    (*a*) 3.0 (*b*) 0.0
    (*c*) 55.0 (*d*) 14.5

80. Which one of the following is not an essential fatty acid?
    (*a*) Oleic acid (*b*) Linoleic acid
    (*c*) Linolenic acid (*d*) Arachidonic acid

81. Milk is a rich source of
    (*a*) Iron (*b*) Iodine
    (*c*) Calcium (*d*) Copper.

82. Which of the following is yellowish in colour?
    (*a*) Buffalo milk (*b*) Goat milk
    (*c*) Human milk (*d*) None of these

83. Which one of the following is not a major constituent of milk?
    (*a*) Ash (*b*) Fat
    (*c*) Lactose (*d*) Vitamins

84. Which one of the following is not a minor constituent of milk?
    (*a*) Enzyme (*b*) Vitamins
    (*c*) Ash (*d*) Phospholipids

85. Which one of the following has the smallest size of fat globules?
    (*a*) Cow milk (*b*) Buffalo milk
    (*c*) Goat milk (*d*) None of these

86. Which one of the following does not furnish any energy to the body?
    (*a*) Fat (*b*) Carbohydrate
    (*c*) Vitamins (*d*) Proteins

87. Which one of the following is not a protein?
    (*a*) Casein (*b*) Globulin
    (*c*) Glycine (*d*) Albumin

88. Acidity of milk is expressed as
    (*a*) Oleic acid (*b*) Lactic acid
    (*c*) Butyric acid (*d*) Citric acid

89. Which one of the following is the heaviest constituent of milk?
    (*a*) Proteins (*b*) Lactose
    (*c*) Casein (*d*) Mineral salts

90. The pH of normal fresh milk is
    (*a*) 6.6 (*b*) 6.8
    (*c*) 7.0 (*d*) None of these

91. Oxytocin remains effective for
    (*a*) 10 to 15 min (*b*) 5 to 7 min
    (*c*) 20 to 30 min (*d*) None of these

92. Which of the following contains the higher number of bacteria?
    (*a*) Fore milk (*b*) Middle milk
    (*c*) Stripping (*d*) None of these

93. Which of the following contains highest percentage of fat?
    (*a*) Fore milk (*b*) Middle milk
    (*c*) Stripping (*d*) None of these

94. The cholesterol content of a single egg-yolk is
    (*a*) 112 mg (*b*) 224 mg
    (*c*) 300 mg (*d*) 448 mg

95. Which of the following contribute to the hardness of milk?
    (*a*) Short chain saturated fatty acids (*b*) Unsaturated fatty acid
    (*c*) Long chain saturated fatty acids (*d*) None of these

96. Which one of the following contains cobalt?
    (*a*) Vitamin B_1 (*b*) Riboflavin
    (*c*) Folic acid (*d*) Vitamin B_{12}

97. Which one of the following is stored in the body?
(*a*) Vitamin B_1 (*b*) Vitamin B_6
(*c*) Vitamin B_{12} (*d*) Vitamin C

98. Which one of the following contains the highest percentage of lactose?
(*a*) Cow milk (*b*) Buffalo milk
(*c*) Goat milk (*d*) Human milk

99. One litre cow milk provides calcium
(*a*) 1 gm (*b*) 2 gm
(*c*) 3 gm (*d*) None of these

100. Maximum clumping of fat globules takes place at
(*a*) 7 °C (*b*) 11 °C
(*c*) 4 °C (*d*) 21 °C

101. Triglycerides of milk fat occur in
(*a*) Fat globules (*b*) Milk serum
(*c*) Fat globules membrane (*d*) None of these

102. Which one of the following is an unsaponificable constituent of milk fat?
(*a*) Triglyceride (*b*) Phospholipids
(*c*) Cholesterol (*d*) None of these

103. Cholesterol content in buffalo ghee is
(*a*) 330 mg % (*b*) 275 mg %
(*c*) 400 mg % (*d*) None of these

104. Melting point of cholesterol is
(*a*) 150 °C (*b*) 114 °C
(*c*) 117 °C (*d*) None of these

105. Vitamin A content in buffalo ghee on an average is
(*a*) 9.5 μg/g (*b*) 15.0 μg/g
(*c*) 7.00 μg/g (*d*) None of these

106. Which one of the following is responsible for the characteristic pleasing flavour of ghee?
(*a*) Lactic acid (*b*) Butyric acid
(*c*) Lactones (*d*) None of these

107. The number of essential amino acids is
(*a*) 6 (*b*) 8
(*c*) 10 (*d*) 12

108. Which one of the following is not a sulphur containing amino acid?
(*a*) Methionine (*b*) Cystine
(*c*) Lysine (*d*) None of these

109. Which one of the following contains the lowest percentage of ash?
(*a*) Cow milk (*b*) Buffalo milk
(*c*) Goat milk (*d*) Human milk

110. Which one of the following is known as vitamin M?
(*a*) Biotin (*b*) Choline
(*c*) Folic acid (*d*) None of these

111. The disease 'Pellagra' in man is caused due to the deficiency of
(*a*) Vitamin B_1 (*b*) Vitamin B_2
(*c*) Ascorbic acid (*d*) None of these

112. 'Fatty liver' is caused due to the deficiency of
(*a*) Cobalamin (*b*) Pantothenic acid
(*c*) Choline (*d*) None of these

113. Adequate heating of milk is indicated by
(*a*) Phosphatase test (*b*) Baudoin test
(*c*) Hansa test (*d*) None of these

114. Which of the following organisms produces red colour in milk?
(*a*) *B. cyanogenes* (*b*) *B. synxanthus*
(*c*) *Serratia marcescens* (*d*) None of these

115. Acidity of ghee is expressed as
(*a*) % lactic acid (*b*) % butyric acid
(*c*) % oleic acid (*d*) None of these

116. The optimum size of container for packaging ghee is
(*a*) 16 kg (*b*) 5 to 10 kg
(*c*) Not more than 4 kg (*d*) None of these

117. The optimum temperature for storing ghee is
(*a*) 5 °C (*b*) 7 °C
(*c*) 21 °C (*d*) None of these

118. The most sensitive method for detecting added water in milk is the determination of
(*a*) Specific gravity (*b*) Refractive index
(*c*) Freezing point (*d*) None of these

119. One I.U. of vitamin A is equal to
(*a*) 0.3 μg of vitamin A acetate (*b*) 0.6 μg of vitamin A acetate
(*c*) 0.9 μg of vitamin A acetate (*d*) None of these

120. The Vieth ratio between lactose, total protein and ash contents of milk is
(*a*) 9 : 13 : 2 (*b*) 13 : 9 : 2
(*c*) 13 : 2 : 9 (*d*) 9 : 13 : 2

121. Addition of cane sugar to milk can be detected by
(*a*) Hansa test (*b*) Cotton's test
(*c*) Holde's test (*d*) None of these

122. The following amount of formalin is sufficient as milk preservative
(*a*) 1 drop/100 ml of milk (*b*) 2 drop/100 ml of milk
(*c*) 4 drop/100 ml of milk (*d*) None of these

123. The density of water is maximum at
(*a*) 15.5 ºC (*b*) 20 ºC
(*c*) 4 ºC (*d*) None of these

124. The amount of cream taken for fat test by Gerber's method is
(*a*) 11 gram (*b*) 10 gram
(*c*) 5 gram (*d*) None of these

125. The specific gravity of Gerber's sulphuric acid is
(*a*) 1.091 (*b*) 1.082
(*c*) 1.815 (*d*) None of these

126. If cream is removed from milk sample and water is added thereafter the specific gravity of the sample will
(*a*) increase (*b*) decrease
(*c*) will become normal (*d*) none of these

127. Boiling point of buffalo milk is
(*a*) 100 ºC (*b*) 117 ºC
(*c*) 100.17 ºC (*d*) 100.55 ºC

128. In India, "milk' means:
(*a*) Cow milk (*b*) Buffalo milk
(*c*) Cow milk or buffalo milk or a mixture thereof
(*d*) None of these

129. Mixed milk means a mixture of :
(*a*) Cow milk, buffalo milk and goat milk (*b*) Milk of all the species
(*c*) Cow and buffalo milk (*d*) None of these

130. There is no formation of cream when the following is allowed to stand overnight?
(*a*) Cow milk (*b*) Buffalo milk
(*c*) Goat milk (*d*) None of these

131. Cow or buffalo body fat is called
(*a*) Lord (*b*) Pork
(*c*) Tallow (*d*) None of these

132. Pig body fat is called
(*a*) Lord (*b*) Pork
(*c*) Tallow (*d*) None of these

133. Presence of sesame oil in genuine ghee can be detected by
(*a*) Nickel test (*b*) Valenta test
(*c*) Holde's test (*d*) Baudoin's test

134. Presence of cotton seed oil in genuine ghee or the genuine ghee from cotton tract areas may be detected by
(*a*) B.N. Ghosh's test (*b*) Boudoin's test
(*c*) Halphen's test (*d*) None of these

135. Addition of any vegetable oil or vegetable fat to genuine ghee can be detected by
(*a*) Nickel test (*b*) Boudoin's test
(*c*) Bomer's phytosterol test (*d*) None of these

136. Which of the following test can be used to detect the adulteration of genuine ghee with Dalda (vanaspati ghee)
(*a*) Bomer's phytosterol test (*b*) Nickel test
(*c*) Holde's test (*d*) Nitric acid test

137. The B.R. Reading of animal body fats range from,
(*a*) 40 to 50 (*b*) 50 to 60
(*c*) 44 to 51 (*d*) None of these

138. R.M. value of a ghee sample indicates:
(*a*) Saturation of fatty acid (*b*) Unsaturation of fatty acid
(*c*) Steam-volatile water-soluble fatty acids (*d*) None of these

139. Iodine value of ghee sample is the indication of
(*a*) Saturation of fatty acid (*b*) Unsaturation of fatty acid
(*c*) Average molecular weight of the fatty acids (*d*) None of these

140. Saponification value indicates
(*a*) Average molecular weight of the fatty acids (*b*) Saturation of fatty acid
(*c*) Steam-volatile water-insoluble fatty acids (*d*) None of these

141. Which of the following in not found in the adulterant fats?
(*a*) Oleic acid (*b*) Linolenic acid
(*c*) Butyric acid (*d*) None of these

142. Which of the following is not found in the milk fat?
(*a*) Butyric acid (*b*) Caproic acid
(*c*) Oleic acid (*d*) Cyclopropenoic acid

143. Cyclopropenoic acid is found in
(*a*) Milk fat (*b*) Animal body fat
(*c*) Cotton seed oil (*d*) None of these

144. Which of the following can reduce methylene blue to colourless?
(*a*) Animal body fat (*b*) Cotton seed oil
(*c*) Milk fat (*d*) None of these

145. Which one of the following is a volumetric method for determining % fat in milk?
(*a*) Gerber's method (*b*) Werner-Schmidt method
(*c*) Adam's paper-coil method (*d*) Roese-Gottlieb method

146. Which one of the following is a gravimetric method for determining fat in milk?
(*a*) Gerber's method (*b*) Babcock method
(*c*) Roese-Gottlieb method (*d*) None of these

147. Which of the following test can not be used for detection of adulteration of genuine ghee with animal body fat?
(*a*) Valenta test (*b*) Carbolic test
(*c*) Opacity test (*d*) Holde's test

148. Which one of the following milk preservative is not an official preservative?
(*a*) Formalin (*b*) Potassium permanganate
(*c*) Hydrofluoric acid (*d*) β-naphthol

149. Which one of the following milk preservative is an official preservative?
(*a*) Salicylic acid (*b*) β-naphthol
(*c*) Mercuric chloride (*d*) Hydrogen peroxide

150. Addition of formalin to milk interferes with the following test:
(*a*) Fat test (*b*) % total solids
(*c*) % Ash (*d*) None of these

151. The specific gravity of amyl alcohol used in the fat test by Gerber's method is
(*a*) 1.081 (*b*) 1.810
(*c*) 0.810 (*d*) 0.081

152. Ashing of milk in muffle furnace is done at:
(*a*) 200°C ± 10°C for 4-5 hrs
(*b*) 300°C ± 10°C for 4-5 hrs
(*c*) 550°C ± 10°C for 4-5 hrs
(*d*) 400°C ± 10°C for 4-5 hrs

153. Which one of the following milk preservative is most suitable for fat test in milk
(*a*) Formalin
(*b*) Potassium di-chromate
(*c*) Hydrogen peroxide
(*d*) None of these

154. The most accurate method of determining the specific gravity of milk is by:
(*a*) Lactometer
(*b*) R.D. Bottle
(*c*) Westphal balance
(*d*) None of these

155. "Zero point" is:
(*a*) Freezing point of milk
(*b*) Freezing point of water
(*c*) Freezing point of ice cream
(*d*) None of these

156. Lactostar (automatic milk analyzer) analyses fat, solid non fat, lactose, protein percentage and freezing point of milk within:
(*a*) 2 minutes
(*b*) 4 minutes
(*c*) 6 minutes
(*d*) None of these

157. The Butyro-Refractometer reading of milk fat ranges from:
(*a*) 45 to 50
(*b*) 40 to 45
(*c*) 50 to 60
(*d*) None of these

158. The Butyro-Refractometer reading of vegetable fat and oils ranges from:
(*a*) Above 50
(*b*) Above 60
(*c*) Above 70
(*d*) None of these

159. Which one of the following does not increase the Butyro-Refractometer reading when added to milk fat?
(*a*) Sesame oil
(*b*) Ground nut oil
(*c*) Coconut oil
(*d*) None of these

160. Dry ice is:
(*a*) Frozen water
(*b*) Frozen milk
(*c*) Frozen carbon dioxide
(*d*) None of these

161. Fishiness flavour in cream and butter is developed due to
(*a*) Aeromonas hydrophila
(*b*) P. fragi
(*c*) P. mephitica
(*d*) Alternaria

162. Freezing point of milk (°C) is
(*a*) – 0.50
(*b*) – 0.54
(*c*) – 0.60
(*d*) – 0.65

163. Average specific gravity of milk is
(*a*) 1.030 (*b*) 1.035
(*c*) 1.040 (*d*) 1.045

164. Refractive index of milk at 20°C is which of the following
(*a*) 1.31 (*b*) 1.32
(*c*) 1.33 (*d*) 1.34

165. Boiling point of milk (°C) is
(*a*) 100.5 (*b*) 110.5
(*c*) 120.5 (*d*) 115.5

166. Which colour of milk is due to which of the following
(*a*) Carotene (*b*) Lactochrome
(*c*) Calcium sensitive a_s casein (*d*) Calcium insensitive k casein

167. Water soluble protein of milk is which of the following
(*a*) α-Lactoalbumin (*b*) β-Lactoglobulin
(*c*) α-Casein (*d*) β-Casein

168. Milk from which cream can not be separated
(*a*) Frozen milk (*b*) Soft curd milk
(*c*) Homogenized milk (*d*) Fermented milk

169. Bacterial count in machine milking in comparison to hand milking is
(*a*) Lower (*b*) Equal
(*c*) Higher (*d*) Zero

170. Melting point of milk fat (°C) is
(*a*) 25-35 (*b*) 30-36
(*c*) 36-40 (*d*) 40-45

171. Pressure required for machine milking (mm of Hg)
(*a*) 15-25 (*b*) 25-40
(*c*) 40-50 (*d*) 50-60

172. Pulsation rate of machine milking (cycle/min)
(*a*) 10-20 (*b*) 30-40
(*c*) 40-120 (*d*) 120-200.

173. Maximum limit of SPC/ml for pasteurized milk is
(*a*) 10,000 (*b*) 20,000
(*c*) 30,000 (*d*) 40,000

174. Maximum limit of SPC/ml for market milk is
(*a*) 1,00,000
(*b*) 2,00,000
(*c*) 3,00,000
(*d*) 4,00,000

175. Lactogenic is also known as
(*a*) Chloretic
(*b*) Galactopaietic
(*c*) Cholacgagoue
(*d*) Mammogenic

176. Which of the following is/are emulsifiers of milk?
(*a*) Milk protein
(*b*) Phospholipids
(*c*) Both (*a*) and (*b*)
(*d*) None of these

177. Which of the following is volumetric method for determination of fat % of milk?
(*a*) Soxhlet method
(*b*) Wast fall balance
(*c*) Richmond sliding method
(*d*) Gerber method

178. Which of the following is method for determination of specific gravity of milk?
(*a*) Wast fall balance
(*b*) Soxhlet method
(*c*) Richmond sliding method
(*d*) Gerber method

179. Which of the following is gravimetric method for determination of fat % of milk?
(*a*) Wast fall balance
(*b*) Soxhlet method
(*c*) Richmond sliding method
(*d*) Gerber method

180. Which of the following is gravimetric method for determination of total solids % of milk?
(*a*) Wast fall balance
(*b*) Soxhlet method
(*c*) Richmond sliding method
(*d*) Gerber method

181. Richmond formula is
(*a*) $TS = 0.25\ L + 1.2$ Fat $+ 0.41$
(*b*) $TS = 0.25\ L + 0.14$ Fat $+ 1.2$(
(*c*) $TS = 1.25\ L + 1.2$ Fat $+ 0.14$
(*d*) $TS = 0.25\ L + 1.2$ Fat $+ 0.14$

182. Curding of milk by admixture of acid in milk is known as
(*a*) Sour curding
(*b*) Sweet milk
(*c*) Acid curding
(*d*) Milk curding

183. Curding of milk by admixture of rennet in milk is known as
(*a*) Sour curding
(*b*) Sweet milk
(*c*) Acid curding
(*d*) Milk curding

184. Ordinary souring of milk is due to citrate produced by
(*a*) Butter
(*b*) Chhana
(*c*) Paneer
(*d*) Cheese

185. Instruments used for measurement of refractive index
(*a*) Babcock bottle
(*b*) Soxhlet apparatus
(*c*) Lactometer
(*d*) Zeiss refractometer

186. Flavouring agent of butter is which of one the following
(*a*) Diacetyl
(*b*) Volatile fatty acid
(*c*) Butter protein
(*d*) All of these

187. Specific heat of milk
(*a*) Higher than water
(*b*) Lower than water
(*c*) Equal to water
(*d*) Independent

188. Which fungi are used for ripening of butter from unpasteurized sweet cream at room temperature?
(*a*) Oidium
(*b*) Oospora
(*c*) Both (*a*) and (*b*)
(*d*) None of these

189. Instrument used for vacuum pasteurization
(*a*) Pasteurized bottle
(*b*) Vacuum creator
(*c*) Flasher
(*d*) Henristassanao bottle

190. Efficiency of pasteurizer is measured by which test
(*a*) Phosphatase test
(*b*) MBR test
(*c*) Reductase test
(*d*) Resazurine test

191. Bacteriological status of milk is judged by which of the following
(*a*) Alizarine test
(*b*) COB test
(*c*) Reductase test
(*d*) Catalase test

192. In starter of the butter and the cheese bacteriophase fed on which of the following
(*a*) Acetate bacteria
(*b*) Propionate bacteria
(*c*) Lactate bacteria
(*d*) Butyrate bacteria

193. Is tetanus toxin appear in milk?
(*a*) Never
(*b*) Some time
(*c*) Depend upon symptom
(*d*) Always

194. Organoleptic test (platform test) is which of the following
(*a*) Smell and appearance of milk
(*b*) Acidity and temperature of milk
(*c*) Sediment and lactometer reading
(*d*) All above

195. Preheating temperature (°C) of milk before pasteurization is
(*a*) 30-35
(*b*) 35-40
(*c*) 40-45
(*d*) 45-50

196. Temperature and time combination for LTLT pasteurization
(*a*) 63°C and 30 min (*b*) 63°C and 30 sec
(*c*) 72°C and 15 sec (*d*) 72°C and 15 min

197. Temperature and time combination for HTST pasteurization
(*a*) 63°C and 30 min (*b*) 63°C and 30 sec
(*c*) 72°C and 15 sec (*d*) 72°C and 15 min

198. Casual agent of which disease is used as index for pasteurization
(*a*) Rabies (*b*) Small pox
(*c*) TB (*d*) Cholera

199. Organism of TB destroyed at what temperature
(*a*) 138°F (*b*) 142°F
(*c*) 143°F (*d*) 144°F

200. Temperature of cream line reduction test
(*a*) 138°F (*b*) 142°F
(*c*) 143°F (*d*) 144°F

201. Which of the following is true?
(*a*) LTLT is batch process (*b*) LTLT is continuous process
(*c*) HTST is both processes (*d*) None of these

202. Uperization temperature and time is (°C and second)
(*a*) 74 and 7 (*b*) 150 and 3/4
(*c*) 72 and 15 (*d*) 63 and 15

203. Test for accessing sweet/soft curding of milk
(*a*) Curd test (*b*) Hill curd test
(*c*) Tension test (*d*) Hill curd tension test

204. Test for mixing of the cow milk and buffalo milk
(*a*) Sediment test (*b*) Hansa test
(*c*) COB test (*d*) Alizarine test

205. Lecithin is a component of which of the following
(*a*) Milk protein (*b*) Milk vitamin
(*c*) Milk lipid (*d*) Milk pigment

206. Flavoured milk contains ______________ fat %.
(*a*) Above 2 % (*b*) 5 %
(*c*) 1-2 % (*d*) 2-5 %

207. Temperature and time for frozen concentrated milk is
(*a*) 4°C and 12 hr
(*b*) 0°C and 12 hr
(*c*) 4°C and 24 hr
(*d*) 0°C and 24 hr

208. Milk gel oxidized if it is pored in container of following metal
(*a*) Bronze
(*b*) Copper
(*c*) Nickel
(*d*) Steel

209. Taint due to improper homogenization is
(*a*) Fishy
(*b*) Oily
(*c*) Rusty
(*d*) None of these

210. Skim milk contains approximately ______________ of water
(*a*) 70 %
(*b*) 80 %
(*c*) 90 %
(*d*) 50 %

211. Nature of fermentation by yeast in milk is
(*a*) Acidic
(*b*) Alcoholic
(*c*) Alkaline
(*d*) Neutral

212. Carotene in cow milk (µg/gm)
(*a*) 26
(*b*) 30
(*c*) 34
(*d*) 40

213. Cow milk is used for making which of the following
(*a*) Butter
(*b*) Cheese
(*c*) Paneer
(*d*) All of these

214. Sheep milk is used for making
(*a*) Butter
(*b*) Cheese
(*c*) Paneer
(*d*) Roquefort cheese

215. Best substitute of human milk is
(*a*) Goat milk
(*b*) Cow milk
(*c*) Buffalo milk
(*d*) Camel milk

216. Minimum fat % required for two stage homogenization is
(*a*) Less than 6%
(*b*) Above 6 %
(*c*) Above 10 %
(*d*) Above 20 %

217. Specific gravity of milk is reduced one hour after milking is known as
(*a*) Ham burg phenomenon
(*b*) Rackangel phenomenon
(*c*) Donnan phenomenon
(*d*) None of these

218. Effect of fat admixture on specific gravity of milk
(*a*) Decreased (*b*) Increased
(*c*) Irregular (*d*) Not affected

219. Which of the following is/are preservatives for milk?
(*a*) Boric acid (*b*) Formalin
(*c*) Sodium carbonate (*d*) All above

220. Byproduct of cheese and cream making is
(*a*) Whey and skim milk (*b*) Whey and butter milk
(*c*) Skim milk and butter (*d*) All of these

221. Enzyme required for cheese formation from milk is
(*a*) Rennin (*b*) Amylase
(*c*) Alkaline phosphatase (*d*) All above

222. Which mineral element is found maximum in cow milk?
(*a*) Sodium (*b*) Calcium
(*c*) Phosphorus (*d*) Iron

223. Which sugar is present in milk?
(*a*) Lactose (*b*) Fructose
(*c*) Glucose (*d*) Sucrose

224. Casein protein is found in which of the following
(*a*) Egg (*b*) Milk
(*c*) Meat (*d*) Cereals

225. Table sugar is
(*a*) Lactose (*b*) Maltose
(*c*) Sucrose (*d*) Fructose

226. Total solids percentage in milk can be calculated by the formula
(*a*) CLR/4 +2.2 F + 1.4 (*b*) CLR/5 +1.2 F +0.14
(*c*) CLR/4 +1.2 F + 1.4 (*d*) CLR +1.2 F + 1.4

227. The milk sugar is
(*a*) Monosaccharide (*b*) Disaccharides
(*c*) Oligosaccharides (*d*) Polysaccharides

228. Which of the following milk is best for infant?
(*a*) Goat (*b*) Buffalo
(*c*) Cow (*d*) Sheep

229. Protein are composed of a series of
(*a*) Nucleic acid (*b*) Amino acid
(*c*) Nucleotide (*d*) None of these

230. National Dairy Development Board (NDDB) is located
(*a*) Punjab (Ludhiana) (*b*) Haryana (Karnal)
(*c*) Madhya Pradesh (Bhopal) (*d*) Gujarat (Anand)

231. The most important factor in determination the milk quality is
(*a*) SNF (*b*) Fat
(*c*) Both above (*d*) Lactose

232. Average total solids in milk is
(*a*) 8.5 % (*b*) 10.2 %
(*c*) 15.0 % (*d*) 12.5 %

233. Average ash percent in milk is
(*a*) 0.8 % (*b*) 1.8 %
(*c*) 2.8 % (*d*) 3.8 %

234. Taste of milk is
(*a*) Sour (*b*) Sweet
(*c*) Above both (*d*) None of these

235. Buffalo milk is not yellow because
(*a*) It contains more fat (*b*) It contains no lactose
(*c*) It contains no carotene (*d*) It contains more calcium

236. The freezing point of milk is about
(*a*) – 029°C (*b*) – 0.10°C
(*c*) – 4.5°C (*d*) – 0.55°C

237. Milk is deficient in which of the following
(*a*) Iron (*b*) Copper
(*c*) Calcium (*d*) Potassium

238. Milk – milk serum =
(*a*) Milk fat (*b*) Skim milk
(*c*) Coagulum (*d*) Casein

239. Milk – milk fat =
(*a*) Milk fat (*b*) Skim milk
(*c*) Coagulum (*d*) Casein

240. Casein is present in milk as which of the following
(*a*) Colloidal suspension (*b*) Emulsion
(*c*) True solvent (*d*) Homogenized

241. Fat is present in milk as which of the following form
(*a*) Colloidal suspension (*b*) Emulsion
(*c*) True solvent (*d*) Homogenized

242. Which constituent is critical for milk yield
(*a*) Milk fat (*b*) Milk casein
(*c*) Milk lactose (*d*) Calcium

243. Which of the following is known as spoon eating milk?
(*a*) Toned milk (*b*) Standard milk
(*c*) Tattoo milk (*d*) Recombined milk

244. Milk viscosity is due to which constituent of milk
(*a*) Casein (*b*) Albumin
(*c*) Globulin (*d*) Phospholipids

245. Which of the following is milk protein?
(*a*) Alpha casein (*b*) Beta casein
(*c*) Gamma casein (*d*) All above

246. Which of the following have maximum casein in its milk?
(*a*) Cow (*b*) Buffalo
(*c*) Human (*d*) Camel

247. Two carbonyl compounds are responsible for flavour in blue cheese.
(*a*) 2-heptanone and 2-nonanone (*b*) Trans-2-hexenal and Benzaldehyde
(*c*) Trans -2-hexenall and trans-2-cis-nonadienal (*d*) None of these

248. The freshly drawn milk turns the red litmus to blue and blue litmus to red due to the presence of
(*a*) Amino acid (*b*) Carbohydrates
(*c*) Lactic acid (*d*) None of these

249. Riboflavin is present in
(*a*) Butter milk (*b*) Skim milk
(*c*) Milk fat (*d*) All of these

250. At which temperature, the specific gravity of milk is determined?
(*a*) 25°C (*b*) 10°C
(*c*) 15°C (*d*) 20°C

251. Specific gravity of milk is

(*a*) >1 (*b*) 1
(*c*) <1 (*d*) 1.5

252. The average freezing point of milk is

(*a*) – 0.545°C (*b*) – 0.535°C
(*c*) – 0.550°C (*d*) – 0.555°C

253. The instrument used for determining of freezing point of milk is known as

(*a*) Cryogenic (*b*) Cryoscope
(*c*) Lactometer (*d*) Psychrometer

254. The thermal conductivity of milk should be

(*a*) 13.5×10^4 (*b*) 2.0×10^5
(*c*) 1.5×10^4 (*d*) None of these

255. Which component of milk imparts immunological properties to milk?

(*a*) Lactobacillus (*b*) Lactoglobulin
(*c*) Amino acid (*d*) Tryptophan

256. On an average, what is the ratio of calcium and phosphorus in milk?

(*a*) 1.43:1 (*b*) 2.43:1
(*c*) 1.7:1.5 (*d*) 1.6:1.4

257. Which vitamin is responsible for eye sight?

(*a*) D (*b*) B
(*c*) A (*d*) C

258. The alternative name of vitamin is

(*a*) Antirichetic (*b*) Calcitonic
(*c*) Keratin (*d*) None

259. What is the daily requirement of milk per capita in India?

(*a*) 280 g (*b*) 256 g
(*c*) 312g (*d*) 250 g

260. One litre of buffalo milk yield energy about

(*a*) 800 calorie (*b*) 900 calorie
(*c*) 1100 calorie (*d*) 1000 calorie

261. One litre of cow milk yield energy about

(*a*) 720 calorie (*b*) 750 calorie
(*c*) 1100 calorie (*d*) 980 calorie

262. The temperature, at which milk is cooled?
(*a*) Less than 10°C (*b*) Above 10°C
(*c*) 10°C (*d*) 25°C

263. Which method of chilling is less expensive, simple and suited for small scale operation?
(*a*) Can cooling (*b*) Can immersion
(*c*) Surface cooler (*d*) Tubular cooler

264. Which operations are performed to clean the milk?
(*a*) Straining (*b*) Filtration
(*c*) Clarification (*d*) All of these

265. When HTST pasteurization is performed, the milk is heated at
(*a*) 70°C (*b*) 60°C
(*c*) 85°C (*d*) 71.5°C

266. The sterilization of milk is done at
(*a*) 115°C for 15 minutes (*b*) 125°C for 10 minutes
(*c*) 130°C for 12 minutes (*d*) 118°C for 8 minutes

267. The process of homogenization was first time employed in 1892 for the manufacturing of
(*a*) Milk (*b*) Margarine
(*c*) RTS (*d*) Fruit juice

268. According to WHO and FAO, the percentage of milk fat in double cream is
(*a*) >18 % (*b*) >28 %
(*c*) >35 % (*d*) >45 %

269. The percentage of lactic acid in special cream grade is
(*a*) 0.1% (*b*) 0.2%
(*c*) 0.3% (*d*) 0.4%

270. Which milk was first made in India?
(*a*) Skim milk (*b*) Cream milk
(*c*) Toned milk (*d*) Double cream milk

271. The conversion of milk into curd takes place with the help of
(*a*) Bacteria (*b*) Molds
(*c*) Yeasts (*d*) None of them

272. Which of the following is capable of to freeze the food at –196°C?
(*a*) LPG (*b*) Frozen CO_2
(*c*) LN_2 (*d*) LH_2

273. Muffle furnace is used for the estimation of
(*a*) Protein
(*b*) Sugar
(*c*) Salt
(*d*) Ash

274. Texture analyzer is employed to determine the ____________ properties of foods.
(*a*) Physical
(*b*) Mechanical
(*c*) Rheological
(*d*) Aerodynamic and hydrodynamic

275. Packaging material for food should be
(*a*) Non toxic
(*b*) Transparent
(*c*) Economically sound
(*d*) All above

276. Which of the following is not considered as a food additives?
(*a*) Vitamins
(*b*) Flavouring agent
(*c*) Emulsifier
(*d*) Antioxidants

277. Sugar and salt is used as
(*a*) Antioxidant
(*b*) Sequestrant
(*c*) Emulsifier
(*d*) Preservative

278. Cottage cheese is an example of
(*a*) High fat
(*b*) Low fat
(*c*) Both
(*d*) none

279. Which of the following is the substitute of butter?
(*a*) margarine
(*b*) enzyme
(*c*) cholesterol
(*d*) ester

280. Dairy products such as cheese and butter milk is
(*a*) Fermented products
(*b*) Main products
(*c*) Byproducts
(*d*) None of these

281. Swiss cheese also known as
(*a*) Brown cheese
(*b*) White cheese
(*c*) Emmental cheese
(*d*) None of these

282. Salt added at a level about 2.5 % of the final butter constitutes flavour and also act as a
(*a*) Preservative
(*b*) antioxidant
(*c*) Catalyst
(*d*) Colouring agent

283. Which bacteria is responsible for conversion of lactose into lactic acid
(*a*) Aspergillus spp.
(*b*) Theujunasis
(*c*) Byssachlamyc fulva
(*d*) Streptococcus lactis

284. Fat percentage in milk cream is
(*a*) 25 (*b*) 40
(*c*) 35 (*d*) 40

285. First time secreted milk from cattle is known as
(*a*) Colustrum (*b*) Clostridium
(*c*) Cholesterol (*d*) None

286. Sterilized canned cream has fat percentage nearly
(*a*) 20% (*b*) 40%
(*c*) 60% (*d*) 80%

287. Frozen cream has fat percentage about
(*a*) 20% (*b*) 40%
(*c*) 60% (*d*) 80%

288. Fruit cream may have fat % approximately
(*a*) 20-25% (*b*) 40-45%
(*c*) 60-65% (*d*) 80-85%

289. Which of the following is responsible for better flavour in cream?
(*a*) Diacetyl (*b*) Propionic acid
(*c*) Acetyl methyl carbinol (*d*) All of the above

290. Separeta refers to
(*a*) Milk without cream (*b*) Milk after separated fat
(*c*) Milk with cream (*d*) All of the above

291. How much percentage of Khoa can be prepared from cow milk?
(*a*) 15.5% (*b*) 18.3%
(*c*) 21.6% (*d*) None

292. Mother dairy in Delhi was established in the year
(*a*) 1974-75 (*b*) 1976-77
(*c*) 1977-78 (*d*) 1978-79

293. Rancidity in cheese is caused by
(*a*) Lipolytic organisms (*b*) Low salts
(*c*) High acidity (*d*) None

294. Condensed milk means
(*a*) Milk with 30 % water (*b*) Milk with 30 % total solids
(*c*) Milk with 9 % fat (*d*) All of the above

295. Condensed milk may be made of
(*a*) Sugar free milk (*b*) Sweet milk
(*c*) Both above (*d*) None of the above

296. Pasty cheese is caused by
(*a*) High moisture and acidity (*b*) Low moisture and acidity
(*c*) High salt percentage (*d*) All of these

297. Evaporated milk may be made from
(*a*) Sugar free milk (*b*) Sweet milk
(*c*) Both above (*d*) None of the above

298. Evaporated milk is homogenized at
(*a*) 45-50°C (*b*) 55-60°C
(*c*) 60-65°C (*d*) 80-82°C

299. Milk powder may be prepared from
(*a*) Milk (*b*) Sepreta
(*c*) Cream (*d*) All of the above

300. Pure ghee should have RM value about
(*a*) >28 (*b*) <24
(*c*) >15 (*d*) None

301. Chhana refers to
(*a*) It is milk product (*b*) It is prepared after curdling of milk
(*c*) The solid particles are separated from paneer water
(*d*) All of the above

302. Which of the following component is found in paneer?
(*a*) Ca (*b*) Vitamin B
(*c*) Protein (*d*) All of the above

303. Export quality chedder paneer contains water about
(*a*) <36% (*b*) >36%
(*c*) <10% (*d*) None

304. Under UHT the milk is heated to
(*a*) 132°C for 1 sec (*b*) 132°C for 1 min
(*c*) 132°C for 1 hour (*d*) None

305. Evaporated milk is accomplished by removing the water upto
(*a*) 50 % (*b*) 60%
(*c*) 80% (*d*) 90%

306. Skim milk contain usually about __________ percent fat.
(*a*) 0.1 (*b*) 1.0
(*c*) 0.9 (*d*) 0.01

307. Moisture content in cow and buffalo milk khoa is respectively about
(*a*) 25.6 and 19.2% (*b*) 25.6 and 25.6%
(*c*) 10.6 and 20.5% (*d*) 35.1 and 18.9%

308. The position of India in milk production globally is
(*a*) First (*b*) Second
(*c*) Third (*d*) Fourth

309. Full name of SNF is
(*a*) Soluble non fat (*b*) Solid non fat
(*c*) Saturated non fat (*d*) Synthetic non fat

310. Rapid heating of cream is accomplished by injecting steam or by a combination of steam injection and evacuation in a process known as
(*a*) Flash pasteurization (*b*) Vat pasteurization
(*c*) Vacreation (*d*) Aseptic processing

311. The most popular of the UHT system are the direct heating methods include
(*a*) Steam injection technique (*b*) Steam infusion technique
(*c*) Bactofugation (*d*) Both (*a*) and (*b*)

312. Steam injection technique of UHT system is referred as
(*a*) Milk injected into steam process (*b*) Steam injected into milk process
(*c*) Vacreation (*d*) None

313. Milk injected into steam process of UHT method is also known as
(*a*) Steam injected technique (*b*) Steam infusion technique
(*c*) Milk injected technique (*d*) Milk infusion technique

314. The efficiency of milk pasteurization or the percentage of reduction of numbers of microorganisms in milk during pasteurization depends on
(*a*) Temperature of pasteurization (*b*) Holding time
(*c*) Total numbers of bacteria
(*d*) The proportion of the total microbial load that are spore-formers or thermoduric organisms
e) All of the above

315. The shelf life of pasteurized milk depends on the
(*a*) Temperature of storage (*b*) Number of microorganisms
(*c*) Types of microorganisms (*d*) All of the above

316. Sealed cans of evaporated milk are processed at about
(*a*) 115 to 118°C for 14-18 min
(*b*) 93 to 100°C for 14-18 min
(*c*) 71.1 to 82°C for 30 min
(*d*) 71 to 100°C for 10-30 min

317. During the manufacturing of processed cheese from cheese, one of the following is used as emulsifiers.
(*a*) Phosphates
(*b*) Citrates
(*c*) Tartrates
(*d*) All of the above

318. Refrigeration temperature are recommended for milk or related products during
(*a*) Storage in the plant and during delivery
(*b*) Retail market
(*c*) Home or restaurant until consumption
(*d*) All of the above

319. Butter should be stored at
(*a*) –17 to –18°C or lower
(*b*) 0 to 5°C
(*c*) –5 to –10°C
(*d*) None

320. Evaporated milk is made by removing about _______ percent of the water from whole milk.
(*a*) 20
(*b*) 40
(*c*) 60
(*d*) 80

321. Among the dairy products prepared in the dry form are
(*a*) Milk, skim milk
(*b*) Cream, ice cream milk
(*c*) Whey, butter milk and malted milk
(*d*) All of the above

322. Most of dry milk is prepared by
(*a*) Roller process with vacuum
(*b*) Roller process without vacuum
(*c*) Spray drying process
(*d*) All of the above

323. Milk is preheated before drying to __________ for the roller process and to __________ for the spray process.
(*a*) 65 to 85°C, 68.8 to 93.3°C
(*b*) 60 to 70°C, 65 to 75°C
(*c*) 50 to 80°C, 60 to80°C
(*d*) None

324. In dry process of milk involve an instantization step, in which the dry powder is
(*a*) Wetted and re-dried
(*b*) Added the steam and dried
(*c*) Dried and wetted
(*d*) None

325. The microbial content of the heat dried dairy product depends on
(*a*) Content of liquid product to be dried
(*b*) Temperature and time of preheating and method of drying
(*c*) Contamination and growth in storage tank, pipes and drying source (air)
(*d*) All of them

326. The high temperature of the roller process of drying milk without vacuum destroyed almost all microorganisms

(*a*) Fungus (*b*) Mold

(*c*) Bacterial spores (*d*) All

327. The use of sorbic or propionic acid or one of their salts is permitted in

(*a*) Cottage cheese (*b*) Yogurt

(*c*) Hard cheese and processed cheese (*d*) All of them

328. The primary objective in adding a preservative to hard cheese or processed cheese is to prevent

(*a*) Surface growth of mold (*b*) Surface growth of bacteria

(*c*) Surface growth of yeast (*d*) All of them

329. Added sugar acts as a preservative and reduces the water activity in

(*a*) Toned milk (*b*) Skim milk

(*c*) Sweetened condensed milk (*d*) Evaporated milk

330. Sodium chloride (common salt) is usually used in the cheese making for

(*a*) Increase digestibility (*b*) Flavour enhancement

(*c*) Controlling the growth of microorganisms (*d*) Both (*b*) and (*c*)

331. Mold spoilage of cheese is usually delayed or prevented when ________ are added or incorporated in the wrapper.

(*a*) Sorbic acid (*b*) Sorbates

(*c*) Propionic acid or propionates (*d*) All of these

332. Milk is an excellent culture medium for much kind of microorganisms because of

(*a*) High in moisture content (*b*) Neutral in pH

(*c*) Rich in microbial foods (*d*) All of the above

333. In raw milk at temperature from 10 to 37°C, which microorganisms are most likely to cause the souring?

(*a*) Streptococcus lactic (*b*) S. faecalis

(*c*) Lactobacillus bulgaricus (*d*) L. thermophilus

334. Butyric acid may be produced in the milk by the action of

(*a*) Coliform bacteria (*b*) Streptococcus spp.

(*c*) Clostridium spp. (*d*) Bacillus spp.

335. The chief gas formers in milk and milk products are

(*a*) Coliform bacteria (*b*) Clostridium spp.

(*c*) Gas forming bacillus spp. (*d*) All of the above

336. The hydrolysis of milk protein by microorganisms usually accompanied by production of bitter flavour caused by some of the peptides release is called
(*a*) Proteolysis (*b*) Hydrogenation
(*c*) Stabilization (*d*) Sterilization

337. The ropiness usually decreases as the acidity of milk or cream
(*a*) Decreases (*b*) Increases
(*c*) Constant (*d*) Increase than decrease

338. Ropiness and sliminess in milk, cream and whey is caused by
(*a*) Enterobacter (*b*) LAB
(*c*) Micrococcus, streptococcus and bacillus (*d*) All of the above

339. The change in milk fat occurs by microbial decomposition are
(*a*) Oxidation of unsaturated fatty acid (*b*) Hydrolysis of the butter fat
(*c*) Combined oxidation and hydrolysis to produce rancidity
(*d*) All of the above

340. Oxidation of the unsaturated fatty acids of milk may results
(*a*) Aldehydes (*b*) Acids
(*c*) Ketones (*d*) Tallow odour and taste
e) All of the above

341. Hydrolysis of the butterfat to fatty acid and glycerol may occur by the enzyme is
(*a*) Lipase (*b*) Lactase
(*c*) Amylase (*d*) Catalase

342. Which one of the following, colour of milk is not changed due to microorganisms.
(*a*) Blue milk (*b*) Yellow
(*c*) Brown milk (*d*) Red milk
e) Green milk

343. Which one of the following produces a bluish gray to brownish colour in milk in pure culture?
(*a*) Pseudomonas syncyanea (*b*) Pseudomonas synxantha
(*c*) Serratia spp. (*d*) Micrococcus roseus

344. Pseudomonas synxantha may cause a _________ colour in the cream layer of milk.
(*a*) Blue (*b*) Yellow
(*c*) Red (*d*) Green

345. In red milk, a red layer at the top of milk is caused by
(*a*) Brevibacterium erythrohenes (*b*) Flavobacterium
(*c*) Geotrichum (*d*) All of the above

346. At refrigeration temperature, proteolysis may be initiated by psychrotophic bacteria such as
(*a*) Pseudomonas
(*b*) Leuconostoc
(*c*) Shiggela
(*d*) Salmonella

347. Spoilage of evaporated milk from leakage, as indicated by the presence of non spore formers may result in gas and swelling caused by
(*a*) Coliform bacteria or yeast
(*b*) Coagulation by streptococci
(*c*) Bitterness caused by cocci
(*d*) All of them

348. The main kind of spoilage in sweetened condensed milk is due to
(*a*) Gas forming fermentating yeasts or more rarely by Coliform bacteria
(*b*) Thickening caused by micrococci which probably produced rennin like enzyme
(*c*) Button's which are mold colonies growing on the milk surface
(*d*) All of the above

349. Cheesiness flavour defect in cream is caused by
(*a*) Lactobacilli
(*b*) Enterobacter
(*c*) Yeast
(*d*) Coliform bacteria

350. Musty flavour in cream is caused by
(*a*) Bacteria
(*b*) Mold and anctinomycetes
(*c*) Pseudomonas
(*d*) None of these

ANSWER KEY (MILK AND MILK PRODUCT PROCESSING)

1.A	2.A	3.B	4.C	5.D	6.A	7.D	8.A	9.B	10.D
11.A	12.B	13.A	14.B	15.C	16.D	17.A	18.B	19.C	20.D
21.A	22.B	23.C	24.D	25.A	26.B	27.C	28.D	29.A	30.A
31.B	32.C	33.D	34.A	35.B	36.C	37.D	38.A	39.B	40.C
41.D	42.A	43.B	44.C	45.A	46.B	47.A	48.A	49.A	50.B
51.D	52.A	53.C	54.B	55.A	56.C	57.B	58.C	59.D	60.A
61.B	62.B	63.C	64.C	65.B	66.B	67.B	68.C	69.C	70.D
71.C	72.B	73.D	74.A	75.D	76.D	77.B	78.C	79.B	80.A
81.C	82.D	83.D	84.C	85.C	86.C	87.C	88.B	89.D	90.A
91.B	92.A	93.C	94.D	95.C	96.D	97.C	98.D	99.A	100.A
101.A	102.C	103.B	104.A	105.A	106.C	107.C	108.C	109.D	110.C
111.D	112.C	113.A	114.C	115.C	116. C	117.C	118.C	119.A	120.B
121.B	122.B	123.C	124.C	125.C	126.C	127.D	128.C	129.C	130.C
131.C	132.A	133.D	134.C	135.C	136.B	137.C	138.C	139.B	140.A
141.C	142.D	143.C	144.B	145.A	146.C	147.D	148.D	149.C	150.A
151.C	152.C	153.B	154.C	155.B	156.A	157.B	158.A	159.C	160.C
161.A	162.B	163.A	164.D	165.A	166.C	167.A	168.C	169.C	170.B
171.B	172.C	173.C	174.A	175.B	176.B	177.D	178.A	179.B	180.C
181.D	182.A	183.B	184.C	185.D	186.A	187.B	188.C	189.B	190.A
191.C	192.C	193.D	194.D	195.B	196.A	197.C	198.C	199.A	200.D
201.A	202.B	203.D	204.B	205.C	206.C	207.A	208.B	209.B	210.C
211.B	212.A	213.B	214.D	215.A	216.B	217.B	218.A	219.D	220.A
221.A	222.B	223.A	224.B	225.C	226.C	227.B	228.A	229.B	230.D
231.C	232.D	233.A	234.B	235.C	236.D	237.A	238.C	239.B	240.A
241.B	242.C	243.C	244.A	245.D	246.B	247.A	248.A	249.B	250.C
251.A	252.A	253.B	254.A	255.B	256.A	257.C	258.A	259.A	260.C
261.A	262.A	263.B	264.D	265.D	266.A	267.B	268.D	269.B	270.C
271.A	272.C	273.D	274.C	275.D	276.A	277.D	278.B	279.A	280.C
281.C	282.A	283.D	284.A	285.A	286.A	287.B	288.B	289.A	290.D
291.B	292.A	293.A	294.D	295.C	296.A	297.A	298.A	299.D	300.A
301.D	302.D	303.A	304.A	305.B	306.A	307.A	308.A	309.B	310.C
311.D	312.B	313.B	314.E	315.D	316.A	317.D	318.D	319.A	320.C
321.D	322.D	323.A	324.A	325.D	326.C	327.D	328.A	329.C	330.D
331.D	332.D	333.A	334.C	335.D	336.A	337.B	338.D	339.D	340.E
341.A	342.E	343.A	344.B	345.A	346.A	347.D	348.D	349.A	350.B

Meat Processing

1. NRC on meat and meat products is located at
   (*a*) Hyderabad
   (*b*) Aligarh
   (*c*) Izzatnagar
   (*d*) Puttur

2. NRC on cold water fisheries is at
   (*a*) Srinagar
   (*b*) Nainital
   (*c*) Dehradun
   (*d*) Shimla

3. Central Institute of Fisheries Technology is established at
   (*a*) Kolkata
   (*b*) Mumbai
   (*c*) Cochin
   (*d*) Chennai

4. Blue revolution is related to
   (*a*) Crops
   (*b*) Oilseeds
   (*c*) Energy source
   (*d*) Fisheries

5. Quality of egg can be judged by
   (*a*) pH
   (*b*) Candling
   (*c*) Annealing
   (*d*) Temperature test

6. Egg shell made up of
   (*a*) Cao
   (*b*) $Ca_3(PO_4)_2$
   (*c*) $CaCO_3$
   (*d*) $Ca(OH)_2$

7. Yellow colour of egg is due to
   (*a*) Carotene
   (*b*) Vitamin B
   (*c*) Xanthophyll
   (*d*) None of these

8. Meat of sheep is known as
   (*a*) Chicken
   (*b*) Pork
   (*c*) Beef
   (*d*) Mutton

9. Buffalo meat contains white fat as
(*a*) WBC
(*b*) RBC
(*c*) Yellow pigment
(*d*) Carotene covered to vitamin A

10. Minimum weight of extra large A grade Egg is
(*a*) 45 g
(*b*) 55 g
(*c*) 60 g
(*d*) 70 g

11. Oiling of egg present the escape of following gas
(*a*) CO_2
(*b*) O_2
(*c*) Water vapour
(*d*) H_2

12. Egg shell contributes about how much percent of egg
(*a*) 11
(*b*) 20
(*c*) 30
(*d*) 35

13. To differentiate between fresh and old eggs which of the following method is used?
(*a*) Breakage of egg
(*b*) Rotation of egg
(*c*) Candling of egg
(*d*) All above

14. In white albumin, how much fat is present?
(*a*) 0 %
(*b*) 17 %
(*c*) 30 %
(*d*) 55 %

15. In yolk, how much protein is present?
(*a*) 0 %
(*b*) 17 %
(*c*) 30 %
(*d*) 50 %

16. Which is major fat in yolk?
(*a*) Phosvitin
(*b*) Conalbumin
(*c*) Ovalbumin
(*d*) Livetin

17. In which albumin, which protein is in maximum amount (54%)
(*a*) Phosvitin
(*b*) Ovalbumin
(*c*) Livetin
(*d*) Conalbumin

18. Major bactericidal protein in albumin is
(*a*) Phosvitin
(*b*) Ovalbumin
(*c*) Livetin
(*d*) Conalbumin

19. Which protein of albumin, binds with biotin
(*a*) Ovamucin
(*b*) Ovamucoid
(*c*) Lysozyme
(*d*) Avidin

20. Energy provide per egg is
(*a*) 75 kcal
(*b*) 90 kcal
(*c*) 100 kcal
(*d*) 150 kcal

21. Fat content per egg is
(*a*) 5.5 gm
(*b*) 6.6 gm
(*c*) 600 IU
(*d*) 20 mg

22. Cholesterol content per egg is about in yolk
(*a*) 240 mg
(*b*) 240 gm
(*c*) 600 IU
(*d*) 2.0 mg

23. Vitamin A content per egg is
(*a*) 540 IU
(*b*) 300 IU
(*c*) 600 IU
(*d*) 450 IU

24. Storage temperature of egg should be
(*a*) 20 °F
(*b*) 30 °F
(*c*) 50-55 °F
(*d*) 90 °F

25. Freezing temperature of chicken is
(*a*) 0 °F
(*b*) -10°F to -20°F
(*c*) -20 °F to -40°F
(*d*) -60°F to -80°F

26. Chilling of carcass (poultry) done at which temperature
(*a*) 0 °F
(*b*) 15 °F
(*c*) 25 °F
(*d*) 30 °F

27. Removal of filoplume from carcass is known as
(*a*) Singing
(*b*) Pinning
(*c*) Picking
(*d*) scalding

28. Removal of feather from scaled bird is known as
(*a*) Singing
(*b*) Pinning
(*c*) Picking
(*d*) Scalding

29. Removal of pin feather from carcass is known as
(*a*) Singing
(*b*) Pinning
(*c*) Picking
(*d*) Scalding

30. Time and temperature combination for bird scalding is
(*a*) 120 -130°F for 20-40 sec
(*b*) 100°F
(*c*) 180 – 190°F for 6-8 sec
(*d*) Any one of these

31. Semi-scalding is done at
    (*a*) 120 -130°F for 20-40 sec
    (*b*) 100°F
    (*c*) 180 – 190°F for 6-8 sec
    (*d*) Any one of these

32. Which is ISI grade of quality of egg?
    (*a*) A to F
    (*b*) 1 to 5
    (*c*) I to V
    (*d*) Any one

33. According ISI, the size of extra large egg
    (*a*) 70 gm
    (*b*) 60 gm
    (*c*) 50 gm
    (*d*) 47 gm

34. According to USDA grade, the size of extra large egg is
    (*a*) 70 gm
    (*b*) 65 gm
    (*c*) 56 gm
    (*d*) 49 gm

35. In good egg, air space should be
    (*a*) 1/8"
    (*b*) 1/16"
    (*c*) 3/36"
    (*d*) 5/16"

36. Egg quality judge by
    (*a*) External appearance
    (*b*) Candling experiment
    (*c*) Broken out examination
    (*d*) All

37. Cocking test include
    (*a*) Chemical test of egg
    (*b*) Biological test of egg
    (*c*) Bacterial test of egg
    (*d*) All

38. Specialties of cold storage for egg is
    (*a*) Temperature 29 - 30°F, humidity 80-90 %
    (*b*) Temperature 30 - 40°F, humidity 60-80 %
    (*c*) Temperature 0 -10°F, humidity 90 %
    (*d*) Temperature 4 -6°F, humidity 100 %

39. Shell treatment of egg is done for
    (*a*) Checking escape of CO_2
    (*b*) Checking escape of O_2
    (*c*) Checking escape of moisture
    (*d*) Checking escape of N_2

40. In shell treatment spraying is done by
    (*a*) Liquid
    (*b*) Coconut oil
    (*c*) Both ((*a*) and ((*b*)
    (*d*) Mustard oil

41. Pasteurizing of egg is done by
    (*a*) Immersing egg in water at 72°F for 30 minutes
    (*b*) Immersing egg in water at 100°F for 15 minutes
    (*c*) Immersing egg in water at 130°F for 30 minutes
    (*d*) None of these

42. Dirty egg can be cleaned by washing with
    (*a*) NaOH (*b*) Warm water
    (*c*) Cold water (*d*) Both ((*a*) and ((*b*)

43. Almost all vitamins are present in eggs except vitamin
    (*a*) A (*b*) B
    (*c*) C (*d*) E

44. The colour of meat is due to the pigment
    (*a*) Heamoglobin (*b*) Myoglobin
    (*c*) Lycopene (*d*) None of these

45. The shell of egg is made of
    (*a*) Calcium carbonate (*b*) Magnesium carbonate
    (*c*) Sodium carbonate (*d*) Calcium propionate

46. Number of layer of outer membrane of egg is made by
    (*a*) Eight (*b*) Ten
    (*c*) Six (*d*) Seven

47. Number of layer of inner membrane of egg is made by
    (*a*) Three (*b*) Ten
    (*c*) Six (*d*) Seven

48. Two carbohydrates are found in meat is ______________ and ______________.
    (*a*) Glucose and fructose (*b*) Glycogen and glucose
    (*c*) Maltose and fructose (*d*) Chitin and lignin

49. Natural emulsifier *i.e.* lecithin found in
    (*a*) Egg (*b*) Meat
    (*c*) Apple (*d*) Cereals

50. Fish protein concentrate contain the high quality protein of about
    (*a*) 30-40% (*b*) 50-59%
    (*c*) 62-67% (*d*) 85-92%

51. Roasted white meat without skin contain fat percent about
    (*a*) 1.8 (*b*) 2.5
    (*c*) 3.5 (*d*) 4.0

52. Composition of an egg is
    (*a*) 70 % water, 7% protein and 11 % fat
    (*b*) 65 % water, 12% protein and 11 % fat
    (*c*) 80 % water, 10% protein and 11 % fat
    (*d*) 62 % water, 10% protein and 11 % fat

53. Meat preservative agents are
    (*a*) Nitrite and nitrate (*b*) Sugar
    (*c*) Acetic acid (*d*) Benzoic acid

54. The major protein of muscles fibre in meat is
    (*a*) Myosin (*b*) Elastin
    (*c*) Gyosin (*d*) None

55. When an animal is well fed, fat percentage between the muscle fibre bundles is called
    (*a*) Fat reduction (*b*) Fat marbling
    (*c*) Fat budding (*d*) Fat bundling

56. The substance added to meat as mild preservative and flavour enhancer is
    (*a*) Common salt (*b*) Sugar
    (*c*) Vinegar (*d*) Brine

57. The chief muscle pigment in meat protein is called
    (*a*) Myosin (*b*) Cyoglobin
    (*c*) Hemoglobin (*d*) Myoglobin

58. The ideal temperature of meat smoke room is
    (*a*) 52°C (*b*) 55°C
    (*c*) 57°C (*d*) 60°C

59. The smoking of meat is done for the period of
    (*a*) 15-20 hrs (*b*) 18-24 hrs
    (*c*) 20-25 hrs (*d*) 30-35 hrs

60. According to USFDA, freezing of meat is done at the temperature of
    (*a*) 0°C (*b*) – 10°C
    (*c*) –15°C (*d*) – 20°C

61. Nutritive value of beef meat is
    (*a*) Protein (18%), fat (23%), water (58%), and ash (0.5%)
    (*b*) Protein (17.5%), fat (22%), water (60%), and ash (0.9%)
    (*c*) Protein (21%), fat (20%), water (55%), and ash (0.9%)
    (*d*) Protein (25%), fat (18%), water (58%), and ash (0.5%)

62. Amount of protein in an egg
(*a*) 5.5 gm (*b*) 6.6 gm
(*c*) 600 IU (*d*) 20 mg

63. The washed poultry chilled to prevent bacterial spoilage and preserve quality at temperature of
(*a*) 32 to 2°C (*b*) – 35 to – 5°C
(*c*) – 40 to – 1°C (*d*) – 20 to – 10°C

64. A grade quality of egg can be maintained as long as
(*a*) 6 month (*b*) 12 month
(*c*) 18 month (*d*) 24 month

65. Pasteurization of egg should be done at the temperature of
(*a*) 60-62°C (*b*) 60-65°C
(*c*) 40-50°C (*d*) 70-75°C

66. The water activity of chilled meat is
(*a*) Very high (*b*) Very low
(*c*) Moderate (*d*) Low

67. Unprotected meat will lose weight by
(*a*) Evaporation (*b*) Respiration
(*c*) Transpiration (*d*) Refrigeration

68. Oxymyoglobin has a colour of
(*a*) Black (*b*) White
(*c*) Red (*d*) Bright red

69. The colour of meat surface will become some what brighter when stored at
(*a*) Lower temperature (*b*) Higher temperature
(*c*) Any temperature (*d*) None

70. The first observation regarding the effect of carbon dioxide on retarding bacterial growth were made in
(*a*) 1982 (*b*) 1882
(*c*) 1873 (*d*) 1802

71. The boxed beef concept was introduced in
(*a*) 1950 (*b*) 1870
(*c*) 1967 (*d*) 1913

72. Shrink bath method was developed by
(*a*) W.R. Grace (*b*) C.M. Davidson
(*c*) V.R. Robert (*d*) G. Moun

73. Vacuum packaged beef of normal pH can be stored at chiller temperature for
(*a*) One month
(*b*) 10 weeks
(*c*) 15 weeks
(*d*) 30 weeks

74. The use of carbon dioxide (CO_2) enriched atmosphere for extending the shelf life of poultry meat was carried out in
(*a*) Germany (1951)
(*b*) USA (1915)
(*c*) USA (1951)
(*d*) Newzeland (1915)

75. Sea foods have a much higher content of
(*a*) Poly unsaturated fat
(*b*) Lipid
(*c*) Protein
(*d*) Vitamin A, D, E, K

76. Heat products sufficiently to inactivate the non proteolytic spores and than keep the product well at
(*a*) 0°C
(*b*) 12°C
(*c*) Below 10°C
(*d*) Below 0°C

77. The partial pressure of oxygen in air is
(*a*) 160 mm Hg
(*b*) 200 mm Hg
(*c*) 150 mm Hg
(*d*) 760 mm Hg

78. The meat is the good culture media for the growth of microorganisms due to
(*a*) High in moisture
(*b*) Nearly neutral pH
(*c*) High in nutrients
(*d*) All of the above

79. After slaughtering, meat undergo undesirable changes occurs in
(*a*) Moisture content
(*b*) Appearance
(*c*) Flavour
(*d*) Both (*b*) and (*c*)

80. Long storage of meat at __________ temperature may allow increase in the number of microorganisms.
(*a*) Drying
(*b*) Freezing
(*c*) Chilling
(*d*) Curing

81. Most meat products are
(*a*) Low acid foods
(*b*) Medium acid foods
(*c*) High acid foods
(*d*) None of them

82. Chemicals added during curing of meat such as
(*a*) Spices
(*b*) Salt
(*c*) Nitrates and nitrites
(*d*) All of the above

83. One of the following ingredients in meat helps to kill spores of anaerobic bacteria by heat and inhibit germination of surviving spores.
(*a*) Salt (*b*) Spices
(*c*) Nitrates (*d*) Nitrites

84. The processing temperature for shelf stable canned cured meats is
(*a*) 65 °C (*b*) 72 °C
(*c*) 98 °C (*d*) 100 °C

85. The non shelf stable cured meats are packed in container upto 22 lb are processed at temperature of about
(*a*) 65 °C (*b*) 72 °C
(*c*) 98 °C (*d*) 100 °C

86. Mostly meat is preserved under low temperature in
(*a*) Freezing (*b*) Chilling
(*c*) Cold storage (*d*) Refrigeration

87. Increasing amounts of CO_2 in the atmosphere inhibits microorganisms but also
(*a*) Hasten the formation of metamyoglobin (*b*) Loss of bloom
(*c*) Loss of natural colour (*d*) All of the above

88. Optimal concentration of CO_2 under chilled condition for most of meat vary from
(*a*) 5-7 % (*b*) 5-15%
(*c*) 10-30% (*d*) Upto 100 %

89. Storage time of meat under chilling can be increased in the presence of ozone in atmosphere varying from
(*a*) 1-2% (*b*) 3-7%
(*c*) 2.5-3.0% (*d*) 10-30%

90. Ozone is an active oxidize agent that may cause tallow flavour to
(*a*) Fat (*b*) Protein
(*c*) Carbohydrates d) β-carotene

91. The fish is most susceptible to
(*a*) Autolysis (*b*) Oxidation
(*c*) Hydrolysis of fat (*d*) Microbial spoilage
(*e*) All of these

92. Rigor mortis is especially important in the preservation of fish to retard
(*a*) Postmortem autolysis (*b*) Bacterial decomposition
(*c*) Hydrolysis of fat (*d*) Both ((*a*) and ((*b*)

93. The best antioxidant used to retard oxidative rancidity of fat or oil of fish is
(*a*) Nordihydroguaiaretic acid (*b*) Ethyl gallet
(*c*) Ascorbic acid (*d*) All of the above

94. To differentiate between fresh and old eggs which of the following method is used?
(*a*) Breakage of egg (*b*) Rotation of egg
(*c*) Candling of egg (*d*) All above

95. Scalding of poultry may be done at the temperature of
(*a*) 60°C for 15 minutes. (*b*) 70°C for 5 second
(*c*) 60°C for 45 second (*d*) 80°C for 2 second

ANSWER KEY (MEAT PROCESSING)

1.A	2. B	3.C	4.D	5.B	6.C	7.C	8.D	9.D	10.C
11.B	12.A	13.A	14.B	15.D	16.D	17.B	18.D	19.D	20.B
21.A	22.A	23.C	24.C	25.C	26.D	27.A	28.C	29.B	30.C
31.A	32.A	33.B	34.A	35.A	36.D	37.B	38.A	39.A	40.C
41.C	42.D	43.C	44.B	45.A	46.C	47.A	48.B	49.A	50.D
51.C	52.B	53.A	54.A	55.B	56.A	57.D	58.C	59.B	60.C
61.B	62.B	63.A	64.A	65.A	66.A	67.B	68.D	69.A	70.B
71.C	72.A	73.B	74.C	75.A	76.C	77.A	78.D	79.D	80.C
81.A	82.D	83.C	84.C	85.A	86.B	87.D	88.C	89.C	90.A
91.E	92.D	93.D	94.C	95.C					

Food Chemistry

1. Food irradiation is used to
   (*a*) Precook foods
   (*b*) Kill insects
   (*c*) Kill microorganisms
   (*d*) Increase nutrient value

2. The calorific value is defined as the amount of heat produced in calories when __________ of substance is completely burnt.
   (*a*) One gram
   (*b*) One mole
   (*c*) 100 grams
   (*d*) 1 kg

3. The heat change when one mole of a liquid is converted into vapour or gaseous state at its boiling point is called
   (*a*) Heat of liquifaction
   (*b*) Heat of fusion
   (*c*) Heat of vaporization
   (*d*) Heat of sublimation

4. The process of sublimation for a solid occurs
   (*a*) At the melting point
   (*b*) At a temperature below its melting point
   (*c*) At a temperature above its melting point
   (*d*) None of these

5. The apparatus used to measure heat change during chemical reactions is called
   (*a*) Polarimeter
   (*b*) Colorimeter
   (*c*) Calorimeter
   (*d*) None of these

6. The boiling point of a liquid is that temperature at which
   (*a*) The vapour pressure of a liquid is equal to the atmosphere pressure
   (*b*) The vapour pressure of a liquid is less than the atmosphere pressure
   (*c*) The vapour pressure of a liquid is greater the atmosphere pressure
   (*d*) The vapour pressure of a liquid is equal to the square root of the atmosphere pressure.

7. Viscosity of a liquid is measure of
   (*a*) Repulsive forces between the liquid molecules
   (*b*) Frictional resistance
   (*c*) Intermolecular forces between the molecules
   (*d*) None of the above.

8. The reciprocal of viscosity is called
   (*a*) Surface tension
   (*b*) Frictional resistance
   (*c*) Fluidity
   (*d*) None of these

9. The dimension in which coefficient of viscosity is expressed are
   (*a*) mass × length × $time^{-1}$
   (*b*) mass × $length^{-1}$ × $time^{-1}$
   (*c*) mass × $length^{-1}$ × time
   (*d*) $mass^{-1}$ × length × $time^{-1}$

10. The SI unit of coefficient of viscosity is
    (*a*) kg m^2 sec
    (*b*) kg m sec^{-1}
    (*c*) kg m^{-1} sec
    (*d*) kg m^{-1} sec^{-1}

11. The fluidity of liquids...............with increase in temperature
    (*a*) Decreases
    (*b*) Remains the same
    (*c*) Increases
    (*d*) None of these

12. The rate of evaporation of a liquid depends upon
    (*a*) Surface area
    (*b*) Temperature
    (*c*) Nature of liquid
    (*d*) All of these

13. The highest temperature at which vapour pressure of liquid can be measured is
    (*a*) Boling point of the liquid
    (*b*) Freezing point of the liquid
    (*c*) Inversion temperature
    (*d*) Critical temperature

14. The units of coefficient of viscosity are
    (*a*) Dyne cm^{-1} sec
    (*b*) Poise
    (*c*) Centi poise
    (*d*) All of these

15. The concentration of a solution is defined as
    (*a*) The amount of solvent present in a given amount of solution
    (*b*) The amount of solute present in a given amount of solution
    (*c*) The amount of solute present in a given amount of solvent
    (*d*) The amount of solvent present in a given amount of solute

16. If n represents the number of moles of a solute and N represents the number of moles of a solvent, the mole fraction of the solvent is given by
    (*a*) $n / (n+N)$
    (*b*) $N / (n+N)$
    (*c*) $(n+N) / n$
    (*d*) $(n+N) / N$

17. A saturated solution is defined as one which is .... with the excess of solid at a particular temperature
    (*a*) in contact
    (*b*) in equilibrium
    (*c*) constant impurities
    (*d*) none of these

18. In super saturated solution, the amount of solute is …………in comparison to saturated solutions.
    (*a*) More (*b*) Less
    (*c*) Equal to (*d*) None of these

19. The solubility generally rises with
    (*a*) Increase in temperature (*b*) Decrease in temperature
    (*c*) Increase in volume of the solvent (*d*) None of these

20. The complete miscible solution can be separated by
    (*a*) A separating funnel (*b*) Evaporation
    (*c*) Fractional distillation (*d*) None of these

21. Sugar dissolves in water due to the formation of
    (*a*) covalent bonds (*b*) ionic bonds
    (*c*) co-ordinate bonds (*d*) hydrogen bonding.

22. The rate of crystallization is the rate at which the
    (*a*) Solid molecules go into the solvent
    (*b*) Solute molecules return to the crystal surface and are deposited on it.
    (*c*) Solute molecules leave the surface of the solution
    (*d*) Solute molecules condense on the surface of the solution.

23. A liquid boils when its vapour pressure becomes equal to
    (*a*) one atmospheric pressure (*b*) zero
    (*c*) very high (*d*) very low

24. Freezing point depression is measured by
    (*a*) Beckmann's method (*b*) Rast's camphor method
    (*c*) Both (*d*) None of these

25. The study of depression in freezing point of a solution is called
    (*a*) Osmotic pressure (*b*) Ebullioscopy
    (*c*) Cryoscopy (*d*) None of these

26. The study of elevation in boiling point of a solution is done under
    (*a*) Osmotic pressure (*b*) Ebullioscopy
    (*c*) Cryoscopy (*d*) None of these

27. A semi-permeable membrane allows the passage of ______________ through it.
    (*a*) Solvent only (*b*) Solute only
    (*c*) Solvent and solute both (*d*) Either solvent or solute

28. The flow of solvent through a semi-permeable membrane from pure solvent to solution is called
(*a*) Diffusion (*b*) Osmosis
(*c*) Effusion (*d*) None of these

29. The flow of solvent molecules can take place from
(*a*) dilute to concentrated solution (*b*) concentrated to dilute solution
(*c*) anyone solution to the other (*d*) none of these

30. The external pressure applied to the solution in order to stop the passage of solvent molecules into solution by a semi-permeable membrane is called
(*a*) osmosis (*b*) osmotic pressure
(*c*) optical pressure (*d*) none of these

31. Two solution are separated by a semi-permeable membrane. The one which is of lower osmotic pressure is called
(*a*) hypotonic (*b*) hypertonic
(*c*) isoosmotic (*d*) isomorphic

32. A hypertonic solution is the one which has ___________ osmotic pressure then the other.
(*a*) Lower (*b*) Equal
(*c*) Higher (*d*) One of these

33. The osmosis taking place from solution to pure water by application of pressure greater than osmotic pressure on the solution is called
(*a*) Equilibrium osmosis (*b*) Irreversible osmosis
(*c*) Reversible osmosis (*d*) Reverse osmosis

34. The osmotic pressure of a solution at a given temperature is directly proportional to
(*a*) Atmospheric pressure (*b*) Concentration
(*c*) Internal energy (*d*) Volume of the solution

35. The osmotic pressure of a solution of a given concentration is directly proportional to
(*a*) Room temperature (*b*) Atmospheric pressure
(*c*) Absolute temperature (*d*) Critical temperature

36. The osmotic pressure is directly proportional to
(*a*) Vapour pressure of the pure solvent (*b*) Lowering of vapour pressure
(*c*) Relative lowering of vapour pressure (*d*) None of these

37. When a bottle of soda water is opened, the partial pressure of carbon dioxide __________
(*a*) decreases (*b*) increases
(*c*) remains the same (*d*) none of these

38. The enzyme which can catalyze the conversion of glucose to ethanol is
(*a*) maltase (*b*) diastase
(*c*) invertase (*d*) zymase.

39. Enzymes are
(*a*) finely divided metals (*b*) transition metal ions
(*c*) auto-catalysts (*d*) proteins

40. The phenomenon of negative catalysis is known as
(*a*) Auto catalysis (*b*) Self catalysis
(*c*) Inhibition (*d*) Enzyme catalysis

41. A colloidal solution consists of
(*a*) A dispersed phase (*b*) A dispersion medium
(*c*) A dispersed phase in a dispersion medium
(*d*) A dispersion medium in a dispersed phase

42. In a colloidal solution, the diameter of dispersed particle is in the range of
(*a*) 10 Å to 100 Å (*b*) 10 Å to 500 Å
(*c*) 10 Å to 1000 Å (*d*) 10 Å to 2000 Å

43. The process of removing ions (or molecules) from a sol by diffusion through a permeable membrane is called
(*a*) ultra-filtration (*b*) dialysis
(*c*) electrophoresis (*d*) osmosis

44. The flocculation and settle down of the discharged sol particles is called
(*a*) peptization (*b*) coagulation
(*c*) osmosis (*d*) diffusion

45. An emulsion is a colloidal solution of a ________ dispersed in another liquid.
(*a*) solid (*b*) liquid
(*c*) gas (*d*) medium.

46. A gel is a colloidal system in which a ________ is dispersed in a ________ medium.
(*a*) solid, liquid (*b*) liquid, solid
(*c*) liquid, liquid (*d*) liquid, gas

47. Milk is an example of
(*a*) sol (*b*) gel
(*c*) emulsion (*d*) true solution

48. The function of an emulsifier is to
(*a*) coagulate a colloidal solution (*b*) stabilize an emulsion
(*c*) stabilize a sol (*d*) electrify a colloidal solution

49. Ice cream is a dispersion of ______________ in cream.
(a) water
(b) water vapour
(c) ice
(d) cooling agent

50. The function of alum used for the purification of water is to
(a) coagulate the sol particles
(b) disperse the sol particles
(c) emulsify the sol particles
(d) absorb the sol particles

51. Smoke is an example of
(a) solid dispersed in solid
(b) solid dispersed in liquid
(c) solid dispersed in gas
(d) gas dispersed in solid

52. Which of the following is not a colloidal solution?
(a) brine solution
(b) fog
(c) smoke
(d) butter

53. Peptization involves
(a) digestion of food
(b) precipitation of colloidal solution
(c) hydrolysis of protein
(d) breaking of a precipitate to colloidal state.

54. The cleansing action of soap is due to
(a) hydrolysis of salt present in soap
(b) ionization of salt present in soap
(c) high molecular mass of soap
(d) emulsification properties of soap.

55. The phenomenon of concentration of molecules of gas or liquid at a solid surface is called
(a) absorption
(b) adsorption
(c) catalysis
(d) none of these

56. The process of adsorption is
(a) exothermic
(b) endothermic
(c) sometimes exothermic, some times endothermic
(d) none of the above

57. Physical adsorption is a ______________ process.
(a) reversible
(b) irreversible
(c) exothermic
(d) none of these

58. Physical adsorption occurs rapidly at ______________ temperature.
(a) low
(b) high
(c) absolute zero
(d) none of these

59. One of the following medium used in desiccated air drying of grains is
(a) Silica gel
(b) $CaCl_2$
(c) CaC_2
(d) Mineral oil

60. The water freed from all ions (cations and anions) is referred to as
(*a*) heavy water
(*b*) concentrated water
(*c*) mineral water
(*d*) demineralized water

61. The process of desorption increases with ________ of pressure.
(*a*) decrease
(*b*) increase
(*c*) sometimes increases, sometimes decreases
(*d*) none of these

62. Adsorption takes place with
(*a*) decrease in enthalpy of the system
(*b*) increase in enthalpy of the system
(*c*) no change in enthalpy of the system
(*d*) none of these

63. In chromatographic analysis, the principle is used
(*a*) Absorption
(*b*) Adsorption
(*c*) Distribution
(*d*) evaporation

64. Which is correct about pH?
(*a*) $pH = -\log(H^+)$
(*b*) $pH = \log 1/(H^+)$
(*c*) $pH = 10^{-pH}$
(*d*) all of these

65. The reaction of an anion or cation with water accompanied by cleavage of O-H bond is called
(*a*) neutralization
(*b*) hydrolysis
(*c*) acidification
(*d*) ionization

66. Hydrolysis is reverse of ________.
(*a*) ionization
(*b*) neutralization
(*c*) acidification
(*d*) saponification

67. The pH value of an aqueous solution is
(*a*) equal to 7
(*b*) equal to 0
(*c*) less than 7
(*d*) more than 7

68. The equation for Lambert's law is
(*a*) $ln(I_0/I) = -bx$
(*b*) $ln(I/I_0) = -bx$
(*c*) $ln(I/I_0) = -\varepsilon Cx$
(*d*) $ln(I/I_0) = \varepsilon bx$

69. The following polysaccharides is composed of β – glycosidic bonds
(*a*) Starch
(*b*) Glycogen
(*c*) Dextrin
(*d*) Cellulose

70. Which is not classified as basic amino acid?
(*a*) Lysine
(*b*) Histidine
(*c*) Valine
(*d*) Arginine

71. Which is not a complex protein?
(*a*) Chromoprotein
(*b*) Protamines
(*c*) Lipoiprotein
(*d*) Nucleo-protein

72. Which one of the following is not a saturated fatty acid?
(*a*) Arachidic acid
(*b*) Arachidonic acid
(*c*) Palmitic acid
(*d*) Stearic acid

73. Which is non-glycerol based lipids
(*a*) Lecithines
(*b*) Steroids
(*c*) Trialein
(*d*) Cephalines

74. The pigments responsible for the red, purple colour of fruit and vegetables is
(*a*) Anthocyanins
(*b*) Xanthophylls
(*c*) Tannins
(*d*) Betalains

75. The enzyme cleave a protein molecule by the hydrolysis of peptide bond is
(*a*) Lipase
(*b*) Protease
(*c*) Rennin
(*d*) None of these

76. One of which is a prominent acid of grapes.
(*a*) Oxalic acid
(*b*) Malic acid
(*c*) Citric acid
(*d*) Tartaric acid

77. Vitamin A is an alcohol which has formula as
(*a*) $C_{20}H_{29}OH$
(*b*) C_2H_5OH
(*c*) $CH_3CH_2CH_2OH$
(*d*) CH_3CH_2OH

78. One of the following yellow pigments found in flour is
(*a*) Anthocyanin
(*b*) Carotenoid
(*c*) Flavanoid
(*d*) Xanthophylls

79. β-carotene is the precursor of the vitamin
(*a*) B
(*b*) C
(*c*) E
(*d*) A

80. One of following enzyme is responsible for brown disclouration of cut fruits.
(*a*) Poly phenol oxidase
(*b*) Amylase
(*c*) Lipase
(*d*) Silage

81. In yellow fruits, which pigment is precursor of vitamin A?
(*a*) Carotenoid
(*b*) Anthocyanin
(*c*) Flavonoid
(*d*) Xanthones

82. The flavours in fruits is largely due one of the following group of compound.
(*a*) Acids (*b*) Alcohols
(*c*) Carbonyl compound (*d*) Easters

83. Benzaldehyde is the flavour component of
(*a*) Almonds (*b*) Cherries
(*c*) Peaches (*d*) All of these

84. The acid which is flavour component of cranberries.
(*a*) 2-methyl-butyric acid (*b*) Butyric acid
(*c*) Acetic acid (*d*) None of these

85. Hesperidin, a flavonoid or flavour glycosides present in peels of orange and lemon is
(*a*) Taste less (*b*) Salty
(*c*) Bitter taste (*d*) Sweet taste

86. Naringin, a flavanone glycoside present in peels of grape fruit
(*a*) Has extremely bitter taste (*b*) Tasteless
(*c*) Sweet taste (*d*) Insipid

87. The lachrymatory factor of onion is produced by rearrangement of unstable flavouring compound formed from precursor 5-1- propenyl Cysteine sulphoxide. Name of the unstable flavouring compound is
(*a*) Propenyl-sulphenic acid (*b*) Two propanol -5-oxide
(*c*) Thiohydroxamic-o-sulphate (*d*) None of these

88. Which one of the following is not essential amino acid?
(*a*) Glycine (*b*) Leucine
(*c*) Isoleucine (*d*) Tryptophan

89. The chemical name of vitamin A is
(*a*) Ascorbic acid (*b*) Ethanol
(*c*) Retinol (*d*) Butanol

90. Enzymes are made up by
(*a*) Nucleic acid (*b*) Protein
(*c*) Vitamin (*d*) Carbohydrates

91. Freezing point of pure water is
(*a*) 0ºC (*b*) –10ºC
(*c*) 100ºC (*d*) 5ºC

92. Hydrogenation of fat brings conversion of
(*a*) fat to vegetables oil (*b*) fat to butter
(*c*) vegetable oil to ghee (*d*) none

93. Best use of enzyme in food industry is
(*a*) Lipase (*b*) Phytase
(*c*) Amylase (*d*) Protease

94. The nutritional value of protein depends on different composition of
(*a*) Fat (*b*) Amino acid
(*c*) Organic acid (*d*) Nucleic acid

95. In 1982 a report entitled 'diet nutrition and cancer' was issued by
(*a*) NAS (*b*) NASA
(*c*) IAS (*d*) WHO/FAO

96. Colour is commonly an index of
(*a*) Decay (*b*) Unripe condition
(*c*) Ripeness or spoilage (*d*) Taste

97. A most important role of thiamine is in utilization of carbohydrate to supply
(*a*) Blood circulation (*b*) Food
(*c*) Both (*a*) and (*b*) (*d*) Energy

98. Enzymes are biological catalysts that promote a wide variety of
(*a*) Biochemical reaction (*b*) Chemical reaction
(*c*) Biological reaction (*d*) Thermodynamic reaction

99. Fats that are liquid at room temperature are called
(*a*) Protein (*b*) Oils
(*c*) Both (*a*) and (*b*) (*d*) None

100. Amino acids are structured unit of
(*a*) Carbohydrate (*b*) Proteins
(*c*) Fats (*d*) Vitamins

101. Formula of vitamin A is
(*a*) $C_{17}H_{20}N_4O_6$ (*b*) $C_{12}H_{18}N_4Cl_2O_5$
(*c*) $C_9H_{17}O_5N$ (*d*) $C_{20}H_{30}O$

102. Calorific value of sucrose per 100 gm is
(*a*) 200 kcal (*b*) 250 kcal
(*c*) 300 kcal (*d*) 398 kcal

103. Which of the following contain no calorie
(*a*) Maltose (*b*) Sucrose
(*c*) Corn syrup (*d*) Aspartame

104. Lipids are
(*a*) Nucleic acid-occurring in living system
(*b*) Fats and their derivatives occurring in living systems
(*c*) Carbohydrates occurring in plants
(*d*) Proteins occurring in animals.

105. The carbohydrates which serves as reserve glucose in body is
(*a*) Sucrose (*b*) Glycogen
(*c*) Collagen (*d*) Protein

106. Mature dry legumes and oil seeds contain protein about
(*a*) 10-15% (*b*) 15-25%
(*c*) 20-40% (*d*) 47-50%

107. Which are an essential constituent of foods?
(*a*) Carbohydrates and Lipids (*b*) Protein
(*c*) Vitamins and Minerals (*d*) Above all

108. Thiamine is
(*a*) Vitamin B_1 (*b*) Vitamin B_2
(*c*) Vitamin B_6 (*d*) Vitamin B_{12}

109. Citrus fruits are rich source of vitamin
(*a*) C (*b*) D
(*c*) A (*d*) B

110. Non reducing sugar is
(*a*) Fructose (*b*) Glucose
(*c*) Sucrose (*d*) Mannose

111. The major portion of fats and oils are made-up of
(*a*) Fatty acid esters of glycerol (*b*) Amino acid
(*c*) Polysaccharides (*d*) Complex protein

112. The term fat and oil indicate whether the material; is liquid or solid. Fats that are liquid at room temperature is called
(*a*) Fat (*b*) Oil
(*c*) Free fatty acid (*d*) Margarine

113. By chemical means hydrogen can be added to an oil to saturate its fatty acid thereby converting it to a solid. This process is termed as
(*a*) Homogenization (*b*) Hydrogenation
(*c*) Purification (*d*) Centrifugation

114. Unsaturated fatty acids are highly reactive with ________ at the point of unsaturation
(*a*) Oxygen (*b*) Carbon dioxide
(*c*) Ozone (*d*) Nitrogen

115. The fatty acid with one or more double bonds can show two type of isomerism namely
(*a*) Positional (*b*) Geometric
(*c*) Symmetric (*d*) Both (*a*) and (*b*)

116. Natural fatty acid contains more than one kind of
(*a*) Triglycerides molecules (*b*) Diglycerides molecules
(*c*) Monoglycerides molecules (*d*) None of these

117. The fats and oils occurs in
(*a*) Fruits and vegetables (*b*) Animals
(*c*) Marine source (*d*) All of these

118. Enzymes are complex of
(*a*) Organic catalysts (*b*) Inorganic catalysts
(*c*) Both (*d*) None

119. Bacterial liquefying α- amylase are mostly used in
(*a*) liquefaction of starch (*b*) liquefaction of lipid
(*c*) liquefaction of protein (*d*) all of the above

120. Which types of amylase are used to form maltose?
(*a*) α-amylase (*b*) β-amylase
(*c*) Both (*d*) None

121. Cereals contain which types of amylase?
(*a*) α-amylase (*b*) β-amylase
(*c*) Both (*d*) None

122. The nature of α-amylase is as
(*a*) Endoenzyme (*b*) Exoenzyme
(*c*) Both (*d*) None

123. Which statement is correct?
(*a*) α and β-amylase both are endoenzyme
(*b*) α and β-amylase both are exoenzyme
(*c*) α -amylase is endoenzyme and β- amylase is exoenzyme
(*d*) None

124. The β-amylase is an exoenzyme that attacks starch from the ____________ ends of the polymers.
(*a*) Reducing sugar
(*b*) Non reducing sugar
(*c*) Invert sugar
(*d*) Crystalline sugar

125. The optimum pH of α-amylase is about
(*a*) 3.5
(*b*) 4.5
(*c*) 5.5
(*d*) 6.5

126. β – amylase produces maltose, it is called the
(*a*) Saccharifying enzyme
(*b*) Sugar fermenting enzyme
(*c*) Both (*a*) and (*b*)
(*d*) None

127. Which enzyme may be important producing soluble organic nitrogen that is utilized during the fermentation?
(*a*) Amylase
(*b*) Lipase
(*c*) Phytase
(*d*) Peptidases

128. Which one of the following are thought as enzyme that split triglycerides?
(*a*) Lipase
(*b*) Peptidases
(*c*) Amylase
(*d*) Phytase

129. Which one of the following is an esterase hydrolyzes phytic acid?
(*a*) Amylase
(*b*) Phytase
(*c*) Lipase
(*d*) Peptidases

130. Highest concentration of lipoxygenase is found in
(*a*) Pigeon pea
(*b*) Ground nut
(*c*) Soybean
(*d*) Chick pea

131. The enzyme that increases the mixing stability of wheat flour dough and alters dough rheology to produce strong dough.
(*a*) Amylase
(*b*) Pectin
(*c*) Lipoxygenase
(*d*) Yeast

132. Which one of the enzyme acts as bleaching agent and a coupled oxidation destroys the yellow pigments in wheat flour?
(*a*) Lipoxygenase
(*b*) Amylase
(*c*) Pectin
(*d*) Yeast

133. Calorific value of foods depends upon the
(*a*) Carbohydrates
(*b*) Proteins
(*c*) Fats
(*d*) All

134. In bomb calorimeter, carbohydrates and fats are completely oxidized to
(*a*) NO_2 and SO_2 (*b*) O_2 and NO_2
(*c*) CO_2 and H_2O (*d*) N_2 and O_2

135. Total bran percent in wheat grain is
(*a*) 11 (*b*) 13
(*c*) 14 (*d*) 15

136. The protein efficiency ratio of wheat protein is about
(*a*) 1.4 to 1.5 (*b*) 1.6 to 1.8
(*c*) 1.1 to 1.3 (*d*) 1.9 to 2.0

137. For removal of temporary hardness of water
(*a*) Alum is added to water (*b*) Lime is added to water
(*c*) Water is boiled (*d*) Water is filtered

138. Toddy and beer contains appreciable vitamin
(*a*) E (*b*) B
(*c*) C (*d*) A

139. In mango juice, availability of carotene is
(*a*) 1600 μg (*b*) 1700 μg
(*c*) 1500 μg (*d*) 1300 μg

140. One of the following have the greater amount of calcium.
(*a*) Apple juice (*b*) Coconut milk
(*c*) Mango juice (*d*) Neera (sweet toddy)

141. Theobromine substance is found in
(*a*) Cocoa (*b*) Coffee
(*c*) Tea (*d*) Chicory

142. Removing the fat from whole milk by using cream separator, this milk is known as
(*a*) Recombined milk (*b*) Homogenized milk
(*c*) Toned milk (*d*) Skimmed milk

143. Alcohol content in beer is about
(*a*) 6.0-7.5 % (*b*) 6.0-10.0 %
(*c*) 3.5-5.0 % (*d*) 2.5-3.0 %

144. Peanuts contains oils about
(*a*) 38-41% (*b*) 25-32%
(*c*) 51-56% (*d*) 42-50%

145. Recommended area for sugar factory establishment should be
(*a*) 10-20 acre (*b*) 20-30acre
(*c*) 40-50 acre (*d*) 50-80 acre

146. The outer shell of sugarcane is hard and is called
(*a*) Pith (*b*) Nodes
(*c*) Inter nodes (*d*) Rind

147. The stalk of sugar cane is built up of a series of internodes, separated from each other by hard partition wall is known as
(*a*) Pith (*b*) Nodes
(*c*) Xylem (*d*) Rind

148. The fair length of stalk of sugar cane should be
(*a*) 5-8 feet (*b*) 15-30 feet
(*c*) 8-12 feet (*d*) 2-6 feet

149. Soft part (pith) of stalk of sugar cane exists ______________ percent juice.
(*a*) 20 (*b*) 30
(*c*) 40 (*d*) 80

150. The part like rind and nodes of stalk of cane contains ___________ percent juice.
(*a*) 20 (*b*) 30
(*c*) 40 (*d*) 80

151. The percentage of sugar (polarization) in the sugar cane varies from
(*a*) 2-4% (*b*) 8-16%
(*c*) 20-25% (*d*) 40-60%

152. The water content in ripe sugar cane varies from
(*a*) 69-75% (*b*) 50-58%
(*c*) 30-43% (*d*) 8-16 %

153. The percentage of water in normal cane juice fall within limits of
(*a*) 69-75% (*b*) 75-88%
(*c*) 30-43% (*d*) 8-16 %

154. The percentage of reducing sugar in normal cane juice varies from
(*a*) 0.3-3.0 % (*b*) 0.5-2.0%
(*c*) 0.5-1.0% (*d*) 0.2-0.6%

155. The percentage of sugar content in sugarcane juice is about
(*a*) 10-21% (*b*) 8-16%
(*c*) 5-10% (*d*) 20-30%

156. The percentage of sugar content in sugarcane is about
(*a*) 10-21% (*b*) 5-10%
(*c*) 8-16% (*d*) 20-30%

157. The pH of sugarcane juice varies from
(*a*) 4.4-5.0
(*b*) 3.5-4.0
(*c*) 7.2-8.1
(*d*) 4.9-5.5

158. The application of heat or addition of chemical (electrolytes) to the cane juice brings about
(*a*) Flocculation
(*b*) Coagulation
(*c*) Clarification
(*d*) Both (*a*) and (*b*)

159. Which are removed from sugarcane juice by lime treatment?
(*a*) Gum and pectins
(*b*) Organic and inorganic acids
(*c*) Colouring matters
(*d*) All of them

160. Sucrose is not soluble in
(*a*) Water
(*b*) Alcohol
(*c*) Chloroform
(*d*) None

161. The density of sucrose is equal to
(*a*) 1.106 g/cm^3
(*b*) 1.256 g/cm^3
(*c*) 1.506 g/cm^3
(*d*) 1.606 g/cm^3

162. The solubility of pure sucrose in water increase with rise in
(*a*) pH
(*b*) acidity
(*c*) temperature
(*d*) pressure

163. The boiling temperature of the sugar solution is ________ than that of water at the same absolute pressure.
(*a*) Higher
(*b*) Lower
(*c*) Remain same
(*d*) None

164. The solubility of sucrose in water upto 40°C is decreased at ________ concentration of salt.
(*a*) Higher
(*b*) Lower
(*c*) Remain same
(*d*) None

165. The specific gravity of sucrose varies from
(*a*) 1.011-1.021
(*b*) 0.953-1.056
(*c*) 1.033-1.106
(*d*) 1.052-1.072

ANSWER KEY (FOOD CHEMISTRY)

1.C	2.A	3.C	4.B	5.C	6.A	7.B	8.C	9.B	10.D
11. C	12.D	13.D	14.D	15.B	16.B	17.B	18.A	19.A	20.C
21.D	22.B	23.A	24.C	25.C	26.B	27.A	28.B	29.A	30.B
31.A	32.C	33.D	34.B	35.C	36.C	37.A	38.D	39.D	40.C
41.C	42.D	43.B	44.B	45.B	46.B	47.C	48.B	49.C	50.A
51.C	52.A	53.D	54.D	55.B	56.A	57.A	58.A	59.A	60.D
61.A	62.A	63.B	64.D	65.B	66.B	67.D	68.B	69.D	70.C
71.B	72.B	73.B	74.A	75.B	76.D	77.A	78.D	79.D	80.A
81.A	82.C	83.D	84.A	85.A	86.A	87.A	88.A	89.C	90.B
91.A	92.C	93.C	94.B	95.A	96.C	97.D	98.A	99.B	100.B
101.D	102.D	103.D	104.B	105.B	106.C	107.D	108.A	109.A	110.C
111.A	112.B	113.B	114.A	115.D	116.A	117.D	118.A	119.A	120.B
121.C	122.A	123.C	124.B	125.B	126.A	127.D	128.A	129.B	130.C
131.C	132.A	133.D	134.C	135.B	136.B	137.C	138.B	139.C	140.D
141.A	142.D	143.C	144.D	145.D	146.D	147.B	148.C	149.D	150.A
151.B	152.A	153.B	154.A	155.A	156.C	157.D	158.D	159.D	160.C
161.D	162.C	163.A	164.B	165.C					

Food Preservation and Microbiology

1. Which fungus is used for lactic acid production?
   (*a*) *Fusarium moniliforme*
   (*b*) *Rhizopus oryzae*
   (*c*) *Aspergillus terreus*
   (*d*) *Aspergillus niger*

2. The microorganism used for meat preservation is
   (*a*) Mesophilic bacteria
   (*b*) Thermophylic bacteria
   (*c*) Psychrophylic microorganisms
   (*d*) Hydrophilic microorganisms

3. pH of acid found in Banana
   (*a*) 5.0
   (*b*) 2.3
   (*c*) 7.3
   (*d*) 8.0

4. pH range of vegetables is around
   (*a*) 5 to 5.5
   (*b*) 5.0 to 7.0
   (*c*) 7.0 to 12.0
   (*d*) 2.0 to 3.0

5. Microorganisms found in milk is classified on the basis of their major characteristics are
   (*a*) Biochemical types
   (*b*) Temperature response
   (*c*) Ability to cause infection and diseases
   (*d*) All of the above

6. Vitamin C is not present in
   (*a*) Banana
   (*b*) Apple
   (*c*) Lemon
   (*d*) Milk

7. Canning is the basic method of food preservation is done at
   (*a*) 100°C for high acidic foods
   (*b*) 121°C for low acidic foods
   (*c*) Both above
   (*d*) None

8. The phosphatase test is done for purpose of
   (*a*) Pasteurized milk testing
   (*b*) Sterilized milk testing
   (*c*) Aseptic processing of milk
   (*d*) Osmotic pressure testing

9. Derivatives of lactic acid is used in the treatments of
(*a*) Tuberoculosis and Asthma
(*b*) Neurological problems
(*c*) Calcium deficiency and anemia
(*d*) Pneumonia and diabetes

10. Which amino acid has two chiral carbons
(*a*) Leucine
(*b*) Isolucine
(*c*) Glycine
(*d*) Valine

11. Riboflavin is obtained from
(*a*) Ashbya gossypii and Eremthecium ashbyii
(*b*) Aspergillus oyyzae and A. niger
(*c*) Rhizopus niveus and rhyzopus stolinerfer
(*d*) None

12. Non protein part of enzyme is called
(*a*) Haloenzymes
(*b*) Prosthetic part
(*c*) Apoenzyme
(*d*) Nanoenzyme

13. Which of the following enzyme was 1st isolated and purified in the form of crystal
(*a*) Urease
(*b*) Pepsin
(*c*) Amylase
(*d*) Ribonuclease

14. Urease enzyme is extracted 1st time from
(*a*) Jack bean
(*b*) Yeast
(*c*) Tomato
(*d*) Jack fruit

15. One of the following, potable water is rich in nitrate caused disease
(*a*) Methaneglobenemia
(*b*) Pneumonia
(*c*) Monomofia disease
(*d*) Loss of appetite

16. Germinating barley seeds are used in preparation of
(*a*) Cheese
(*b*) Wine
(*c*) Beer
(*d*) Lactic acid

17. Whey is the residual part of cheese formation (curd) used for the production of
(*a*) Acetic acid
(*b*) Citric acid
(*c*) Lactic acid
(*d*) Gluconnic acid

18. Wonder drug is also known as
(*a*) Penicillin
(*b*) Streptomycin
(*c*) Insulin
(*d*) Griseoflavin

19. Industrial production of ethanol from starch is brought about by a certain species of
(*a*) Saccharomyces
(*b*) Penicillium
(*c*) Azatobacter
(*d*) Lactobacillus

20. Rennin used in cheese manufacturing is an
    (*a*) Antibiotic (*b*) Alkoid
    (*c*) Enzyme (*d*) Inhibitor

21. Shikonin is the name of
    (*a*) Vitamin (*b*) Enzyme
    (*c*) Steroids
    (*d*) A red dye used in the manufacturing of cosmetics

22. Which organism is useful in the preparation of Roquefort cheese ?
    (*a*) Aspergillus (*b*) Penicillium
    (*c*) Rhizopus (*d*) Yeast

23. Which one of these participates in the manufacturing of dextrin ?
    (*a*) Leuconostoc (*b*) Pseudomonas
    (*c*) Lactobacillus (*d*) Mucor

24. Maximum quantity of alcohol is obtained from
    (*a*) Maize (*b*) Sugarcane
    (*c*) Wood (*d*) Tapioca

25. The first acid to be produced by fermentation was
    (*a*) Lactic acid (*b*) Citric acid
    (*c*) Acetic acid (*d*) Malic acid

26. Which enzymes are used for converting corn starch into high fructose syrup?
    (*a*) Glucoisomerase (*b*) Glucoamylase
    (*c*) Amylase (*d*) All of above

27. The main advantages of waxing on fruits are
    (*a*) Reduced moisture loss (*b*) Improved appearances
    (*c*) Reduce spoilage due attack of microorganisms
    (*d*) All of the above

28. Fungi can produced by the following items
    (*a*) Antibiotics (*b*) Vitamins
    (*c*) Acid (*d*) All

29. The sour taste of vinegar is due to
    (*a*) Acetic acid (*b*) Lactic acid
    (*c*) Formic acid (*d*) Citric acid

30. Canning is the method of
    (*a*) Spoilage of foods (*b*) Preservation of foods
    (*c*) Making plant free from disease (*d*) Isolating bacteria from soils

31. Enzymes are basically
(*a*) Vitamin
(*b*) Protein
(*c*) Minerals
(*d*) Fat

32. Golden rice is rich source of
(*a*) Vitamin C
(*b*) Vitamin A
(*c*) Vitamin D
(*d*) Vitamin B_{12}

33. High lysine content is found in
(*a*) Wheat
(*b*) Maize
(*c*) Rice
(*d*) Bajara

34. Clove is the part of
(*a*) Flower
(*b*) Thalamus of flower
(*c*) Auxiliary bud
(*d*) Seed

35. Cryopreservation is done at
(*a*) 273 ºC
(*b*) –196 ºC
(*c*) 0 ºC
(*d*) 273 ºK

36. Geocarpic fruit is
(*a*) Onion
(*b*) Water melon
(*c*) Ground nut
(*d*) Carrot

37. Lactic acid is produced from one of the following raw material
(*a*) Whey
(*b*) Curd
(*c*) Cheese
(*d*) Butter

38. Pulses not have the amino acid
(*a*) Methionine, Glycine
(*b*) Cysteine, Methionine
(*c*) Lysine, Cysteine
(*d*) Tryptophan, Lysine

39. Millets and cereals not have the amino acid
(*a*) Tryptophan, Lysine
(*b*) Cysteine, Methionine
(*c*) Lysine, Glycine
(*d*) Cysteine, Tryptophan

40. The enzyme is capable of lysising (dissolving) the cell wall of bacteria.
(*a*) Lysozyme
(*b*) Lipase
(*c*) Rennin
(*d*) Protease

41. The factors, by which the growth of microorganisms are influenced.
(*a*) Moisture
(*b*) Oxygen
(*c*) Food supply
(*d*) All of the above

42. Which one of the following group of bacteria can not tolerate relatively low and high temperature?
(*a*) Psychrophytic (*b*) Mesophyllic
(*c*) Thermodiuric (*d*) Thermophyllic

43. The first and most important change occurs in milk due to bacterial activity is making of
(*a*) Lactic acid (*b*) Citric acid
(*c*) Butyric acid (*d*) Malic acid

44. Botulism is an example or type of
(*a*) Food intoxication (*b*) Food infection
(*c*) Both of these (*d*) None of these

45. The microorganism responsible for botulism in food intoxication is
(*a*) Streptococcus botulinum (*b*) Salmonella botulinum
(*c*) Clostridium botulinum (*d*) All of these

46. In botulism, the toxin produced in food which is ingested by *Clostridium botulinum* is a
(*a*) Mycotoxin (*b*) Aflatoxin
(*c*) Neurotoxin (*d*) Haemotoxin

47. Clostridium botulinum is
(*a*) Aerobic bacteria (*b*) Anaerobic bacteria
(*c*) Facultative anaerobic (*d*) Facultative aerobic

48. Which of the following bacterium are not involved in spoilage of the finished cheese?
(*a*) Cladosporium (*b*) Penicillium pubervulum
(*c*) PM nigra (*d*) Pseudomonas fragae

49. Sauerkraut is a fermented product made from the
(*a*) Tomato (*b*) Cabbage
(*c*) Cauliflower (*d*) Barley

50. The best quality sauerkraut should contain __________ percent salt by weight.
(*a*) 3.0-4.0 (*b*) 2.25-2.5
(*c*) 2.0-2.25 (*d*) 1.0-2.0

51. Slimy or ropy kraut is caused by
(*a*) *Lactobacillus plantarum* (*b*) *Lactobacillus brewis*
(*c*) *Leuconostoc* (*d*) *Lactobacillus aceti*

52. Blackening of pickles is caused due to the formation of
(*a*) Ferrous sulphate (*b*) Ferric sulphate
(*c*) Lead tannate (*d*) None of these

53. Blackening of pickles is also caused due to black pigmented bacteria which is
(*a*) *Bacillus nigrificans* var. subtilis
(*b*) *Pseudomonas nigrificans* var. subtilis
(*c*) *Leuconostoc*
(*d*) *None of these*

54. Among the following which bacteria is rod-shaped?
(*a*) Streptococcus
(*b*) Micrococcus
(*c*) Sarcina
(*d*) Capsulated

55. Which bacteria can transmit food borne infection and also capable of causing human diseases?
(*a*) Clostridium perfringens
(*b*) Salmonella
(*c*) Shigella
(*d*) All of above

56. Name of the bacteria causes neurotoxicity is
(*a*) Clostridium botulinum
(*b*) Campylobacter jejuni
(*c*) Shigella dysenterial
(*d*) Bacillus cercus

57. Name of important food borne parasite is
(*a*) Tape worm
(*b*) Trichinosis nematode
(*c*) Flat worm
(*d*) Bug

58. The temperature which double the chemical reaction takes place in food products is
(*a*) 10°C
(*b*) 20°C
(*c*) 30°C
(*d*) 40°C

59. How much concentration of nitrogen and oxygen in air is?
(*a*) 79% and 20%
(*b*) 69% and 30%
(*c*) 50% and 50%
(*d*) 75% and 23%

60. Among the following which are destroyed by sun light?
(*a*) Riboflavin
(*b*) Vitamin A and C
(*c*) Food colour
(*d*) All of above

61. What is the optimum temperature for the growth of microorganisms?
(*a*) 12-14°C
(*b*) 16-32°C
(*c*) 16-18°C
(*d*) 16-38°C

62. What is the important feature of 'Psychrtrophs'?
(*a*) Can grow at 4°C
(*b*) Are parasites
(*c*) Can grow down to 0°C
(*d*) None of these

63. The higher humidity produce
(*a*) Lower population of microorganisms
(*b*) Higher population of microorganisms
(*c*) Average population of microorganisms
(*d*) None of these

64. The term 'water activity' is related to
(*a*) Absolute Humidity
(*b*) Equilibrium Relative humidity
(*c*) Action of water on food
(*d*) All of these

65. How an acid affect the microorganism when combined with heat
(*a*) Destruction of microorganisms
(*b*) Provide good condition for growth of microorganism
(*c*) Increase the aerobic reaction
(*d*) Denaturates of their food protein

66. Smoke acts as
(*a*) Good preservative
(*b*) Poor preservative
(*c*) Very poor preservative
(*d*) None of these

67. Smoke used as colour enhancer in
(*a*) Meat processing
(*b*) Milk processing
(*c*) Vegetable processing
(*d*) Cereal processing

68. What is the main role of FDA?
(*a*) Ensure how much chemical to be used in food
(*b*) The temperature at which food is kept
(*c*) How to increase the growth of food
(*d*) Both (*a*) and (*b*)

69. In which food product, sorbic acid is used as inhibitor to control the growth of mould.
(*a*) Mushroom
(*b*) Cheese
(*c*) Leafy vegetables
(*d*) Fruits

70. Heat preservation of food is generally done in the form of
(*a*) Dehydration/drying
(*b*) Pasteurization/sterilization
(*c*) Canning
(*d*) All the above

71. Among the following, the most heat resistance pathogens found in food.
(*a*) Clostridium botulinum
(*b*) Bacillus stearothermophilus
(*c*) Micrococcus
(*d*) Both (*a*) and (*b*)

72. Heat energy is transferred by
(*a*) Conduction
(*b*) Convection
(*c*) Radiation
(*d*) All of the above

73. In natural convection, the heated portion of food becomes
(*a*) Lighter in density
(*b*) Higher in density
(*c*) Higher in mass
(*d*) Both (*a*) and (*c*)

74. At which temperature the best quality juice is retained after pasteurization.
(*a*) 63°C for 30 min
(*b*) 88°C for 1 min
(*c*) 100°C for 12 sec
(*d*) 121°C for 2 sec

75. HTST stands for
(*a*) Hot temperature short time
(*b*) High temperature short time
(*c*) High time short temperature
(*d*) None of them

76. GMP stands for
(*a*) Good manufacturing practices
(*b*) Gene modified products
(*c*) Gas modified packaging
(*d*) None of them

77. Aseptic packaging is generally used in
(*a*) Food pasteurization
(*b*) Food sterilization
(*c*) Food fermentation
(*d*) Food sanitization

78. What should be the temperature of commercial and house hold refrigerator?
(*a*) 6-8°C
(*b*) 4.5-7°C
(*c*) 6-10°C
(*d*) 5-7°C

79. Air blast freezers are typically operated at temperature of _________ with forced air velocity is of _________.
(*a*) –30 to – 45°C and 10-15 m/s
(*b*) – 10 to – 35°C and 10-15 m/s
(*c*) –2 to – 10°C and 10-15 m/s
(*d*) 30 to 45°C and 10-15 m/s

80. The principal requirements of effective refrigerator storage are
(*a*) Controlled temperature
(*b*) Air circulation
(*c*) Humidity control
(*d*) Modification of gas atmosphere
(*e*) All of these

81. The freezing rate will be faster when
(*a*) The temperature difference between food and refrigerant is greater
(*b*) The thinner the food piece or greater the heat transfer rate of the food package
(*c*) The greater the velocity of refrigerated air
(*d*) The more intimate the contact between the food and cooling medium
(*e*) All of the above

82. The term 'food dehydration' generally refers
(*a*) To artificial drying under controlled condition
(*b*) To remove the water upto predetermined level
(*c*) To dry food upto bone dry condition
(*d*) Both (*a*) and (*c*)

83. Dehydrated foods will be
(*a*) Dry
(*b*) Hygroscopic
(*c*) Bone dry
(*d*) Having high relative humidity

84. In the principal cereals, the percentage of carbohydrate is about
(*a*) 10-14%
(*b*) 58-72%
(*c*) 8-13%
(*d*) 2-5 %

85. IRRI is related to_______________ and located at_______________.
(*a*) Rice, Mexico
(*b*) Rice, Philippines
(*c*) Rape seed, Bharatpur
(*d*) Rice, Rome

86. The protein content is higher in
(*a*) Cereals
(*b*) Pulses and legumes
(*c*) Oil seeds
(*d*) Fruit and vegetables

87. Peanut is a
(*a*) Cereal
(*b*) Pulse
(*c*) Oil seed
(*d*) Legume

88. The water content is comparatively
(*a*) Lower in milk, higher in apple
(*b*) Higher in milk, lower in apple
(*c*) Similar in milk as well as in apple
(*d*) None of these

89. Among the following which pigment is termed as flavonoid
(*a*) Chlorophyll
(*b*) Carotenoid
(*c*) Anthocyanin
(*d*) Anthoxanthens
(*e*) Both (*c*) and (*d*)

90. Betalains are
(*a*) Green water soluble pigment
(*b*) Green water in soluble pigment
(*c*) Red water in soluble pigment
(*d*) Red water soluble pigment

91. Flavonoids are
(*a*) pH sensitive
(*b*) pH resistant
(*c*) tending toward deeper yellow in alkaline medium
(*d*) both (*a*) and (*c*)

92. The mild stimulant used in cola flavour is
(*a*) Opium
(*b*) Caffeine
(*c*) Nicotine
(*d*) Solanine

93. Starches are
(*a*) Sweet
(*b*) Readily soluble in cold water
(*c*) They form paste and gel in hot water
(*d*) They not occur in cereals.

94. Fats are
(*a*) Gradually softer on heating
(*b*) Not form emulsion with water and air
(*c*) as lubricant in foods
(*d*) Both (*a*) and (*c*)

95. Lecithin are
(*a*) Structurally like fats
(*b*) Contain phosphoric acid
(*c*) They have a polar and a non polar end
(*d*) All of the above

96. Lipase enzyme are used as
(*a*) Chill proofing of buns and related products
(*b*) Classification of fruit juices
(*c*) Removal of glucose from egg white before drying
(*d*) Improvement in whipping quality of dried egg white and flavour production in cheese and chocolate.

97. Which vitamin is also called α-tocopherol?
(*a*) Vitamin E
(*b*) Vitamin B
(*c*) Vitamin C
(*d*) Vitamin K

98. Among the following which one is known as nicotinamide in the United Kingdom
(*a*) Riboflavin
(*b*) Folacin
(*c*) Niacin
(*d*) Biotin

99. Vitamin B_6 is the name given to closely related substances such as
(*a*) Pyridoxine
(*b*) Pyridoxal
(*c*) Pyridoxamine
(*d*) All of the above

100. Those foods, do not spoil unless handled carelessly are called
(*a*) Highly perishable
(*b*) Semi perishable
(*c*) Perishable
(*d*) Non-perishable

101. Those foods are properly handled and stored; they will remain unspoiled for a fairly long period are known as
(*a*) Highly perishable
(*b*) Semi perishable
(*c*) Perishable
(*d*) Non-perishable

102. The group includes important daily foods that spoils readily unless special preservatives methods are used and termed as
(*a*) Highly perishable
(*b*) Semi perishable
(*c*) Perishable
(*d*) Stable foods

103. The flour and sugar is classified as
(*a*) Highly perishable
(*b*) Semi perishable
(*c*) Perishable
(*d*) Stable foods

104. Leafy vegetables is recommended as
(*a*) Highly perishable
(*b*) Semi perishable
(*c*) Perishable
(*d*) Non-perishable

105. A dry food such as bread is most likely to be spoiled by
(*a*) Bacteria
(*b*) Yeast
(*c*) Molds
(*d*) Virus

106. The moist or neutral foods such as milk, meat, fish and egg ordinarily are spoiled by
(*a*) Molds
(*b*) Yeast
(*c*) Bacteria
(*d*) Virus

107. In which process, microorganisms are not destroyed?
(*a*) Drying and dehydration
(*b*) Irradiation
(*c*) Sterilization
(*d*) Freezing

108. Which one of the process is the reverse of freezing?
(*a*) Drying
(*b*) Thawing
(*c*) Blanching
(*d*) Aseptic processing

109. In growth of microbial culture, the phase in which, there is no growth or even a decline in numbers of microorganisms is called
(*a*) Lag phase
(*b*) Exponential phase
(*c*) Death phase
(*d*) Positive acceleration phase

110. The phase in which the rate of growth is continuously increasing is termed as
(*a*) Lag phase
(*b*) Exponential phase
(*c*) Positive acceleration phase
(*d*) Logarithmic phase

111. The phase in which the rate of multiplication is most rapid and is constant is known as
(*a*) Lag phase
(*b*) Exponential phase
(*c*) Death phase
(*d*) Survival phase

112. The phase during which no cell division occurs but remaining cells survive on endogenous nutrients is called
(*a*) Lag phase
(*b*) Logarithmic phase
(*c*) Survival phase
(*d*) Death phase

113. In special conditions, removal of microorganisms from food may be accomplished by
(*a*) Filtration
(*b*) Centrifugation or sedimentation or clarification
(*c*) Washing or trimming
(*d*) All of these

114. The foods contain pH above 5.3 are known as
(*a*) Low acid foods (*b*) Medium acid foods
(*c*) Acid foods (*d*) High acid foods

115. One of the following is on example of low acid foods.
(*a*) Peas (*b*) Asparagus
(*c*) Pears (*d*) Cape goose berry

116. The pH of high acidic foods should be
(*a*) Above 5.3 (*b*) 3.7 and below
(*c*) 5.3 – 4.5 (*d*) 4.5-3.7

117. Tomato, pears and pine apples are acid foods contain the pH value between
(*a*) 5.3-4.5 (*b*) 3.7 and below
(*c*) 4.5-3.7 (*d*) Above 5.3

118. Medium acid foods like spinach and pumpkin having pH between
(*a*) 3.7 and below (*b*) 4.5-3.7
(*c*) 5.3-4.5 (*d*) Above 5.3

119. Most of the molds and their spores are killed by moist heat at
(*a*) 60ºC in 5 to 10 min. (*b*) 60ºC in 15 to 20 min.
(*c*) 50ºC in 5 to 10 min. (*d*) None of these

120. The thermal death time of bacterial spore of *Clostridium botulinum* at 100ºC is
(*a*) 10-20 min (*b*) 50-100 min
(*c*) 100-330 min (*d*) 20-50 min

121. Generally TDT curve is plotted on
(*a*) Simple graph paper (*b*) Semi log paper
(*c*) Log-log paper (*d*) Plain paper

122. The symbol 'D' in thermo-bacteriology is indicated for
(*a*) Dehydration (*b*) Drying
(*c*) Diffusion (*d*) Decimal reduction time

123. The time of heating at a specified temperature required to destroy 90 percent of organisms in a population is known as
(*a*) F-value (*b*) D-value
(*c*) Z-value (*d*) F_o- value

124. The D-value, decimal reduction time decreased the surviving population by
(*a*) One log cycle (*b*) Two log cycle
(*c*) Ten log cycle (*d*) None of these

125. The time in minutes required to destroy the organisms in a specified medium at 250°F (121.1°C) is referred as

(*a*) F-value (*b*) Z-value
(*c*) D-value (*d*) None of these

126. The degree Fahrenheit required to reduce the TDT tenfold is called

(*a*) F-value (*b*) Z-value
(*c*) D-value (*d*) None of these

127. Heat penetration into container for sterilization is dependent on

(*a*) Size, shape and material of container
(*b*) Temperature of food and retort
(*c*) Consistency of can content and size and shape of foods
(*d*) All of the above

128. The monograph method is most rapid for the estimation of

(*a*) Thermal process time (*b*) Drying time
(*c*) Freezing time (*d*) Blanching time

129. The heat treatment that kills part but not all of microorganisms below 100°C temperature is referred as

(*a*) Sterilization (*b*) Pasteurization
(*c*) Heat at about 100°C (*d*) Blanching

130. Temperature and time combination in ultra pasteurization method employed for milk is

(*a*) 62.8°C for 30 min (*b*) 71.7°C for 15 sec
(*c*) 137.8°C for at least 2 sec (*d*) 121.1°C for 30 min

131. The average heat treatment for carbonated juice would be

(*a*) 65.6°C for 30 min (*b*) 62.8°C for 30 min
(*c*) 60°C for 30 min (*d*) 85 to 87.8°C for 30 sec

132. Thermal treatment for apple juice in can and bulk form would be done at

(*a*) 65.6°C for 30 min (*b*) 62.8°C for 30 min
(*c*) 60°C and 85 to 87.8°C for 30 to 60 sec (*d*) none

133. Dried fruits usually are pasteurized in the packaged at

(*a*) 65.6°C to 85°C for 30-90 min (*b*) 72°C for 15 sec
(*c*) 68.2°C for 30 min (*d*) 85 to 87.8°C for 30 sec

134. The process of hermetically sealing of foods in container and sterilizing by heat treatment for long period of storage is known as

(*a*) Canning (*b*) Aseptic packaging
(*c*) Dehydration (*d*) Irradiation

135. The father of canning is
(*a*) Spllanzani
(*b*) Nicolas Appert
(*c*) Robert Kotch
(*d*) Louise Pasteur

136. Tomato juice may be pasteurized to kill spores of *Bacillus coagulaus* before canning at
(*a*) 121 to 132ºC
(*b*) 121ºC
(*c*) 137.8ºC
(*d*) 71.7ºC

137. Carbon dioxide under pressurize packaged food inhibit various microorganism like
(*a*) Aerobic bacteria and molds
(*b*) Yeasts
(*c*) LAB
(*d*) None

138. Mold infection in bread can be prevented by adding of
(*a*) Salts
(*b*) Sugar
(*c*) Sodium benzoate
(*d*) Sodium and calcium propionate

139. Ropiness of bread can be eliminated by adding of
(*a*) Citrate
(*b*) Sorbate
(*c*) Pectate
(*d*) Propionate

140. The basic purpose of smoking in foods is
(*a*) Adding desired flavour
(*b*) Preserving the food
(*c*) Improvement in colour
(*d*) Tenderizing action on meat
(*e*) All of the above

141. Smoking temperature for meat vary between
(*a*) 20-30ºC
(*b*) 30-42ºC
(*c*) 43-71ºC
(*d*) 71-85ºC

142. Wood smoke used for food contains a large number of volatile compounds that may have
(*a*) Bacteriostatic effect
(*b*) Bacteriocidal effect
(*c*) Fungicidal effect
(*d*) Both (*a*) and (*b*)

143. The rate of germicidal action of smoke increases with
(*a*) Its concentration
(*b*) Temperature
(*c*) Above both
(*d*) None of these

144. Wood smoke is more effective against vegetative cells than
(*a*) Bacterial spore
(*b*) Fungi
(*c*) Yeast and molds
(*d*) None of these

145. The concentration of mycostatic material from wood smoke necessary to prevent mold growth increases with in the
(*a*) Humidity of storage
(*b*) Temperature of storage
(*c*) Gas composition of storage
(*d*) Oxygen concentration

146. Mustard flour and the volatile oil of mustard are very effective against
(*a*) *Saccharomyces cerevisiae* (*b*) *Clostridium botulinum*
(*c*) *Aspergillus* (*d*) *Salmonella*

147. One of the following is most effective against yeast
(*a*) Volatile oil of mustard (*b*) Oil of cinnamon and cloves
(*c*) Oil of thyme and bay leaves (*d*) None of them

148. Halogen kill organism during washing of foods or equipments by
(*a*) Oxidation (*b*) Injury to cell membranes
(*c*) Direct combination with cell proteins (*d*) All of the above

149. Hypochlorite treated water in food processing plants is used for
(*a*) Drinking (*b*) Processing
(*c*) Cooling (*d*) Washing
(*e*) All of the above

150. An oxidizing agent like hydrogen peroxide (H_2C_2) is employed as a preservative along with
(*a*) Water (*b*) Humidity
(*c*) Heat (*d*) Acid

151. One of the following bacteria are destroyed in the processing of sugar by a combination of heat and hydrogen peroxide
(*a*) Thermoduric (*b*) Thermophiles
(*c*) Osmophilic (*d*) Saccharolytic

152. One of the following is used as an inert gas over foods that should not be exposed to air.
(*a*) Hydrogen dioxide (*b*) Hydrogen peroxide
(*c*) Nitrogen (*d*) Sulphur dioxide

153. Antifungal antibiotics include
(*a*) Grisefulvin (*b*) Pimaricin
(*c*) Fulcine and actidione (*d*) Rimacidin and nystalin
(*e*) All of the above

154. Antimycotics permitted in food packaging materials (under specified conditions) are
(*a*) Na and Ca propionate (*b*) Na benzoate
(*c*) Sorbic acid and sorbates (*d*) Methyl and propyl parabens
(*e*) All of these

155. Which one antibiotic has been used in Europe to suppress anaerobes in cheese and cheese products?
(*a*) Nisin (*b*) Natamycin
(*c*) Neomycin (*d*) Bacitracin

156. Natamycin is effective against
(*a*) Bacteria (*b*) Vegetative spore
(*c*) Yeast and molds (*d*) None of above

157. Microbial destruction without the generation of high temperature suggest the term
(*a*) Cold sterilization (*b*) Oxidation
(*c*) Chemical sterilization (*d*) None of the above

158. The factors that influence the effectiveness of UV rays on foods are
(*a*) Time (*b*) Intensity
(*c*) Penetration (*d*) All of the above

159. The quantity or dose of irradiation actually absorbed by an organism or products thus is expressed by the product of
(*a*) Time and intensity (*b*) Time and specific gravity
(*c*) Mass and intensity (*d*) Moisture content and intensity

160. The UV rays affect only __________ of most irradiate food directly to the UV lamp.
(*a*) Inner surface (*b*) Outer surface
(*c*) Front side of UV rays (*d*) Whole product

161. One of the following is most resistant against the rays of UV lamp.
(*a*) Bacteria (*b*) Mold
(*c*) Yeast (*d*) None

162. The resistance of mold in comparison to bacteria exposed to UV rays is more
(*a*) 2 to 5 times (*b*) 5 to 15 times
(*c*) 1 to 3 times (*d*) Equal

163. Which one ionizing radiation is used to preserve food economically?
(*a*) *x*-rays (*b*) γ-rays
(*c*) β-rays (*d*) photon

164. The quantity of γ-rays or *x*-radiations which produces one electrostatic unit of electric charge of either sign in a cubic centimeter of air under standard condition is termed as
(*a*) Roentgen (*r*) (*b*) Roentgen-equivalent- physical (rep)
(*c*) 1 rad (*d*) 1MeV

165. The unit of radiation dosage, being equivalent to the absorption of 100 erg per gram of irradiated material is called
(*a*) rad (*b*) Mrad
(*c*) Krad (*d*) MeV

166. The radiation dose like 1Krad is equivalent to
(*a*) 100 rad (*b*) 1000 rad
(*c*) 10000 rad (*d*) 100000 rad

167. An electron volt (eV) is the energy gained by a/an __________ in moving through a potential difference of 1 volt.
(*a*) Neutron (*b*) Electron
(*c*) Proton (*d*) Photon

168. A MeV is the measurement of
(*a*) Intensity of irradiation (*b*) Absorbed energy
(*c*) Dose of irradiation (*d*) All of these

169. A rep is a measure of the __________ that is effective within the foods.
(*a*) Intensity of irradiation (*b*) Dose of irradiation
(*c*) Absorbed energy (*d*) None of these

170. Radiation sterilization which would imply high dose treatment, with the resulting product being shelf life stable is also known as
(*a*) Radurization (*b*) Radicidation
(*c*) Radappertization (*d*) Picowaved

171. 'Commercially sterile' means
(*a*) Kill all microorganisms but some are unable to grow during storage
(*b*) Completely destruction of all microorganisms
(*c*) Partially killing of microorganisms by thermal treatment
(*d*) Any one (*b*) or (*c*)

172. Pressurized packaged liquids or paste are packed under pressure of a propellant gas is called
(*a*) Aerosols (*b*) Tetra pack
(*c*) Ultra pack (*d*) Vacuum pack

173. Carbon dioxide under pressure inhibit various microorganism including
(*a*) Aerobic bacteria and molds (*b*) Anaerobic bacteria and molds
(*c*) Yeast and molds (*d*) None of these

174. Low temperature preservation are used to
(*a*) Retard biochemical reaction (*b*) Retard of enzymatic activity
(*c*) Slow down or stop the growth and activity of microorganisms
(*d*) Above all

175. Which one of the following prevents the growth of most food borne microorganisms?
(*a*) Freezing (*b*) Refrigeration
(*c*) Commercial refrigeration (*d*) All of the above

176. Refrigeration temperature is used to
(*a*) Prevent the growth of microorganisms (*b*) Slow down the growth rates
(*c*) Increase the growth rate (*d*) No change in growth rate

177. Commercial refrigeration temperature retards the growth of various food borne pathogens at
(*a*) Lower than 5 to 7.2ºC (*b*) More than 7.2ºC
(*c*) 10ºC (*d*) 121ºC

178. *Clostridium botulinum* type E which has a minimum temperature for growth of about
(*a*) 0ºC (*b*) –3.3ºC
(*c*) 3.3ºC (*d*) 10ºC

179. The use of temperature above or below freezing, although the application of mechanical refrigeration is implied in
(*a*) Cold storage (*b*) Freezing
(*c*) Refrigeration (*d*) Commercial sterile

180. Most commercial storage freezer are used
(*a*) At or below –18ºC (*b*) Above –18ºC
(*c*) At 18ºC (*d*) At –10ºC

181. A high alcohol yielding variety of yeast used to produce industrial alcohol, wine and distilled liquors is
(*a*) *S. cerevisiae var. ellipsoideus* (*b*) *S. uvarum*
(*c*) *S. fragilis* (*d*) *S. lactis*

182. Yeast grow in soy sauce with its high content of salt of
(*a*) 18 % (*b*) 20 %
(*c*) 3 % (*d*) 1 %

183. Osmophilic yeasts grow well in food which has
(*a*) High concentration of sugar or salt (*b*) Low concentration of sugar or salt
(*c*) Normal concentration of sugar or salt (*d*) All of the above

184. The range of temperature for growth of most yeasts are, in general to that molds, with the optimum around ________ and the maximum about__________.
(*a*) 25-30ºC and 35-47ºC (*b*) 10-15ºC and 25-30ºC
(*c*) 15-20ºC and 30-35ºC (*d*) None of these

185. Enzymatic and microbial changes due to chilling in food is
(*a*) Prevented (*b*) Not prevented
(*c*) Slowed considerably (*d*) Both (*b*) and (*c*)

186. The chilling temperature for any food is selected on the basis of
(*a*) Kind of food (*b*) Time of storage
(*c*) Condition of storage (*d*) All

187. The banana should be kept in the refrigerator; it keeps best at temperature of
   (*a*) 13.3 to 16.7°C (*b*) 8 to 10°C
   (*c*) 0 to 8°C (*d*) 10 to 20°C

188. Too low relative humidity, what would be happen in foods?
   (*a*) Loss of moisture and weight (*b*) Wilting and softening of the vegetables
   (*c*) Shrinkage of fruits (*d*) All of these

189. Changes in humidity as well as in temperature during low temperature storage of foods may cause
   (*a*) Sweating or precipitation of moisture (*b*) Shrinkage of fruits
   (*c*) Drying of foods (*d*) None of these

190. In the presence of optimal concentration of CO_2 or O_3 in CAP with chilling storage,
   (*a*) A food will remain unspoiled for a longer period
   (*b*) A higher RH can be maintained without harm to keeping quality of certain foods
   (*c*) A high storage temperature can be used without shortening the keeping time of the food than is possible with ordinary chilling storage
   (*d*) All of the above

191. The rate of freezing of food depends upon
   (*a*) Method employed and freezing temperature (*b*) Circulation of air or refrigerant
   (*c*) Size and shape of package (*d*) Kind of food
   (*e*) All of the above

192. Temperature and time combination in sharp freezing for foods are as
   (*a*) –15 to –29°C and 3-72 hr (*b*) –17.8 to – 45.6°C and 2hr
   (*c*) –17.8 to –34.4°C and 10 hr (*d*) None

193. Quick freezing of food is accomplished by
   (*a*) Direct immersion of food or packaged foods in a refrigerant
   (*b*) Indirect contact with refrigerant
   (*c*) Air blast
   (*d*) All of the above

194. Optimum temperature of sharp freezing used to frozen the food is
   (*a*) –23.3°C (*b*) – 45.6°C
   (*c*) –17.8°C (*d*) 34.4°C

195. In indirect contact method of quick freezing, the temperature of refrigerant is
   (*a*) –23.3°C (*b*) –17.8 to –45.6 °C
   (*c*) –17.8 to –34.4°C (*d*) None

196. Air blast freezing method freeze the food at the temperature of
(*a*) –17.8 to –34.4°C
(*b*) –15 to –29°C
(*c*) –23.5°C
(*d*) –17.8 to – 45.6°C

197. The freezing method in which half moisture is removed from food before freezing is called
(*a*) Freeze drying
(*b*) Dehydro-freezing
(*c*) Dehydration
(*d*) Osmotic dehydration

198. The advantages of quick freezing over slow freezing are
(*a*) Smaller ice crystal formation
(*b*) Shorter period of solidification
(*c*) Quick prevention of microbial growth
(*d*) Rapid slow down of enzyme activity
(*e*) All of these

199. When ice crystal of frozen food evaporate from an area at the surface, the defect is called
(*a*) Freezer burn
(*b*) Chilling injury
(*c*) Freezing injury
(*d*) Cold shock

200. The pink or reddish liquid that comes from meat during thawing is called
(*a*) Drip or bleeding
(*b*) Metacryotic liquid
(*c*) Leakage
(*d*) Cold shock

201. The liquid oozing out from fruits or vegetables on thawing is referred as
(*a*) Drip
(*b*) Sweating
(*c*) Leakage
(*d*) Seepage

202. Rapid cooling of cells from an optimal temperature to 0°C can also results in death referred as
(*a*) Freezer burn
(*b*) Chilling injury
(*c*) Cold shock
(*d*) Freezing injury

203. Best quality dried product of coffee, milk, tea and fruits puree is obtained by using the one following drier
(*a*) Drum drier
(*b*) Rotary drier
(*c*) Spray drier
(*d*) Freeze drier

204. Liquid food should be comparatively dried at
(*a*) low temperature and under vacuum
(*b*) high temperature and under vacuum
(*c*) low temperature and high RH
(*d*) high temperature and high RH

205. Case hardening in food takes place by
(*a*) Freezing
(*b*) Improper drying
(*c*) Chilling
(*d*) Drying

206. The factors that affect the drying of any food products are
   (*a*) Temperature and RH of hot air
   (*b*) Velocity of air
   (*c*) Drying time and drying method
   (*d*) All of the above

207. Before drying, fruits are treated by SO_2 with the amount of
   (*a*) 100-200 ppm (*b*) 350-600 ppm
   (*c*) 600-905 ppm (*d*) 1000-3000 ppm

208. Most foods are packed just after drying from the protection against
   (*a*) Moisture (*b*) Microbial Contamination
   (*c*) Infestation with insect (*d*) All of the above

209. The predominant kinds of organisms in dry milk are
   (*a*) Thermoduric streptococci (*b*) Micrococci
   (*c*) Sporeformers (*d*) All of these

210. The foods which contain 20 to 40 percent moisture and have non refrigerated shelf stability is referred as
   (*a*) Flesh product (*b*) Protective product
   (*c*) Colloid (*d*) Intermediate moisture foods

211. Which one of the following is not a intermediate moisture foods.
   (*a*) Jam (*b*) Jelly
   (*c*) RTS (*d*) Honey

212. The effect of water activity (a_w) on microorganisms in foods is influenced by
   (*a*) pH and temperature (*b*) oxygen level
   (*c*) nutritional concentration (*d*) all of the above

213. The water activity (a_w) of most intermediate moisture foods would be between
   (*a*) 0.10-0.30 (*b*) 0.32-0.45
   (*c*) 0.46-0.72 (*d*) 0.75-0.85

214. The water activity (a_w) of foods usually adjusted by
   (*a*) Drying (*b*) Additives (sugar or salt)
   (*c*) Combination of drying and additives (*d*) All

215. Those foods additives which are specifically added to prevent the deterioration or decomposition of a food have been referred to as
   (*a*) Physical preservatives (*b*) Chemical preservatives
   (*c*) Biological preservative (*d*) None

216. The additives in food are added when deterioration may be caused by
(*a*) Microorganism
(*b*) Food enzymes
(*c*) Purely Chemical reaction
(*d*) All of the above

217. The inhibition of growth and activity of microorganisms is the one of the main purpose of the use of
(*a*) Pasteurization
(*b*) Blanching
(*c*) Chemical preservatives
(*d*) None of these

218. Preservative may inhibit microorganism by interfering with their
(*a*) cell membrane
(*b*) enzyme activity
(*c*) genetic mechanism
(*d*) all of the above

219. The effectiveness of chemical preservatives depend upon
(*a*) Concentration of chemical
(*b*) Kind and number of organisms
(*c*) Time and temperature
(*d*) Physical-chemical characteristics of substrate in which organism is found
(*e*) All of these

220. Physio-chemical characteristic of substrate means
(*a*) Moisture and pH
(*b*) Kind and amount of solute
(*c*) Surface tension
(*d*) Colloids and other protective substance
(*e*) All of the above

221. A chemical preservative have a wide range of antimicrobial activity and should be
(*a*) Non toxic and economic
(*b*) No effect on flavour, taste or aroma of original food
(*c*) Should not encourage the development of resistant strain
(*d*) All

222. Antimicrobial preservative added to foods can be classified on the basis of
(*a*) Those added preservative not defined as such by low
(*b*) Substances generally recognized as safe for addition to foods
(*c*) Chemical considered to be food additives
(*d*) Chemical proved safe and approved by the USFDA
(*e*) All of these

223. Preservatives added to inhibit or kill microorganisms may be classified on their
(*a*) Chemical composition
(*b*) Mode of action
(*c*) Specificity or effectiveness
(*d*) Legality
(*e*) All of the above

224. Sugar, salt, oil, spices and essence are classified into
(*a*) Preservative class I
(*b*) Preservative class II
(*c*) Both above
(*d*) None

225. One of the following organic acids is used in syrup, drinks, jam and jelly as substitute for fruit flavour and preservation
(*a*) Lactic acid
(*b*) Citric acid
(*c*) Benzoic acid
(*d*) Acetic acid

226. Lactic acid or acetic acid are used in foods along with
(*a*) Brine
(*b*) Sugar syrup
(*c*) Corn syrup
(*d*) None

227. Na or Ca propionate is used extensively in the
(*a*) Cheese and spreads
(*b*) Baked items
(*c*) Dairy confectionary foods
(*d*) All of the above

228. The basic purpose of Na or Ca propionate to be used in food for
(*a*) Prevention of mold growth
(*b*) Prevention of rope development
(*c*) Regulation of acidity
(*d*) Both (*a*) and (*b*)

229. The effectiveness of Na or Ca propionate decreases with
(*a*) increase in pH
(*b*) increase in acidity
(*c*) decrease pH
(*d*) increase relative humidity

230. The effectiveness of sodium benzoate increases with
(*a*) increase in pH
(*b*) increase in acidity
(*c*) constant pH
(*d*) increase relative humidity

231. The pH at which sodium benzoate is most effective against growth of bacteria is
(*a*) 2.5 – 4.0
(*b*) 4.0 – 4.5
(*c*) 5.0 – 6.5
(*d*) Above 7.0

232. Sorbic acid and its salts are used as direct antimicrobial additives in food in the form of
(*a*) Spray
(*b*) Dip
(*c*) Coating on packaging material
(*d*) All of these

233. Sorbic acid and its salts are known to inhibit the
(*a*) Yeast
(*b*) Molds
(*c*) Bacteria
(*d*) Both (*a*) and (*b*)

234. Sorbets are more effective than sodium benzoate at the pH value above
(*a*) 2.0
(*b*) 4.0
(*c*) 6.0
(*d*) 7.0

235. One of the following is not added in fruits juices.
(*a*) Propionate (*b*) Benzoate
(*c*) Sorbate (*d*) Sulpher dioxide

236. One of the following is mostly used in pickles and sauce/ketchup
(*a*) Calcium Propionate (*b*) Sodium benzoate
(*c*) Potassium sorbate (*d*) Acetic acid

237. Which antimicrobial preservative is not added in jam and jelly?
(*a*) Propionate (*b*) Benzoate
(*c*) Sorbate (*d*) Acetate

238. The combination of nitrates and nitrites used in curing of
(*a*) Pickles (*b*) Cheese
(*c*) Meat (*d*) Egg

239. Nitrite can react with secondary and tertiary amines to form nitrosamine which is known as
(*a*) Carcinogenic (*b*) Oxidation
(*c*) Flocculation adsorption (*d*) None

240. The problem of possible carcinogenic nitrosamine may be greatest in
(*a*) Bacon (*b*) Beef
(*c*) Pork (*d*) Mutton

241. Sulphur dioxide are used in food to prevent
(*a*) Microbial load
(*b*) Prevent enzymatic and non enzymatic reaction
(*c*) Disclouration of some foods
(*d*) All of these

242. Ethylene oxide and propylene oxide used for
(*a*) Sterilizing packaging material (*b*) Fumigation of ware house
(*c*) Cold sterilization of surgical items (*d*) All of these

243. The FDA restricted the used of ethylene oxide in
(*a*) Spices (*b*) Processed natural seasoning
(*c*) Both (*a*) and (*b*) (*d*) None

244. Effect of sugar and salt on water activity of foods
(*a*) Decreased (*b*) Increased
(*c*) Remain same (*d*) Increased then decreased

245. High sugar concentration as preservatives is used in
(*a*) Sweetened condensed milk (*b*) Fruit in syrup
(*c*) Jam, jelly and candies (*d*) All of the above

246. Ethanol is most germicidal in the concentration between
(a) 5-10% (b) 10-25%
(c) 25-70% (d) 70-95%

247. Flavour extract e.g., vanilla and lemon extract are preserved along with
(a) Ethanol (b) Formaldehyde
(c) Glycerol (d) Propylene glycol

248. The effectiveness of NaCl in foods varies directly with its
(a) Concentration (b) Temperature
(c) Salinity (d) Both (a) and (b)

249. Most bacteria, yeast and molds grow best in the temperature range of about
(a) 10-20°C (b) 30-40°C
(c) 16-38°C (d) Above 40°C

250. Bread is prepared by
(a) Yeast (b) Bacteria
(c) Fungus (d) Molds

251. Most of the bacteria are killed at temperature range of
(a) 80-90°C (b) 82-93°C
(c) 95-100°C (d) 100-105°C

252. The greatest drawback at present to the use of x-rays in food preservation is
(a) Low efficiency (b) High cost
(c) High penetration (d) Both (a) and (b)

253. Gamma rays have good penetration, but their effectiveness decreases exponentially with
(a) Depth (b) Volume
(c) Mass (d) Area

254. The bactericidal efficiency of a given dose of irradiation to food depends upon
(a) Kind, species and number of organisms or species
(b) Composition and physical state of food
(c) The presence or absence of oxygen
(d) All of the above

255. Low level irradiation (1kGy) can be used to kill insects and to inhibit spoilage in
(a) Fresh fruits and vegetables (b) Dried vegetables
(c) Pickles (d) Stored grains

256. Dry or dehydrated vegetables (herbs and spices) can be irradiated to kill insects and bacteria by the dose of
(*a*) 0.8 kGy
(*b*) 10 kGy
(*c*) 30 kGy
(*d*) 30 Gy

257. Radiation dose (kGy) for inhibiting spouting of potato, onions and garlic is used about
(*a*) 0.05-0.15
(*b*) 0.08-0.15
(*c*) 0.25-0.35
(*d*) 0.1-1.0

258. Meat foods, sea foods and baked foods are irradiated by the dose of
(*a*) 8-30 kGy
(*b*) 1-10 kGy
(*c*) 20-70 kGy
(*d*) 1-4 kGy

259. Bacteria in wheat flour include
(*a*) Spores of bacillus
(*b*) Coliform bacteria
(*c*) Achromyces, micrococcus, alcaligenes and other
(*d*) All of the above

260. Species of *Fusarium* and *Pencillium* are the dominant mold which contaminate to the
(*a*) Corn meal
(*b*) Corn flour
(*c*) Corn sweetener
(*d*) Both (*a*) and (*b*)

261. During slicing of bread, contamination may take place from the microorganism by
(*a*) Air
(*b*) Knives
(*c*) Wrapper
(*d*) All

262. Spores of bacteria able to cause ______________ in bread will survive in the baking process.
(*a*) Ropiness
(*b*) Bitterness
(*c*) Mustiness
(*d*) All of the above

263. *Aspergillus flavus* and *A. parasiticus* molds are responsible to produce a toxin in food is known as
(*a*) Aflatoxin
(*b*) Mycotoxin
(*c*) Neurotoxin
(*d*) Enterotoxin

264. A storage temperature for dry food products is recommended about
(*a*) 1-3°C
(*b*) 4.4-7.2°C
(*c*) 10-15°C
(*d*) 37°C

265. The moisture content of corn at which aflatoxin formation by mold take place
(*a*) 15%
(*b*) 20%
(*c*) More than 20%
(*d*) None

266. One of the following chemicals can be reduce the mold growth in high moisture corn.
(*a*) Ammonia (2%) and propionic acid (1%) (*b*) SO_2 (2%) and citric acid (0.25%)
(*c*) Sodium benzoate (0.06%) (*d*) All of the above

267. The major factor involved in the spoilage of stored grain by mold including
(*a*) Microbial load
(*b*) pH, temperature and moisture above 12-13%
(*c*) Physical damage
(*d*) All of the above

268. One of the following molds spoiled the cereals grains.
(*a*) Aspergillus (*b*) Penicillium and fusarium
(*c*) Mucor and rhyzopus (*d*) All of the above

269. A moisture content of wheat flour is less than ________ percent has been reported to prevent the growth of all microorganisms.
(*a*) 15% (*b*) 13%
(*c*) 17% (*d*) 10%

270. Optimum moisture content of wheat flour, at which mold grows very well.
(*a*) 10% (*b*) 13%
(*c*) 15% (*d*) 17%

271. Population of both mold and bacteria in flour increased if moisture content is over than
(*a*) 10% (*b*) 13%
(*c*) 15% (*d*) 17%

272. Ropinèss of bread is caused by a mucoid variant of
(*a*) *Bacillus subtilis or B. licheniformis* (*b*) *Serratia marcescens*
(*c*) *Monilia (Neurosporca) sitophila* (*d*) *Geotrichum aurantiacum*

273. The red colour on bread results the growth of pigmented bacteria usually
(*a*) *Geotrichum aurantiacum* (*b*) *Bacillus subtilis*
(*c*) *Serratia marcescens* (*d*) *Monilia (Neurospora) sitophila*

274. Chalk like spots has been blamed on the growth of yeast like fungi is
(*a*) *Endomycopsis fibuligera* (*b*) *Trichosporon variable*
(*c*) *Aspergillus* (*d*) Both (*a*) and (*b*)

275. The deterioration of breads, cakes, pies and other bakery products due to mostly by physical changes during holding without microorganisms is called
(*a*) Physical injury (*b*) Mechanical shock
(*c*) Staling (*d*) Proofing

276. Approximate bacterial load in raw cane juice is between
(*a*) $10^2 - 10^8$ per gram
(*b*) $10^2 - 10^4$ per ml
(*c*) $10^4 - 10^8$ per gram
(*d*) $10^4 - 10^8$ per ml

277. Raw sugar contain number of bacteria in the range of
(*a*) $10^2 - 10^4$ per gram
(*b*) $10^2 - 10^4$ per ml
(*c*) $10^4 - 10^8$ per gram
(*d*) $10^4 - 10^8$ per ml

278. Bacterial counts in maple sap are usually less than
(*a*) 1000 per ml
(*b*) 10000 per gram
(*c*) 10000 per ml
(*d*) 100000 per gram

279. The chief sources of microorganisms in honey are
(*a*) Nectar of flower
(*b*) Honey bees
(*c*) Both above
(*d*) None

280. An enzyme of honey contain a bacteriostatic as well as a lytic effect on most of the gram positive bacteria is
(*a*) Apoenzyme
(*b*) Lysozyme
(*c*) Lipase
(*d*) Peroxidase

281. The main groups of bacteria present during maturation of nectar to honey are
(*a*) Gluconobacter
(*b*) Lactobacillus
(*c*) Salmonella
(*d*) Both (*a*) and (*b*)

282. Liquid sugar with sugar content as high as 67 to 72ºBrix will support the growth of yeasts like
(*a*) Saccharomyces
(*b*) Candida
(*c*) Rhodotorula
(*d*) All of the above

283. The percentage of sugar content in honey with the range of
(*a*) 75-90%
(*b*) 70-80%
(*c*) 20-30%
(*d*) 40-60%

284. The honey is rich source of
(*a*) Glucose and fructose
(*b*) Glucose and levulose
(*c*) Glucose and maltose
(*d*) Sucrose and levulose

285. The pH value of honey lies between
(*a*) 1.2-2.2
(*b*) 2.2-3.0
(*c*) 3.2-4.2
(*d*) 5.0-5.6

286. Chief source of spoilage of honey is
(*a*) Osmophilic yeast
(*b*) Osmophilic bacteria
(*c*) Mesophilic molds
(*d*) All of the above

287. The critical moisture content of honey for the initiation of yeast growth has been placed at
(*a*) 15%
(*b*) 17%
(*c*) 21%
(*d*) 25%

288. Smallest and largest enzyme, respectively is
(*a*) Urease and zymase
(*b*) Catalase and isomerase
(*c*) Peroxidase and catalase
(*d*) Amylase and pectinase

289. Spoilage of fresh fruits and vegetables usually occurs during
(*a*) Harvesting
(*b*) Transportation and Storage
(*c*) Processing
(*d*) All of the above

290. Pre-cooling and refrigeration during storage of the fresh fruits and vegetables will ________ the growth of microorganisms.
(*a*) Increase
(*b*) Decrease
(*c*) No effect
(*d*) None

291. Washing of foods with detergent or germicidal solution will ________ microbial load.
(*a*) Reduce
(*b*) Increase
(*c*) No effect
(*d*) None

292. Added ingredients such as sugar and starch to foods may add spoilage organisms especially spores of
(*a*) Thermophiles
(*b*) Psychotrophs
(*c*) Thermoduric
(*d*) Halophiles

293. Adequate control of temperature and RH of environment of foods will ________ growth of microorganisms.
(*a*) Increase
(*b*) Reduce
(*c*) Constant
(*d*) None

294. Contamination from equipment at the processing plant can be reduced by adequate
(*a*) Cleaning
(*b*) Sanitization
(*c*) Both (*a*) and (*b*)
(*d*) Sanitation

295. Cooling of fruits and vegetables before normal cold storage after harvesting by use of cold water is referred as
(*a*) Hydro cooling
(*b*) Air cooling
(*c*) Chilling
(*d*) Refrigeration

296. The blanching of foods at (86 to 98°C) brings about a great reduction in the number of microorganisms as much as
(*a*) 20-40%
(*b*) 50-60%
(*c*) 80-90%
(*d*) 90-99%

297. The kind of bacteria in vegetables most likely to grow on thawing will depends upon
(*a*) Temperature (*b*) Elapsed time
(*c*) Moisture content (*d*) Both (*a*) and (*b*)

298. Most bacteria multiple best at the temperature of 16-38°C, is called
(*a*) Psychrophilic (*b*) Mesophilic
(*c*) Thermophilic (*d*) All of the above

299. Under favourable conditions, bacteria can ____________ their number after every 30 minutes.
(*a*) Five times (*b*) Triple
(*c*) Double (*d*) Remain same

300. Bacteria can grow under aerobic and anaerobic conditions is called
(*a*) Aerobic (*b*) Anaerobic
(*c*) Facultative (*d*) Putrefactive

301. The temperature at which thermophilic bacteria can grow very well is
(*a*) 52-62°C (*b*) 66-82°C
(*c*) 62-72°C (*d*) 80-100°C

302. 'Neurotoxicity' is caused by which pathogenic microorganisms.
(*a*) Salmonella sps (*b*) Shigella sps
(*c*) Clostridium botulinum (*d*) Clostridium perfringens

303. Microorganisms cause food deterioration and found in intestinal tracts of human and animal.
(*a*) Vibrio cholera (*b*) Bacillus cereus
(*c*) Salmonella sps (*d*) Escherichia coli

304. Most of bacteria are killed at the temperature range of
(*a*) 66-82°C (*b*) 50-70°C
(*c*) 82-93°C (*d*) 93-110°C

305. The acid effective against the growth of several molds
(*a*) H_2SO_4 (*b*) HCl
(*c*) Sorbic acid (*d*) Acetic acid

306. The water activity equals to
(*a*) $T/100$ (*b*) RH × 100
(*c*) $T \times 100$ (*d*) RH/100

307. Most bacteria requires water activity in the range of
(*a*) 0.50-0.60 (*b*) 0.60-0.80
(*c*) 0.90-1.00 (*d*) 1.80-2.00

308. An ice cream mix inoculated with typhoid bacteria still retained over bacteria per milliliter after one year frozen is
(*a*) 2,00,000 (*b*) 5,00,000
(*c*) 6,00,000 (*d*) 10,00,000

309. If sufficient acid is present, to gain sterility in food requires
(*a*) 93°C temperature for 15 min (*b*) 66°C temperature for 2 min
(*c*) 100°C temperature for 40 min (*d*) 82°C temperature for 10 min

310. The moderate temperature ranges over which most of foods are handled.
(*a*) 5-10°C (*b*) 20-50°C
(*c*) 30-60°C (*d*) 10-38°C

311. *Entamoeba histolytica* is a parasite contaminant of food responsible for
(*a*) Bacterial dysentery (*b*) Fungal dysentery
(*c*) Amoebic dysentery (*d*) Histolic dysentery

312. Diarrhoea is caused by
(*a*) Vibrio cholera (*b*) Bacillus cereus
(*c*) Yersinia enterocolitica (*d*) Shigella sps

313. To destroying insect eggs in flour, a centrifuge type machine used named as
(*a*) Centoleter (*b*) Mycoleter
(*c*) Insectoleter (*d*) Entoleter

314. Food spoilage can be reduced by
(*a*) Pasteurization and sterilization (*b*) Canning
(*c*) Blanching (*d*) All of the above

315. The factor influencing the kind and rate of spoilage of fish is/are
(*a*) Kind of fish (*b*) Condition of catching of fish
(*c*) Contaminants
(*d*) Temperature and use of an antibiotic ice or dip
(*e*) All of them

316. Salt fish are spoiled by salt tolerant or halophilic bacteria of genus
(*a*) Serratia (*b*) Micrococcus and bacillus
(*c*) Alcaligene and pseudomonas (*d*) All of the above

317. The eggs has various type of protecting itself from known as
(*a*) Microbial invasion (*b*) Water and RH
(*c*) Temperature (*d*) All of the above

318. The shell and the thin surface layer of proteinaceous material of egg is known as
(*a*) Cuticle or bloom
(*b*) Pericarp
(*c*) Seed coat
(*d*) Chalaza

319. The rate of physical and chemical changes in the eggs depends on the
(*a*) Time and temperature of holding
(*b*) Relative humidity
(*c*) Composition of the atmosphere about the eggs
(*d*) All of the above

320. The frozen eggs should be stored at
(*a*) 0 to 5°C
(*b*) 10 to 12°C
(*c*) –10 to 0°C
(*d*) –17.8 to – 20.5°C

321. An especially black colouration in the yolk and a dark colour in white is caused by
(*a*) Pseudomonas
(*b*) Acromonas
(*c*) Proteus melanovogenes
(*d*) Alcaligenes

322. Pink rot in eggs are caused by the strain of
(*a*) Pseudomonas
(*b*) Alcaligenes
(*c*) Flavobacterium
(*d*) Paracolobactrum

323. Pin spot molding in egg is caused due to
(*a*) Mold growth
(*b*) Bacterial growth
(*c*) Moisture
(*d*) Low temperature

324. In storage atmosphere of high humidity, a variety of molds may cause
(*a*) Pin spot molding
(*b*) Superficial fungal spoilage
(*c*) Fungal spoilage
(*d*) Off flavour development

325. Fishy flavours are produced in egg by certain strains of
(*a*) Escherichia coli
(*b*) Enterobacter cloacae
(*c*) Streptomyces
(*d*) None

326. The coliform bacteria, anaerobes and yeast can cause gassiness and off-flavour in which milk products.
(*a*) Fermented milk
(*b*) Cheese
(*c*) Ice cream
(*d*) Both (*a*) & (*b*)

327. The centrifugal procedure used for removing bacteria from milk is known as
(*a*) Bactofugation
(*b*) Centrifuge
(*c*) Ultra filtration
(*d*) Homogenization

328. The presence of heat resistant microorganisms or spores may be potential source of spoilage in
(*a*) Cheese (*b*) Butter
(*c*) Evaporated and sweetened condensed milk (*d*) Double toned milk

329. Pasteurization should be accomplished without deleteriously affecting
(*a*) Flavour and appearance (*b*) Nutritional properties
(*c*) Creaming of the milk (*d*) All of these.

330. The heat treatment of cream also destroys ___________ that may cause deterioration of butter during storage.
(*a*) Lipase (*b*) Amylase
(*c*) Lactase (*d*) PPO

331. Spoilage of heated foods may have
(*a*) Physical cause (*b*) Chemical cause
(*c*) Biological cause (*d*) Both (*b*) and (*c*)

332. The most important kind of chemical spoilage of canned foods is
(*a*) Hydrogen swell (*b*) Oxygen swell
(*c*) Sulphur dioxide swell (*d*) None

333. Hydrogen swell in canned foods are favoured by
(*a*) Increasing acidities of foods and Increasing temperature of storage
(*b*) Imperfection in the tinning and lacquering of the interior of the can
(*c*) A poor exhaust
(*d*) Presence of soluble sulphur and phosphorus compounds
(*e*) All of them

334. The defect caused by interaction between the steel base of the can and the contained food include
(*a*) Discolouration of the inside of the can and foods
(*b*) Production of off flavour in foods
(*c*) Cloudiness of liquors or syrup and loss in the nutritive value
(*d*) Corrosion or perforation of the metals
(*e*) All of the above

335. Acid foods such as fruits are processed at temperature of _______, which killing all vegetative cells of bacteria, yeasts and molds and their spores and some bacterial spores.
(*a*) 72°C (*b*) 80°C
(*c*) 100°C (*d*) 121°C

336. A __________ has flat ends, one of which will become convex when the side of the can is struck sharply or the temperature of contents is increased.
(*a*) Flats (*b*) Flipper
(*c*) Soft swell (*d*) Breather

337. A __________ has both ends of can bulged, but one or both ends will stay concave if pushed in or if a swollen ends in pushed in, an opposite flat end will pop out.
(*a*) Flipper (*b*) Springer
(*c*) Hard swell (*d*) Soft swell

338. When both ends of the can bulged, but the gas pressure is low enough to permit the ends to be dented by manual pressure is called
(*a*) Breather (*b*) Hard swell
(*c*) Springer (*d*) Soft swell

339. A __________ has such high gas pressure from within that the ends are too hard to dented by hand.
(*a*) Soft swell (*b*) Hard swell
(*c*) Flat (*d*) Flipper

340. A __________ is a can with a minute leak that permits air to move in or out but does not necessarily allow microorganisms to enter.
(*a*) Springer (*b*) Flipper
(*c*) Breather (*d*) None

341. The __________ of canned foods is important in determining the heat process necessary for their sterilization.
(*a*) pH (*b*) acidity
(*c*) water content (*d*) optical density

342. The classification of kind of spoilage of canned foods is based on the kinds of changes produced in the food *i.e.*
(*a*) Putrefaction (*b*) Acid formation
(*c*) Gas formation (*d*) All of the above

343. The most important kinds of biological spoilage of commercial canned foods are
(*a*) Flat sour spoilage (*b*) TA spoilage
(*c*) Putrefaction (*d*) All of the above

344. A fourth important kind of spoilage caused by action of acid of foods on iron of can results in a
(*a*) Flat sour spoilage (*b*) TA spoilage
(*c*) Hydrogen swell (*d*) Putrefaction

345. The ends of the can of food remain flat during souring, or the development of lactic acid in the food by the flat sour bacteria is known as
(*a*) Flat sour spoilage
(*b*) TA spoilage
(*c*) Hydrogen swell
(*d*) Putrefaction

346. The flat sour spoilage occurs chiefly in low acid foods, such as peas or corn and is caused by species of
(*a*) Bacillus
(*b*) Lactobacillus
(*c*) Clostridium
(*d*) All

347. The ability of B. coagulans in flat sour spoilage to grow in tomato juice depends on
(*a*) No. of spores present
(*b*) The availability of oxygen
(*c*) pH
(*d*) All

348. TA spoilage is the short form of
(*a*) Thermophilic anaerobe
(*b*) Thermal allowance
(*c*) Thermal absorption
(*d*) None

349. TA spoilage is caused due to
(*a*) Clostridium botulinum
(*b*) Clostridium thermosaccharolyticum
(*c*) B. coagulant
(*d*) B. stearothermophilus

350. Sulfide or sulphur stinker spoilage in low acid food is caused by
(*a*) Dusulfotomaculum nigrificans
(*b*) Clostridium thermosaccharolyticum
(*c*) B. stearothermophilus
(*d*) None of them

351. Mesophilic clostridium species spoiled the canned foods when the pH is
(*a*) 2.5-4.5
(*b*) 4.5
(*c*) Above 4.5
(*d*) Below 2.5

352. Home canned foods heated to 100°C may be spoiled by the saccharolytic bacteria with production of
(*a*)`Butyric acid
(*b*) Carbon dioxide
(*c*) Hydrogen
(*d*) All of the above

353. Under spoilage of mesophilic clostridium species, the putrefactive anaerobic produces CO_2 and hydrogen causing the can to
(*a*) Flat
(*b*) Swell
(*c*) Burst
(*d*) Breather

354. The presence of viable non spore forming bacteria in heated processed canned foods indicates
(*a*) Flipper
(*b*) Springer
(*c*) Swell
(*d*) Leakage of container

355. Canned fruits, jam, jellies, fruit juices, syrups and sweetened condensed milk spoiled by fermentative yeast with swelling of can by the production of
(*a*) Hydrogen (*b*) Carbon dioxide
(*c*) Sulphur dioxide (*d*) Hydrogen sulfide

356. Jam, jellies, marmalade and fruit butters will permit mold growth when sugar concentration is as high as
(*a*) 60 % (*b*) 65 %
(*c*) 70 % (*d*) 75 %

357. Strain of Aspergillus, Penicillium and Citromyces found growing in jelly and canned fruits are able to grow in sugar concentration upto
(*a*) 65.0 % (*b*) 67.5 %
(*c*) 70.0 % (*d*) 72.5 %

358. One of the following is pectin fermenting molds in bottled and canned fruits and have caused spoilage.
(*a*) Aspergillus (*b*) Penicillium
(*c*) Byssochlamys fulva (*d*) Citromyces

359. Low acid canned foods with a pH above 5.3 are subjected to
(*a*) Flat sour spoilage (*b*) Putrefaction
(*c*) TA spoilage (*d*) Both (*a*) and (*b*)

360. Canned foods with a pH between 5.3 and 4.5 are likely to undergo
(*a*) Flat sour spoilage (*b*) TA spoilage
(*c*) Sulfide spoilage (*d*) Putrefaction

361. High acid canned foods with a pH between 4.5 and 3.7 usually are spoiled by a
(*a*) Special flat sour bacterium (*b*) Saccharolytic anaerobe
(*c*) TA spoilage (*d*) Both (*a*) and (*b*)

362. High acid canned foods with a pH below 3.7 ordinarily do not undergo spoilage by microorganisms but in cans may become
(*a*) CO_2 swell (*b*) H_2 swell
(*c*) NH_3 swell (*d*) H_2S swell

363. Spoilage in canned meats and fish without gas production with souring and changes in colour and texture may be caused by
(*a*) Bacillus sps. (*b*) Homo-fermentative LAB (faecalis)
(*c*) Both (*a*) and (*b*) (*d*) None

364. One of the following species can produce bitterness, acidity and curdling in canned milk, cream and evaporated milk.
(*a*) Bacillus sps.
(*b*) Lactobacillus sps.
(*c*) Clostridium sps.
(*d*) Streptococcus sps.

365. The spoilage of fatty foods (fats or oils) due to oxidation, chemical or microbiological is termed as
(*a*) Oxidative rancidity
(*b*) TA spoilage
(*c*) Halogenation
(*d*) Maceration

366. Any change in fats or oils that is accomplished by undesirable flavour is known as
(*a*) Spoilage
(*b*) Rancidity
(*c*) Hydrolysis
(*d*) Adsorption

367. The change in fat or oil or fatty foods resulting from hydrolysis, or by lipase originally present or by those from microorganisms causing
(*a*) Oxidative rancidity
(*b*) Hydrolytic rancidity
(*c*) Ketonic rancidity
(*d*) Flavour reversion

368. Hydrolytic rancidity in fatty foods is due to
(*a*) Hydrolysis
(*b*) Lipase
(*c*) Microorganisms
(*d*) All of the above

369. The appearance of objectionable flavour from less oxidation than is needed to produce rancidity is called
(*a*) Auto oxidation
(*b*) Ketonic rancidity
(*c*) Flavour reversion
(*d*) Oxidative rancidity

370. The oxidation of fats and oils may be catalyzed by
(*a*) Metals and rays
(*b*) Moisture
(*c*) Microorganisms
(*d*) All of the above

371. Some of the pigments produced by microorganisms are fat soluble and therefore can diffuse into fat producing discoloration ranging through
(*a*) Yellow
(*b*) Red
(*c*) Purple
(*d*) Brown
(*e*) All of the above

372. The types of spoilage of mayonnaise and similar dressing is/are
(*a*) Separation of the oil or water from emulsion
(*b*) Oxidation and hydrolysis of the oils by chemical or biological action
(*c*) Growth of microorganisms to produce gas, off-flavour, or other defects
(*d*) All of the above

373. The decomposition of salad dressing and related products can be caused by
(*a*) Bacteria (*b*) Yeast
(*c*) Molds (*d*) All of the above

374. Molds can grow on salad dressing if air is available and are favored by the addition of
(*a*) Starch or pectin (*b*) Pectin
(*c*) Edible salt (*d*) None

375. The yeasts are most likely causes of spoilage of soft drinks for such beverages are acid and contain sugar
(*a*) Toruplosis and candila (*b*) Saccharomyces
(*c*) Zygosaccharomyces (*d*) All of the above

376. Types of spoilage of soft drinks is
(*a*) Cloudiness (*b*) Ropiness
(*c*) Gassiness (*d*) Both (*a*) and (*b*)

377. The cloudiness in soft drink takes place due to the growth of various
(*a*) Yeast (*b*) Mold
(*c*) Bacteria (*d*) Both (*a*) and (*c*)

378. Ropiness in soft drink is caused by capsulated bacteria belongs to
(*a*) Bacillus sps. (*b*) Lactobacillus sps.
(*c*) Clostridium sps. (*d*) Streptococcus sps

379. Treatment of spices with __________ greatly reduces their content of microorganisms.
(*a*) Sulphur dioxide (*b*) Propylene oxide
(*c*) Carbon dioxide (*d*) Nitrogen oxide

380. Which salt is used in foods
(*a*) Solar salt (*b*) Mined or rock salt
(*c*) Welled salt (*d*) All of the above

381. One of the following microorganisms may produce mycotoxin on nuts such as aflatoxin production in pea nut.
(*a*) Yeast (*b*) Mold
(*c*) Bacteria (*d*) Parasite

382. Butterfat and meat fat becomes "tallowy" as the result of
(*a*) Oxidation (*b*) Dehydration
(*c*) Frying (*d*) Freezing

383. A mixed lactic acid culture is used in the manufacturing of
(*a*) Cultured butter milk (*b*) Yogurt
(*c*) Several types of cheese (*d*) All of the above

384. Spray dried or lyophilized culture of *Propionibacterium freudenreichii* are added to milk used in the manufacture of __________ to improve the flavour.
(*a*) Chedder cheese
(*b*) Cottage cheese
(*c*) Swiss cheese
(*d*) All of the above

385. Activity of culture for food fermentation depends on
(*a*) Condition of mother or intermediate culture
(*b*) Culture medium
(*c*) Incubation time and temperature
(*d*) All of the these

386. The food fermentation may take place by
(*a*) Yeast
(*b*) Molds
(*c*) Bacteria
(*d*) Combination of the microorganisms
(*e*) All of the above

387. The bread, beer, wine and distilled liquors is fermented products made by with the help of
(*a*) Yeast
(*b*) Mold
(*c*) Bacteria
(*d*) All

388. Fermented milk is produced by
(*a*) Yeast fermentation
(*b*) Mold fermentation
(*c*) Bacterial fermentation
(*d*) All of the above

389. Yeast and bacteria are involved in the manufacturing of __________ from sugar bearing materials.
(*a*) Vinegar
(*b*) Sugar
(*c*) Syrup
(*d*) Alcohol

390. Molds are important in the preparation of
(*a*) Pickles
(*b*) Sauerkraut
(*c*) Cheese and oriental foods
(*d*) Vinegar

391. Dough is usually leavened by __________ which ferment the sugars in dough and produce mainly CO_2 and alcohol.
(*a*) Bread yeast
(*b*) Bacterial culture
(*c*) Mold
(*d*) Citric acid

392. Dough conditioner or yeast foods added in dough to improve the dough characteristics are
(*a*) $KBrO_3$
(*b*) KIO_3
(*c*) CaO_2
(*d*) $(NH_4)_2 S_2O_8$
(*e*) All of them

393. The rate of gas production by yeast in dough is increased by adding
(*a*) More yeast
(*b*) Sugar or amylase bearing malt
(*c*) Dough conditioner within limit
(*d*) All of them

394. In the sponge method of bread making, some of the ingredients are mixed at ________ temperature and allowed to ferment to the desired maturity.
(*a*) 23-24°C (*b*) 25-30°C
(*c*) 30-34°C (*d*) 35-45°C

395. In the straight dough method of bread making, all the ingredients are mixed at ________ temperature.
(*a*) 23-24°C (*b*) 24-25°C
(*c*) 26-28°C (*d*) 30-32°C

396. The fermentation room where the dough is held for most of the leavening process is usually held at the temperature of about
(*a*) 20°C (*b*) 25°C
(*c*) 27°C (*d*) 30°C

397. Leavening of dough takes place by
(*a*) Bread yeast (*b*) Microorganisms
(*c*) Chemicals (*d*) All of the above

398. Thermal treatment in baking process serves as to
(*a*) kill the yeast (*b*) inactivate the enzyme
(*c*) expand the gas (*d*) set the structure of the loaf
(*e*) All of the above

399. "Set" of bread under baking process, ________ gives structural support in the dough, but ________ support the structure of baked bread.
(*a*) Starch, gluten (*b*) Gluten, starch
(*c*) Yeast, salt (*d*) Yeast, sugar

400. The amount of salt by weight is added to shredded cabbage for manufacturing of sauerkraut.
(*a*) 1.0-1.5% (*b*) 1.5-2.0%
(*c*) 2.25-2.5% (*d*) 2.5-3.5 %

401. Dextran, a gummy poly saccharide is a neutral glucan made by *Leuconostoc mesenteroides* from
(*a*) Molasses or refined sucrose medium (*b*) Cereals waste
(*c*) Starch (*d*) Dextrose

402. Dextran is used as ________ for sugar syrup, ice cream and confections.
(*a*) Antioxidant (*b*) Stabilizer
(*c*) Gelling agent (*d*) Preservatives

403. The protein, a biocatalysts called
(*a*) Protein enhancer (*b*) Dough conditioner
(*c*) Enzyme (*d*) Preservative

404. The principal mold used in citric acid production is
(*a*) Aspergillus niger
(*b*) Mucor
(*c*) Penicillum
(*d*) Rhyzopus

405. In food indusıries, citric acid is used as
(*a*) Flavouring agent
(*b*) Adjust the pH
(*c*) To prevent discolouration or browning
(*d*) Synergists with antioxidants for oils
(*e*) All of them

406. The food borne illness caused by the presence of a bacterial toxin formed in the food is referred as
(*a*) Bacterial food intoxication
(*b*) Bacterial food infection
(*c*) Both
(*d*) None

407. The food borne illness caused by the entrance of bacteria into the body through ingestion of contaminated foods and the reaction of the body to their presence or to their metabolites is known as
(*a*) Bacterial food intoxication
(*b*) Bacterial food infection
(*c*) Both
(*d*) None

408. Botulism is a disease caused by the ingestion of food containing the neurotoxin produced by
(*a*) Clostridium botulinum
(*b*) Clostridium perfringens
(*c*) Salmonella
(*d*) Escherichia coli

409. Staphylococcal intoxication caused by a toxin in the food by
(*a*) Clostridium botulinum
(*b*) Salmonella sps
(*c*) Shigella
(*d*) Staphylococcus aureus

410. Mycotoxins are
(*a*) Fungal metabolites
(*b*) Bacterial metabolites
(*c*) Plant parasites
(*d*) Enzyme

411. The syndrome resulting the ingestion of toxin in a mold containing food is referred as
(*a*) Mycotoxicosis
(*b*) Aflatoxin
(*c*) Neurotoxin
(*d*) Botulism

412. Aflatoxin is resulting of
(*a*) Aspergillus flavus
(*b*) Alternaria
(*c*) Fusarium
(*d*) Helminthosporium

413. Optimal conditions for the production of aflatoxin would be
(*a*) a_w (0.85), temperature (25-40°C)
(*b*) a_w (0.65), temperature (30-35°C)
(*c*) a_w (0.90), temperature (20-25°C)
(*d*) a_w (0.75), temperature (10-30°C)

414. Food borne disease take place in human by the consumption of
(*a*) Solid foods
(*b*) Milk
(*c*) Water or other beverage
(*d*) All of the above

415. The strength of the sewage or food waste containing organic matter is expressed in term of
(*a*) BOD (*b*) COD
(*c*) DO (*d*) TDS

416. FAO is primary concerned with food production through improved method of
(*a*) Production (*b*) Processing and preservation
(*c*) Distribution of foods (*d*) All of the above

417. WHO activities are more related to the
(*a*) Health of consumer (*b*) Maintenance of food wholesomeness
(*c*) Processing and preservation (*d*) Both (*a*) and (*b*)

418. The joint FAO/WHO Food Standard Commission is a forum for the cooperation among nations to develop or agree upon various international standards for
(*a*) Agriculture industry (*b*) Health industry
(*c*) Food industry (*d*) Sugar industry

419. The maximum acceptable number of microorganisms or specific types of microorganisms, as determined by prescribed method, in a food being purchased by a firm or agency for it own use is known as
(*a*) Microbiological standard (*b*) Microbiological specification
(*c*) Microbiological guideline (*d*) Microbiological hazard

420. That part of a law or administrative regulation designating the maximum acceptable number of microorganisms or of specific types of microorganisms, as determined by prescribed methods, in a food produced, packed or stored or imported into the area of jurisdiction of an enforcement agency is called
(*a*) Microbiological standard (*b*) Microbiological specification
(*c*) Microbiological guideline (*d*) Microbiological hazard

421. That level of bacteria in final products or in a shipped product that requires identity and correction of causative factors in a current and future production, or in handling after production.
(*a*) Microbiological standard (*b*) Microbiological specification
(*c*) Microbiological guideline (*d*) Microbiological hazard

422. The chilling of foods is accomplished by the use of
(*a*) Cold water or ice (*b*) Mechanical refrigeration
(*c*) Vacuum cooling (*d*) All

ANSWER KEY (FOOD PRESERVATION AND MICROBIOLOGY)

1.A 2. C 3.A 4.B 5.D 6.D 7.C 8.A 9.C 10.B
11. A 12.B 13.A 14.A 15.A 16.C 17.C 18.A 19.A 20.C
21.D 22.B 23.A 24.B 25.A 26.D 27.D 28.D 29.A 30.B
31.B 32.B 33.C 34.A 35.B 36.C 37.A 38.B 39.A 40.A
41.D 42.B 43.A 44.A 45.C 46.C 47.B 48.D 49.B 50.B
51.A 52.A 53.A 54.A 55.D 56.A 57.B 58.B 59.A 60.D
61.D 62.C 63.B 64.B 65.A 66.B 67.A 68.A 69.B 70.D
71.D 72.D 73.D 74.D 75.B 76.A 77.B 78.B 79.A 80.E
81.E 82.D 83.C 84.B 85.B 86.B 87.D 88.C 89.E 90.D
91.D 92.B 93.C 94.D 95.D 96.D 97.A 98.C 99.D 100.D
101.B 102.C 103.D 104.A 105.C 106.C 107.D 108.B 109.A 110.C
111.B 112.C 113.D 114.A 115. A 116. B 117.C 118.C 119.A 120.C
121.B 122.D 123.B 124.A 125.A 126.B 127.D 128.A 129.B 130.C
131.A 132.C 133.A 134.A 135.B 136.A 137.A 138.D 139.D 140.E
141.C 142.D 143.C 144.A 145.A 146.A 147.A 148.D 149.E 150.C
151.B 152.C 153.E 154.E 155.A 156.C 157.A 158.D 159.A 160.B
161.B 162.B 163.B 164.A 165.A 166.B 167.B 168.A 169.C 170.C
171.A 172.A 173.A 174.D 175.A 176.B 177.A 178.C 179.A 180.A
181.A 182.A 183.A 184.A 185.D 186.D 187.A 188.D 189.A 190.D
191.E 192.A 193.D 194.A 195.B 196.A 197.B 198.E 199.A 200.A
201.C 202.C 203.C 204.A 205.B 206.D 207.D 208.D 209.D 210.D
211.C 212.D 213.D 214.D 215.B 216.D 217.C 218.D 219.E 220.E
221.D 222.E 223.E 224.A 225.B 226.A 227.D 228.D 229.A 230.B
231.A 232.D 233.D 234.B 235.A 236.D 237.D 238.C 239.A 240.A
241.D 242.D 243.C 244.A 245.D 246.D 247.A 248.D 249.C 250.A
251.B 252.D 253.A 254.D 255.A 256.C 257.A 258.C 259.D 260.D
261.D 262.A 263.A 264.B 265.C 266.A 267.D 268.D 269.B 270.C
271.D 272.A 273.C 274.D 275.C 276.D 277.A 278.C 279.C 280.B
281.D 282.D 283.B 284.B 285.C 286.A 287.C 288.C 289.D 290.B
291.A 292.A 293.B 294.C 295.A 296.D 297.D 298.B 299.C 300.C
301.B 302.C 303.D 304.C 305.C 306.D 307.C 308.C 309.A 310.D
311.C 312.C 313.D 314.D 315.E 316.D 317.A 318.A 319.D 320.D
321.C 322.A 323.A 324.B 325.A 326.D 327.A 328.C 329.D 330.A
331.D 332.A 333.E 334.E 335.C 336.B 337.B 338.D 339.B 340.C
341.B 342.D 343.D 344.C 345.A 346.A 347.D 348.A 349.B 350.A
351.C 352.D 353.B 354.D 355.B 356.C 357.B 358.C 359.D 360.B
361.D 362.B 363.C 364.A 365.A 366.B 367.B 368.D 369.C 370.D
371.E 372.D 373.D 374.A 375.A 376.D 377.D 378.A 379.B 380.D
381.B 382.A 383.D 384.C 385.D 386.E 387.D 388.C 389.A 390.C
391.A 392.E 393.D 394.A 395.C 396.C 397.D 398.E 399.B 400.C
401.A 402.B 403.C 404.A 405.E 406.A 407.B 408.A 409.D 410.A
411.A 412.A 413.A 414.D 415.A 416.D 417.D 418.C 419.B 420.A
421.C 422.D

Food Processing and Engineering

1. Dimension of viscosity of liquid is
   (*a*) $M^{-1}L^{-1}T^{-1}$ (*b*) $ML^{-1}T$
   (*c*) $M^{-1}LT^{-1}$ (*d*) $ML^{-1}T^{-1}$.

2. Milk, having a viscosity μ is flowing through a circular tube of diameter *D*. The Reynolds number thus can be expressed as *DG*μ, where *G* is
   (*a*) Mass flow rate (*b*) Mass velocity
   (*c*) Volumetric flow rate (*d*) Linear velocity.

3. A food gel requires a threshold level of stress to initiate the flow, and then it follows a linear stress–strain relationship. This kind of fluid is known as
   (*a*) Pseudoplastic (*b*) Dilatant
   (*c*) Bingham plastic (*d*) Bingham pseudoplastic.

4. Finned tube heat exchanger is used for heating air. It helps in
   (*a*) Increasing the thermal conductivity of air
   (*b*) Increasing the heat transfer by natural convection
   (*c*) Decreasing the pressure drop in the heat exchanger
   (*d*) Getting more surface area for heat transfer.

5. Rayleigh number is the product of
   (*a*) Nusselt number and Prandtl number (*b*) Grashof number and Reynolds number
   (*c*) Reynolds number and Prandtl number (*d*) Grashof number and Prandtl number

6. Unsteady state unidirectional heat flow in a slab can be expressed as
   $\partial T/\partial\theta = \alpha\,(\partial^2 T/\partial x^2)$, where α is
   (*a*) Thermal conductivity (*b*) Specific heat
   (*c*) Thermal diffusivity (*d*) Overall heat transfer coefficient

7. The difference between the values of initial and equilibrium moisture content of a food is known as
   (*a*) Unbound moisture content (*b*) Bound moisture content
   (*c*) Free moisture content (*d*) Critical moisture content.

8. LSU drier is basically a
   (*a*) Cocurrent flow drier
   (*b*) Counter current drier
   (*c*) Cross-flow drier
   (*d*) Continuous flow mixing type flow batch drier

9. Drying of fruit pulp can be accomplished by a
   (*a*) Tray drier
   (*b*) Fluidized bed drier
   (*c*) Drum drier
   (*d*) Spray drier

10. If ρ_b and ρ_t are the bulk and true density of grain respectively, then the void fraction in a grain bed can be expressed as
    (*a*) $1 - \rho_b / \rho_t$
    (*b*) ρ_b / ρ_t
    (*c*) $\rho_b / \rho_t - 1$
    (*d*) $\rho_t / \rho_b - 1$

11. Temperature profile in a wall under steady state conditions has been observed to be non-linear. The reason for non-linearity could be
    (*a*) Heat generation within the wall
    (*b*) Thermal conductivity being a function of temperature
    (*c*) Both of the above
    (*d*) None of the above

12. In a dairy farm the 'manger' is meant for
    (*a*) Breading the cattle
    (*b*) Feeding the cattle
    (*c*) Rearing the calf
    (*d*) Storing the feed for longer period.

13. Unit for measurement of vacuum is
    (*a*) kgf / cm^2
    (*b*) Torr
    (*c*) mm of Hg
    (*d*) none of the above

14. Conduction heat transfer is quantified by
    (*a*) Fourier's law
    (*b*) Laplace law
    (*c*) Buke-Plummer equation
    (*d*) Blake-Kozney equation

15. Specific heat, coefficient of viscosity and thermal conductivity are related to
    (*a*) Thermal diffusivity
    (*b*) Prandtl number
    (*c*) Schmidt number
    (*d*) Froude number

16. Heat transfer coefficient in natural convection increases with increasing
    (*a*) Biot number
    (*b*) Nusselt number
    (*c*) Grashof number
    (*d*) None of the above

17. Constant rate of drying is directly proportional to
(*a*) Convective heat transfer coefficient (*b*) Latent heat of vapourization
(*c*) Wet bulb temperature (*d*) None of the above

18. Freeze drying time is directly proportional to the .... of the material being dried.
(*a*) Thickness (*b*) Square of the thickness
(*c*) Cube of thickness (*d*) Fourth power of thickness

19. Decimal reduction time in microbial destruction is inversely proportional to
(*a*) Z value (*b*) Universal gas constant
(*c*) Initial concentration (*d*) Reaction rate

20. Butter must contain________ % fat.
(*a*) 60 (*b*) 70
(*c*) 80 (*d*) 90

21. The particle density of an agricultural produce is 1.95 g/cc. the porosity of the bulk is 36%. The bulk density of the produce is
(*a*) 1.00 (*b*) 1.25
(*c*) 1.50 (*d*) 1.75

22. The temperature range for ultra-high-temperature (UHT) sterilization of milk is
(*a*) 90 to 100°C (*b*) 100 to 115°C
(*c*) 135 to 150°C (*d*) None of the above

23. Losses of nutrient during heating of food products are guided by____________order reaction.
(*a*) Zero (*b*) First
(*c*) Second (*d*) None of the above

24. The unit of absolute viscosity is
(*a*) Volt (*b*) Ampere
(*c*) Poise (*d*) Stoke

25. A dimensionless number representing the ratio of inertial to viscous forces in fluid flow is better known as
(*a*) Reynolds number (*b*) Froude number
(*c*) Prandtl number (*d*) Stanton number

26. Air at 40°C and 50% rh has a wet bulb depression of 10°C. If the relative humidity decreases to 40%, the wet bulb depression will
(*a*) Increase (*b*) Decrease
(*c*) Remain same (*d*) Follow no definite trend

27. The dimensions of pressure are:
(*a*) ML^2T^{-2} (*b*) MLT^{-2}
(*c*) $ML^{-1}T^{-2}$ (*d*) ML^2T^{-3}

28. Fundamental dimensions of coefficient of viscosity are
(*a*) L^2T^{-1} (*b*) MLT^{-1}
(*c*) $ML^{-1}T$ (*d*) $ML^{-1}T^{-1}$

29. The silo pits are storage structures primarily for the purpose of :
(*a*) Sanitation (*b*) Grain storage
(*c*) Animal fodder (*d*) Manures and fertilizers

30. A 100 kg of clean bold paddy would yield about 20-22 kg of
(*a*) Husk (*b*) Bran
(*c*) Germ (*d*) White rice

31. Solid particles of different densities from a mixture could be separated by
(*a*) Sink and float method (*b*) Differential settling method
(*c*) Both 'a' and 'b' (*d*) None of these

32. Moisture content of a mass of paddy was reported to be 25 % (db). The value on wet basis will be
(*a*) 33% (*b*) 30%
(*c*) 20% (*d*) 15%

33. The equipment employed for separating the solid particles from mixture of solid and gas is at faster rate is
(*a*) Gravity separator (*b*) Cyclone separator
(*c*) Centrifugal decanter (*d*) None of these.

34. The process of the removal of solid particles from a fluid by passing through a filtering medium is called
(*a*) Cleaning (*b*) Washing
(*c*) Filtration (*d*) Scrubbing

35. In a ball mill or pebble mill, most of the size reduction is achieved by:
(*a*) Shearing (*b*) Cutting
(*c*) Crushing (*d*) Impact

36. The amount of heat required to raise the temperature of a given object by unit value is known as
(*a*) Specific heat (*b*) Thermal conductivity
(*c*) Thermal capacity (*d*) Volumetric heat capacity

37. The mass of air contained in one cubic meter at m.s.l. is :
(*a*) 12 kg (*b*) 7.5 kg
(*c*) 1.2 kg (*d*) 0.75 kg

38. It is proposed to concentrate orange juice by boiling off excess water. The relevant unit operation representing this process is known as
(*a*) Distillation (*b*) Evaporation
(*c*) Drying (*d*) Crystallization

39. If a hammer grinder is operated at the recommended speed, its capacity depends on
(*a*) Rate at which ground feed is removed (*b*) Size of the opening in the screen
(*c*) Rate of transporting the material near the mill
(*d*) Width of the hopper

40. In batch type of pasteurizer milk is heated for 30 minutes to a temperature of
(*a*) 49ºC (*b*) 61ºC
(*c*) 71ºC (*d*) 79ºC

41. Psychrometric chart is used to determine the following property of air:
(*a*) Relative humidity (*b*) Dew point temperature
(*c*) Both the above (*d*) None of the above

42. Henderson's equilibrium moisture content equation is
(*a*) $1 - Rh = \exp(-CTM^N)$ (*b*) $1 - Rh = \exp(-CTM)^N$
(*c*) $1 - Rh = \exp(CTM^N)$ (*d*) None of the above three

43. Separation of broken seeds on the basis of length may be done by using
(*a*) Air screen cleaner (*b*) Scalper
(*c*) Indented cylinder separator (*d*) Debearder

44. The equilibrium moisture content is zero at:
(*a*) 100 percent moisture content (*b*) 100 percent relative humidity
(*c*) Zero relative humidity (*d*) Infinite relative humidity

45. Which one of the following principles is different from the remaining ones?
(*a*) Fick's law (*b*) Fourier law's of heat conduction
(*c*) Wein's law (*d*) Ohm's law

46. The average calorific value of paddy husk is about :
(*a*) 2100 kcal per kg (*b*) 2600 kcal per kg
(*c*) 3000 kcal per kg (*d*) 3500 kcal per kg

47. Oil from oil cake is extracted by
(*a*) Expeller (*b*) Ghani
(*c*) Rotary mill (*d*) Solvent extraction

48. Milling refers to
(*a*) Size reduction (*b*) Separation
(*c*) None of the above (*d*) Both of the above

49. Dryer is designed by considering the
    (*a*) Air factor
    (*b*) Dryer factor
    (*c*) Grain factor
    (*d*) All of the above

50. The constant of proportionality in relating stress and strain for elastic bodies is known as
    (*a*) Young's modulus
    (*b*) Bulk modulus
    (*c*) Rigidity modulus
    (*d*) None of the above

51. The heart of the refrigeration system is
    (*a*) Condenser
    (*b*) Evaporator
    (*c*) Water pump
    (*d*) Compressor

52. The value of angle of repose of paddy is
    (*a*) 30°
    (*b*) 35°
    (*c*) 45°
    (*d*) 25°

53. Butter is made from the milk cream containing
    (*a*) 5 percent fat
    (*b*) 33 percent fat
    (*c*) 66 percent fat
    (*d*) 99 percent fat

54. HTST pasteurization means
    (*a*) High temperature simple tank
    (*b*) High time short temperature
    (*c*) High temperature short time
    (*d*) High time small temperature

55. Ghee contains:
    (*a*) 50 percent fat
    (*b*) 75 percent fat
    (*c*) 80 percent fat
    (*d*) 98 percent fat

56. The boiling point of NH_3 refrigerant is:
    (*a*) 10 °C
    (*b*) 0 °C
    (*c*) –3 °C
    (*d*) –33.3 °C

57. The UHT processing of milk is
    (*a*) Unit hydro treatment
    (*b*) Ultra high temperature treatment
    (*c*) Ultra high time temperature
    (*d*) None of these

58. Homogenizer is used for
    (*a*) making milk fat free
    (*b*) making milk solid free
    (*c*) breaking the bond between solid and liquid milk
    (*d*) breaking the milk particles into uniform size

59. The food is preserved by
    (*a*) Cooling
    (*b*) Heating
    (*c*) Boiling
    (*d*) All of the above

60. Indented cylinder grader separates the material on the basis of
(*a*) Width difference
(*b*) Length difference
(*c*) Thickness difference
(*d*) None of these

61. Safe moisture range for packaging of crop seeds in vapour proof container is
(*a*) 8 to 9 per cent
(*b*) 10 per cent
(*c*) 11 to 12 per cent
(*d*) None of the above

62. The relative humidity will be reduced to half its original value if the temperature of air is increased by
(*a*) 5ºC
(*b*) 11ºC
(*c*) 15ºC
(*d*) 20ºC

63. Bukhari type structures are generally used for storage of wheat, gram, paddy, maize and sorghum. The capacity varies from:
(*a*) 0.5 to 1.0 tonnes
(*b*) 3.5 to 18.0 tonnes
(*c*) 18.5 to 28.0 tonnes
(*d*) 20 to 25 tonnes

64. In a godown, extra space for alleyways for inspection and disinfection of stacks is provided which is generally about
(*a*) 30 %
(*b*) 20 %
(*c*) 5%
(*d*) 1%

65. The efficiency of a cyclone separator increases by
(*a*) Increasing the air inlet velocity
(*b*) Decreasing the size of the particles
(*c*) Reducing the size of the separator
(*d*) Reducing the air outlet diameter

66. Froude number is expressed as
(*a*) V^2 / vgL
(*b*) V / gL
(*c*) V^2 / gD
(*d*) V / gL^2

67. The dimension of calorie is
(*a*) $ML^2\theta^2$
(*b*) $ML\theta^{-2}$
(*c*) $ML^{-2}\theta^{-1}$.
(*d*) $ML\theta^{-1}$

68. A body whose emmisivity is less than one is called
(*a*) Black body
(*b*) Gray body
(*c*) Opaque body
(*d*) Hollow body painted black inside

69. Temperature yielding maximum storage life of food materials decreases in the order
(*a*) Potato < Fish < Milk < Orange
(*b*) Fish < Milk < Orange < Potato
(*c*) Fish < Potato < Orange < Milk
(*d*) Milk < Fish < Orange < Potato

70. Sterilization of milk is carried out to extent of destroying
(*a*) Only pathogenic organisms
(*b*) Only spoilage organisms
(*c*) All types of organisms
(*d*) Only heat resistant organisms

71. Type of moisture that can be removed by common drying techniques is
    (*a*) Equilibrium moisture
    (*b*) Total moisture
    (*c*) Free moisture
    (*d*) Bound moisture

72. Size reduction of grain is caused by impact in
    (*a*) Attrition mill
    (*b*) Roller mill
    (*c*) Hammer mill
    (*d*) Jaw crusher

73. Convective heat transfer coefficient of a body is dependent partly upon its
    (*a*) Temperature
    (*b*) Composition (type of materials)
    (*c*) Shape and size
    (*d*) None of the above

74. Natural convection is associated with
    (*a*) Prandtl number only
    (*b*) Reynolds number only
    (*c*) Grashof number only
    (*d*) Nusselt number only

75. Thermal death time (TDT) curve is plotted between
    (*a*) Time Vs Temperature
    (*b*) Temperature Vs [illegible] (time)
    (*c*) Time Vs log (numbers of survivors)
    (*d*) None of the above

76. Larger decimal reduction time implies
    (*a*) microorganisms more heat resistant
    (*b*) microorganisms less heat resistant
    (*c*) microbial growth reduces with time
    (*d*) no effect on microorganisms

77. Psychrometric chart is made for
    (*a*) Any pressure
    (*b*) One atmosphere
    (*c*) Below one atmosphere
    (*d*) Above one atmosphere

78. Psychrometric relations developed are valid at only
    (*a*) Any pressure
    (*b*) One atmosphere
    (*c*) Below one atmosphere
    (*d*) Above one atmosphere

79. Time of freezing of food is longer than that for thawing. The reason for this is
    (*a*) Water has higher thermal conductivity than ice
    (*b*) Water has higher density than ice
    (*c*) Water has lower thermal conductivity than ice
    (*d*) Water has higher thermal diffusivity than ice

80. Nusselt number in convective heat transfer is equivalent to the following in convective mass transfer
    (*a*) Schmidt number
    (*b*) Prandtl number
    (*c*) Sherwood number
    (*d*) Archemedes number

81. A disc type centrifuge has milk inlet at radius R_m, cream outlet at radius R_1 and skim milk outlet at radius R_2. In order to have a very high separation of milk fat globules from milk, the position of R_m should be

(*a*) Close to R_1
(*b*) Close to R_2
(*c*) Midway between R_1 and R_2
(*d*) none of the above

82. The relationship between friction factor *f* and Reynolds number Re of fluid flowing in laminar regime through a circular pipe can be expressed as : (*a* and *b* are constant)

(*a*) $f = a + b$ (Re)
(*b*) $f = a$ / Re
(*c*) $f = a + \text{Re}^b$
(*d*) $f = a + b$ / Re

83. A bed of granular material consisting of particles having diameter *D* and density ρ is to be fluidized. If the depth of bed is *L* and bed porosity is *p*, the pressure drop of a fluid flowing through the bed will be: (*g* is acceleration due to gravity)

(*a*) $D\rho\,(1 - p)g$
(*b*) $L\rho\,(1 - p)g$
(*c*) $D\rho\, pg$
(*d*) $L\rho\, pg$

84. A refrigerant vapour is to be condensed inside a shell and tube heat exchanger using water as the cooling medium. The water should flow through

(*a*) Shell side of the heat exchanger
(*b*) Tube side of the heat exchanger
(*c*) Any of (*a*) or (*b*) above
(*d*) None of the above

85. The effect of temperature (°K) on the viscosity μ of a liquid food may be expressed as

(*a*) $\mu = AT^{-B}$
(*b*) $\mu = Ae^{BT}$
(*c*) $\mu = Ae^{B/T}$
(*d*) $\mu = AT^{B}$

86. Storage of grain that requires protection form solar heat should use the following as roofing material:

(*a*) Galvanized iron
(*b*) Asphalt
(*c*) Aluminium
(*d*) Asbestos cement

87. Most cereals foods contain mainly

(*a*) Protein
(*b*) Fat
(*c*) Starch
(*d*) Vitamins

88. Solid food materials are generally

(*a*) Elastic
(*b*) Viscoplastic
(*c*) Viscoelastic
(*d*) Plastic

89. Fluids which become more fluid, viscosity decreases, with time as they are stirred are known as

(*a*) Pseudoplastic
(*b*) Dilitant
(*c*) Thixotropic
(*d*) Rheopectic

90. Whenever the refrigeration system has to charged the usual practice is to feed the suction line with the

(a) Compressor stopped
(b) Compressor running
(c) Condenser in operation
(d) Condenser not in operation

91. Convective heat transfer coefficient is expressed by ________________ number

(a) Prandtl
(b) Grashof
(c) Nusselt
(d) None of the above

92. Difference between a pasteurizer and sterilizer is only in ____________

(a) Heating agent used
(b) Temperature attained
(c) Design of equipment
(d) None of the above

93. The dimension of thermal diffusivity is____________

(a) MLT^{-2}
(b) $M^{-1}LT$
(c) $M^{o}LT$
(d) None of the above

94. In the parallel flow heat exchanger, the mean temperature difference is given by the following expression

(a) $\Delta t_i - \Delta t_e$
(b) $(\Delta t_i - \Delta t_e)/2$
(c) $(\Delta t_i - \Delta t_e)/ \ln(\Delta t_i/\Delta t_e)$
(d) $(\Delta t_i + \Delta t_e)/ \ln(\Delta t_i/\Delta t_e)$

95. Viscosity of a gas

(a) Increases with increasing temperature
(b) Decreases with decreasing temperature
(c) Does not depend on the temperature
(d) First increases and then decreases

96. Discolouration of any food is primarily associated with the reasons involving:

(a) Lipids, water and protein
(b) Carbohydrate, water and protein
(c) Water and carbohydrate

97. The temperature at which the water in a fruit will freeze is dependent on the amount of following present in the fruit:

(a) Sugar
(b) Fibre
(c) Oil
(d) Water

98. Pascal is a unit of

(a) displacement
(b) temperature
(c) pressure
(d) viscosity

99. When heat flows in steady state through two metal plates separated by a negligibly small air gap between them, the rate of heat flow will be governed by

(a) Thermal diffusivity of air
(b) Thermal diffusivity of metal plates
(c) Thermal conductivity of air
(d) Thermal conductivity of metal plates

100. Diffusion is
(*a*) Mixing of two portions of fluids
(*b*) Mixing of a gas in two containers at different pressures
(*c*) Mixing of two portions of a fluid at different temperatures
(*d*) Mixing of unlike fluids

101. The unit of Universal gas constant is.
(*a*) Erg/°K (*b*) Kcal/°K
(*c*) Dynes/°K (*d*) W/m°K

102. Central Food Technological Research Institute (CFTRI) is premier research institute under
(*a*) ICAR (*b*) CSIR
(*c*) ICMR (*d*) UGC

103. The general gas equation is
(*a*) $PV = RT$ (*b*) $PV = mRT$
(*c*) $PV = (RT)^m$ (*d*) $PV = R/T$

104. Food spoilage occurs due to
(*a*) Bacteria (*b*) Moulds
(*c*) Yeast (*d*) All of the above

105. Specific gravity separator is used for separating seeds on the basis of
(*a*) Same size different specific gravity (*b*) Same size and same specific gravity
(*c*) Different size and same specific gravity
(*d*) Different size and different specific gravity

106. A cyclone separator is used for separating
(*a*) Solid particles from liquids (*b*) Liquid particles from gases
(*c*) Fine particles from solid (*d*) All of the above

107. The thermocouple is a common device used for the measurement o
(*a*) Air flow (*b*) Wind velocity
(*c*) Pressure (*d*) Temperature

108. Mode of heat transfer in solids is
(*a*) Conduction (*b*) Convection
(*c*) Radiation (*d*) All of the above

109. Which of following methods of determining moisture content of food grain is not an indirect method?
(*a*) Electrical resistance method (*b*) Dielectric method
(*c*) Chemical method (*d*) Air oven method

110. Conversion of 20 percent grain moisture from wet basis to dry basis would give
(*a*) 12 percent
(*b*) 16 percent
(*c*) 20 percent
(*d*) 25 percent

111. Optimum moisture content at harvest (percentage wet basis) for paddy should be
(*a*) 20-22
(*b*) 12
(*c*) 10
(*d*) 14

112. Paddy parboiling before milling is
(*a*) Essential
(*b*) Desirable
(*c*) Undesirable
(*d*) Optional treatment

113. Approximate quantity of paddy husk obtainable from one tonne of clean bold paddy is
(*a*) 0.24 to 0.26 tonne
(*b*) 0.20-0.22 tonne
(*c*) 0.28 to 0.30 tonne
(*d*) 0.32 to 0.34 tonne

114. The LSU type drier was developed in
(*a*) India
(*b*) UK
(*c*) USA
(*d*) Canada

115. In thin layer drying of food grains, the thickness of the layer is limited to
(*a*) 10 cm
(*b*) 20 cm
(*c*) 30 cm
(*d*) 40 cm

116. The moisture content of paddy at the time of milling should be
(*a*) 9 to 10 %
(*b*) 11 to 12 %
(*c*) 13 to 14 %
(*d*) 15 to 18 %

117. In dry milling of pulses, prior to treatment with oil, the following operation is done
(*a*) Pitting
(*b*) Grading
(*c*) Conditioning
(*d*) Drying

118. In the traditional dal mills the yield of split pulses is
(*a*) 65 to 75 %
(*b*) 75 to 80 %
(*c*) 80 to 85 %
(*d*) Above 85%

119. In a HTST pasteurizer, milk is heated to
(*a*) 61 °C
(*b*) 71 °C
(*c*) 81 °C
(*d*) 91 °C

120. Operation of Hammer mill is an example of dynamic force application by
(*a*) Crushing
(*b*) Impact
(*c*) Shearing
(*d*) Cutting

121. Holding time under High Temperature Short Time Milk Pasteurizer is
(*a*) 1 min. (*b*) 45 sec.
(*c*) 30 sec. (*d*) 15 sec.

122. Commonly used unit of refrigeration is approximately equivalent to
(*a*) 220 kcal/min (*b*) 100 kcal/min
(*c*) 80 kcal/min (*d*) 50 kcal/min

123. Number of fractions that a single screen can make are
(*a*) 4 (*b*) 3
(*c*) 2 (*d*) 1

124. Separation of grain by screening is done on the basis of
(*a*) Size (*b*) Colour
(*c*) Texture (*d*) None of the above

125. The relationship between the equilibrium moisture content and relative humidity for biological materials has been given by the person named
(*a*) Hicks (*b*) Pearson
(*c*) Jensson (*d*) Henderson

126. In which region of the country, Kothar type structures are used
(*a*) Eastern (*b*) Western
(*c*) Northern (*d*) Southern

127. Among the following materials, the specific heat is highest for
(*a*) Sand (*b*) Glass
(*c*) Iron (*d*) Water

128. On the Celsius scale, the absolute zero temperature is
(*a*) 0 ºC (*b*) –32 ºC
(*c*) –100 ºC (*d*) –273.15 ºC

129. Which of these materials has the least thermal conductivity?
(*a*) Concrete (*b*) Water
(*c*) Glass (*d*) Air

130. The most commonly used refrigeration system now –a –days is
(*a*) Ice refrigeration (*b*) Vapour compression refrigeration
(*c*) Steam jet refrigeration (*d*) Absorption refrigeration

131. Maximum moisture content (wet basis) for safe storage of wheat is
(*a*) 20 % (*b*) 18 %
(*c*) 13 % (*d*) 8 %

132. Recommended temperature for safe storage of grapes is
(*a*) 4 to 5 °C (*b*) 10 to 15 °C
(*c*) –1 to 10 °C (*d*) 20 to 25 °C

133. Mature green tomato may be kept green for 5 to 6 weeks at 13°C in an atmosphere having O_2 content of
(*a*) 3 % (*b*) 10 %
(*c*) 15 % (*d*) 21 %

134. Diffusion coefficient has a dimension of ________
(*a*) MLT^{-2} (*b*) ML^2T^{-1}
(*c*) $M^oL^3T^{-1}$ (*d*) None of them

135. An ideal temperature for storage of potato is
(*a*) –5 °C (*b*) 0 °C
(*c*) 4 °C (*d*) 10 °C

136. The recommended relative humidity (%) for safe storage of potato is
(*a*) 85-90 (*b*) 60-80
(*c*) 50-60 (*d*) 40-50

137. In air tight grain storage, insects are killed when the O_2 level falls to about
(*a*) 2 % (*b*) 10 %
(*c*) 15 % (*d*) 20 %

138. In bag storage structure, the clear distance between the stack is kept as
(*a*) 1 m (*b*) 2 m
(*c*) 3 m (*d*) 4 m

139. Plant capacity required for apple cooling with total heat load of 1,44,000 kcal in 24 hours is
(*a*) 0.5 tonnes (*b*) 1.0 tonnes
(*c*) 1.5 tonnes (*d*) 2.0 tonnes

140. Which of the following chiefly aims to reduce the oxygen level?
(*a*) Metal bin storage (*b*) Silo
(*c*) Air tight storage (*d*) All of the above

141. The velocity of a fluid in a pipe of diameter *D* is *V*, the pipe is connected to another pipe of diameter 2*D*. Reynolds number in the pipe of diameter *D* in relation to the pipe of diameter 2*D* is
(*a*) Double (*b*) Four times
(*c*) Half (*d*) Same

142. The Jenssen formula is used to determine the pressure in a
(*a*) Shallow bin (*b*) Deep bin
(*c*) Any type of bin (*d*) Dam

143. Rankine's earth pressure coefficient is given by
(*a*) (1 + cos Φ) / (1 + sin Φ)
(*b*) (1 – sin Φ) / (1+ cos Φ)
(*c*) (1 + sin Φ) / (1 – sin Φ)
(*d*) (1 – sin Φ) / (1+ sin Φ)

144. 1000 kg of a wet solid is dried from 60 to 20 % moisture by weight. The moisture removed in kg is
(*a*) 520
(*b*) 200
(*c*) 400
(*d*) 500

145. 1000 kg of a solute containing 40% by weight of dissolved salts is cooled to a temperature at which the solubility of the salt is 20 kg/100 kg of solvent. The mass of salt in kg crystallizing out will be
(*a*) 200
(*b*) 280
(*c*) 300
(*d*) 400

146. The ratio of molecular diffusivity of heat to molecular diffusivity of masses called
(*a*) Lewis number
(*b*) Schmidt number
(*c*) Sherwood number
(*d*) Prandtl number

147. A composite wall of two materials *A* and *B* of rectangular cross section has the temperature of the outside surfaces of *A* and *B* maintained at 900°C and 200 °C respectively. Thickness of *A* is twice of *B* and the thermal conductivity of *A* is only half of that of *B*. The temperature at the junction of *A* and *B* will be (in °C)
(*a*) 350
(*b*) 760
(*c*) 340
(*d*) 433.3

148. A falling film evaporator is recommended for concentrating a
(*a*) Heat sensitive solution
(*b*) Viscous liquid
(*c*) Foaming solution
(*d*) Scaling solution

149. A large part of droplets in a spray drier can be expected to have the size of
(*a*) 80 microns
(*b*) 5 microns
(*c*) 40 microns
(*d*) 150 microns

150. Temperature at which ice cream hardening room operates is
(*a*) –50 °C
(*b*) –30 °C
(*c*) –10 °C
(*d*) 0 °C

151. The most effective evaporator is a ....type
(*a*) Plate
(*b*) Rising film
(*c*) Falling film
(*d*) Shell and tube

152. The energy required in grinding large solid particles is inversely proportional to the ____________ of the particles
(*a*) Diameter
(*b*) Density
(*c*) Strength
(*d*) Shape

153. Kg of steam/kg of water vapour removed in an evaporator is called
(*a*) Efficiency (*b*) Effectiveness
(*c*) Steam use ratio (*d*) Specific steam consumption

154. Diffusion coefficient has a dimension of ________
(*a*) MLT^{-2} (*b*) ML^2T^{-1}
(*c*) $M^oL^3T^{-1}$ (*d*) None of them

155. If dried grains are stored in a silo for a long period of time and the bottom layer of the grain is spoiled due to accumulation of moisture then the season for this spoilage is that the grain is ________than the outside air.
(*a*) warmer (*b*) cooler
(*c*) more wet (*d*) drier

156. Plank's equation describing freezing of food was derived from a few assumptions was ___________ not one of these assumptions.
(*a*) Pseudo-steady state condition (*b*) A definite freezing point
(*c*) Freezing starts at the freezing point (*d*) None of the above

157. A two fluid nozzle is used in a
(*a*) Spray dryer (*b*) Falling film evaporator
(*c*) Drum dryer (*d*) Tray dryer

158. Coefficient of performance of a refrigerator is given by
(*a*) Heat removed by the evaporator divided by the heat rejected by the condenser
(*b*) Heat removed by the evaporator divided by the compressor work
(*c*) Heat rejected by the condenser divided by the compressor
(*d*) None of the above

159. In thermal processing of food destruction of microorganisms is governed by a ____________ order equation.
(*a*) Zero (*b*) First
(*c*) Second (*d*) Third

160. Pasteurization of milk is to be carried out at a minimum temperature of
(*a*) 135°C (*b*) 90°C
(*c*) 71°C (*d*) 62°C

161. Homogenization of milk must have 90 % of fat globules smaller than ________in diameter.
(*a*) 1 μ (*b*) 2 μ
(*c*) 3 μ (*d*) 4 μ

162. Loss of nutrients during thermal processing of food is guided by ________ order equation.
(*a*) Zero (*b*) First
(*c*) Second (*d*) Third

163. Decimal reduction time for any thermal process is inversely proportional to a function of the ____________ of the process.
(*a*) temperature (*b*) time
(*c*) pressure (*d*) none of the above

164. The maximum length of a stanchion barn to house 72 cows should be
(*a*) 30 m (*b*) 40 m
(*c*) 50 m (*d*) 100 m

165. From the management point of view, the farm stead should be located
(*a*) near the centre of the farm (*b*) on the one side of the farm
(*c*) near a well in the farm (*d*) none of the above

166. Convective heat transfer coefficient of a body is dependent partly upon its
(*a*) temperature (*b*) composition (type of material)
(*c*) shape and size (*d*) none of the above

167. The viscosity of water at ambient temperature is about
(*a*) 10^{-5} Pa s (*b*) 10^{-3} Pa s
(*c*) 10^{-1} Pa s (*d*) 10 Pa s

168. Shedd's chart is used for calculation of
(*a*) Energy consumption (*b*) Temperature measurement
(*c*) Static pressure drop (*d*) Flow measurement

169. Dry ice is the name of
(*a*) Liquid N_2 (*b*) Solidified H_2O
(*c*) Solidified CO_2 (*d*) Liquid CO_2

170. Ratio of Rankine and Kelvin temperature scale is
(*a*) 0.5 (*b*) 1.0
(*c*) 1.5 (*d*) 1.8

171. The nature of curve between shear stress (*y*-axis) and rate of shear (*x*-axis) for Bingham liquid is
(*a*) Straight line (*b*) Parabola
(*c*) Zigzag line (*d*) None of these

172. The equation $-dC/dt = kC$ represents a ____________ order kinetic reaction.
(*a*) Zero (*b*) First
(*c*) Second (*d*) Third

173. Temperature dependent of the reaction rate constant in a first order reaction can be described by
(*a*) Newton's law (*b*) Kick's law
(*c*) Arrhenius law (*d*) None of these

174. The negative sign in Fourier's equation is inserted to satisfy following law of thermodynamics
(*a*) Zero (*b*) First
(*c*) Second (*d*) None of these

175. In a parallel flow heat exchanger, if hot fluid is cooled from 10°C, the LMTD would be theoretically equal to
(*a*) 40 °C (*b*) 50 °C
(*c*) 60 °C (*d*) 70 °C

176. In heat exchanger, ____________ % cut baffles are generally used
(*a*) 10 (*b*) 20
(*c*) 25 (*d*) 30

177. Surface heat transfer coefficient would be higher for
(*a*) Still air (*b*) Moving air
(*c*) High viscosity liquids (*d*) Condensing steam

178. Tao's chart can be used for computation of
(*a*) Freezing time (*b*) Drying time
(*c*) Boiling time (*d*) None of these

179. Rittinger's law is applicable in the process of
(*a*) Sieve analysis (*b*) Size reduction
(*c*) Freezing (*d*) Drying

180. In thermal processing, decimal reduction time refers to the time required to destroy following percent of microorganisms
(*a*) 50 (*b*) 60
(*c*) 90 (*d*) 100

181. The instrument characteristics identifying repeatability is known as
(*a*) Accuracy (*b*) Repeatability
(*c*) Precision (*d*) None of these

182. Ribbon mixers are generally used for mixing of
(*a*) Powder and liquid (*b*) Liquid and liquid
(*c*) Powder and particles (*d*) None of these

183. Soxhlet apparatus is used for
(*a*) Liquid-solid extraction (*b*) Liquid–liquid extraction
(*c*) Solid-solid extraction (*d*) None of these

184. Plank's equation is basically used for the calculation of
(*a*) Drying time (*b*) Freezing time
(*c*) Boiling time (*d*) None of these

185. The major forces acting in cyclone separator is/are
(*a*) Gravity and centrifugal (*b*) Gravity
(*c*) Centrifugal (*d*) Centripetal

186. The heart of the homogenizer is known as
(*a*) Pump (*b*) Body
(*c*) Source of power (*d*) None of these

187. A sensible cooling of air ____________ relative humidity of air.
(*a*) decrease (*b*) increases
(*c*) remains constant (*d*) none of these

188. At 100 % RH, wet bulb temperature is ______________ dew point.
(*a*) same as (*b*) more than
(*c*) less than (*d*) none of these

189. Centigrade and Fahrenheit scales have the same reading at
(*a*) 40 °C (*b*) –40 °C
(*c*) 14 °C (*d*) –14 °C

190. For sterilization of bottles, the bottles should be rinsed with
(*a*) Caustic soda (*b*) Acetic acid
(*c*) H_2SO_4 (*d*) Plain water

191. It is difficult to heat air by steam because air has very
(*a*) High thermal conductivity (*b*) Low thermal conductivity
(*c*) Both (*a*) and (*b*) (*d*) None of these

192. During fruit juice concentration by evaporating steam, economy can be improved by
(*a*) Single effect (*b*) Double effect
(*c*) Multiple effect (*d*) None of these

193. Reynolds number is the ratio of __________ force to _______force.
(*a*) Inertia, viscous (*b*) Gravity, centrifugal
(*c*) Both (*a*) and (*b*) (*d*) None of these

194. Hard water when heated forms scales of
(*a*) Carbonates (*b*) Silicates
(*c*) Phosphates (*d*) None of these

195. Thermal diffusivity for gases is generally ________ than those of liquids
(*a*) more (*b*) less
(*c*) both (*a*) and (*b*) (*d*) none of these

196. For a perfect black body, the value of absorpivity, reflectivity and transmissivity will be
(*a*) 1, 0, 1 (*b*) 0, 0, 1
(*c*) 1, 0, 0 (*d*) 1, 1, 0

197. Intermediate moisture foods are characterized by moisture content between
(*a*) 10-20 % (*b*) 20-40 %
(*c*) 40-50 % (*d*) 60-70 %

198. Highly perishable fruits are pre-cooled to
(*a*) 2-6 °C (*b*) 6-8 °C
(*c*) 8-10 °C (*d*) 10-12 °C

199. EMC of a food product placed in a stream of drying air depends upon
(*a*) Temperature and relative humidity (*b*) Velocity and temperature
(*c*) Velocity and relative humidity (*d*) None of these

200. The range of speed of disc atomizer of a spray drier is
(*a*) 1000-2000 rpm (*b*) 2000-4000 rpm
(*c*) 4000-5000 rpm (*d*) 6000-20000 rpm

201. The value of universal gas constant (R) is
(*a*) 283 J/kg °K (*b*) 283 J/kg °C
(*c*) 283 kJ/kg °K (*d*) 233 J/kg °C

202. Specific heat is the amount of heat necessary to change the temperature of a unit weight of a substance by
(*a*) 10 °C (*b*) 1 °C
(*c*) 11 °C (*d*) 100 °C

203. The permissible pH value of potable water is
(*a*) 5.5-6.5 (*b*) 7.0-8.5
(*c*) 4.5-5.0 (*d*) 8.5-9.5

204. The hardness limit for potable water ranges between
(*a*) 10-20 ppm (*b*) 30-50 ppm
(*c*) 75-115 ppm (*d*) 120-200 ppm

205. The maximum permissible nitrite content for drinking water is
(*a*) 20 ppm (*b*) 10 ppm
(*c*) 5 ppm (*d*) Zero ppm

206. The total count of bacteria in potable water should not exceed
(*a*) 100 per cc
(*b*) 150 per cc
(*c*) 200 per cc
(*d*) 500 per cc

207. In safe drinking water, the BOD should be
(*a*) 20
(*b*) 30
(*c*) 60
(*d*) Nil

208. In one litre drinking water, the coliform count should not exceed
(*a*) 10
(*b*) 20
(*c*) 30
(*d*) 40

209. The most commonly used solvent in the solvent extraction is
(*a*) Butane
(*b*) n-hexane
(*c*) Propane
(*d*) none of these

210. The basis for measuring the thermodynamic property, the temperature is given by
(*a*) Zeroth law of thermodynamics
(*b*) First law of thermodynamics
(*c*) Second law of thermodynamics
(*d*) Third law of thermodynamics

211. The working of constant volume thermometer is based on the principle of
(*a*) Newton's law
(*b*) Charles's law
(*c*) Boyle's law
(*d*) Gay-Lussac's law

212. Within elastic limits
(*a*) Load is less
(*b*) Load is gradually applied
(*c*) Load is static
(*d*) Deformation is proportional to load
(*e*) Deformation is permanent

213. A body which is permanently deformed is said to have undergone
(*a*) Elastic deformation
(*b*) Limit of elastic deformation
(*c*) Plastic deformation
(*d*) Uniform deformation
(*e*) None of the above

214. According to Hooke's law
(*a*) Stress is proportional to strain
(*b*) Stress/strain is constant
(*c*) Average stress is proportional to average strain
(*d*) Within elastic limits average stress is proportional to average strain
(*e*) None of the above

215. A body which does not contain voids or empty spaces is known as
(*a*) Continuous body
(*b*) An isotropic body
(*c*) Heterogeneous body
(*d*) Crystalline body
(*e*) None of the above

216. The limiting load beyond which the material no longer behaves elastically is known

(*a*) Breaking load
(*b*) Limiting load
(*c*) Load bearing capacity
(*d*) Plastic limit
(*e*) Elastic limit

217. The ability of a material to absorb energy when deformed elastically and to return it when unloaded is called

(*a*) Hardness
(*b*) Fatigue strength
(*c*) Creep
(*d*) Toughness
(*e*) Resilience

218. The ability of a material to absorb energy in the plastic range is known as

(*a*) Hardness
(*b*) Fatigue strength
(*c*) Creep
(*d*) Toughness
(*e*) Resilience

219. Thermosetting plastics

(*a*) Permanently set with heat and can not be deformed when again subjected to heat
(*b*) Soften on application of heat and can be molded again
(*c*) Are produced on synthesis basis
(*d*) None of the above

220. Precision is

(*a*) The repeatability of the result of a measurement process
(*b*) Agreement of the result of a measurement with true value of the measured quantity
(*c*) The ability of a measuring device to detect small differences in a quantity being measured
(*d*) The ability of an instrument to reproduce same reading under identical conditions
(*e*) Error of judgment in reading of observation

221. The degree of perfection used in instruments, the methods and the observations, is known as

(*a*) Precision
(*b*) Accuracy
(*c*) Efficiency
(*d*) Least count
(*e*) Error

222. The accuracy depends upon

(*a*) Precision of instrument
(*b*) Precision of method
(*c*) Good planning
(*d*) All of the above
(*e*) None of the above

223. If a measuring tape is too long as compared to standard, the error will be known as

(*a*) Instrumental error
(*b*) Personnel error
(*c*) Natural error
(*d*) Manufacturing error
(*e*) Superficial error

224. Natural error in measurements may be due to
(*a*) Humidity
(*b*) Temperature
(*c*) Wind
(*d*) Gravity
(*e*) Any of the above

225. In work study, the symbol $\Rightarrow$ implies
(*a*) Delay
(*b*) Transport
(*c*) Operation
(*d*) Inspection
(*e*) Temporary storage

226. In work study, the symbol $\square$ implies
(*a*) Delay
(*b*) Transport
(*c*) Operation
(*d*) Inspection
(*e*) Temporary storage

227. In work study, the symbol D implies
(*a*) Delay
(*b*) Transport
(*c*) Operation
(*d*) Inspection
(*e*) Temporary storage

228. In work study, the symbol $\bigcirc$ implies
(*a*) Delay
(*b*) Transport
(*c*) Operation
(*d*) Inspection
(*e*) Temporary storage

229. In work study, the symbol $\bigtriangledown$ implies
(*a*) Delay
(*b*) Transport
(*c*) Operation
(*d*) Inspection
(*e*) Temporary storage

230. In break even analysis, the total cost consists of
(*a*) Fixed cost
(*b*) Variable cost
(*c*) Fixed cost+ variable cost
(*d*) Fixed cost+ variable cost + profits

231. PERT stands fro
(*a*) Project Estimation and Return Times
(*b*) Planning Estimation, Review Techniques
(*c*) Project Evaluation and Review Techniques
(*d*) Precise Estimation and Review Techniques

232. PERT analysis is based on
(*a*) Optimistic time
(*b*) Pessimistic time
(*c*) Most likely time
(*d*) All of the above

233. PERT is
(*a*) Based on one time estimates
(*b*) Synthesis in concepts
(*c*) Used for repetitive works
(*d*) Built of activities orientated program
(*e*) All of the above

234. The assumption in PERT is
(*a*) A project is always behind schedule
(*b*) Cost of project is always more than the estimated cost
(*c*) A project can be subdividual into a set of predictable, independent activities
(*d*) Activities are complex in nature and time taken by them are indeterminate
(*e*) None of the above

235. At 'break even point'
(*a*) Constant expenses = profits
(*b*) Total sales = variable expenses
(*c*) Variable expenses – profits = total sales
(*d*) None of the above

236. SIMO charts are used in
(*a*) Method study
(*b*) Micro-motion study
(*c*) Process analysis
(*d*) Layout analysis
(*e*) Gathering statistical data about accidents

237. The standard deviation is
(*a*) An average variation
(*b*) Same as the meridian
(*c*) One half of the mode
(*d*) A measure of dispersion
(*e*) One third of mean

238. The mode is
(*a*) Same as median
(*b*) The arithmetic average of the distribution
(*c*) The geometric average of the distribution
(*d*) The score with least frequency of distribution
(*e*) The score with greater frequency of distribution

239. The distribution is used when
(*a*) Sample size is large
(*b*) Sample size is small
(*c*) Sample size is small and the population standard deviation is not known
(*d*) Sample size is large and the population variation is not known
(*e*) None of the above

240. The degree of fermentation in brewing process can be measured by
(*a*) Conductivity meter
(*b*) pH meter
(*c*) Bolometer
(*d*) Smoke meter
(*e*) Turbidity meter

241. A piezometer is used to measure
(*a*) Very low pressure (*b*) Pressure differential
(*c*) Atmospheric pressure (*d*) High pressure
(*e*) Pressure above and below atmospheric pressure

242. Thermocouple are generally used for temperature measurement upto
(*a*) 250 ºC (*b*) 500 ºC
(*c*) 1000 ºC (*d*) 1600 ºC
(*e*) 2600 ºC

243. The function of reference electrode in a pH meter is to
(*a*) Measure average pH value (*b*) Produce a constant voltage
(*c*) Provide temperature compensation (*d*) Produce a constant current
(*e*) None of the above

244. Which of the following will be the most alkaline solution?
(*a*) pH 1 (*b*) pH 4
(*c*) pH 7 (*d*) pH 10
(*e*) pH 14

245. The pH value of pure water sould be
(*a*) 0 (*b*) 7
(*c*) 10 (*d*) 14

246. Which of following crop produce maximum quantity of oil
(*a*) Til (*b*) Mustard
(*c*) Ground nut (*d*) Soybean

247. Object of pre-cooling is
(*a*) to fresh the product (*b*) to improve the quality
(*c*) to remove the field heat (*d*) none

248. In Tea, which of the following beverage substance is present?
(*a*) Caffeine (*b*) Thiobromine
(*c*) Piperine (*d*) None

249. The most processed product from areca nut is
(*a*) Kottapak (*b*) Kalipa
(*c*) Nuli (*d*) All

250. In preservation of foods, how much quantity of colour is permitted?
(*a*) 250 ppm (*b*) 300 ppm
(*c*) 200 ppm (*d*) 500 ppm

251. In freezing, which principle is followed?
(*a*) Condensation
(*b*) Crystallization
(*c*) Drying
(*d*) All

252. For inactivation of microbial growth, which method is followed?
(*a*) Freezing
(*b*) Sulphuring
(*c*) Blanching
(*d*) none

253. Cryo-preservation is associated with
(*a*) Liquid N_2
(*b*) Liquid O_2
(*c*) Liquid NH_3
(*d*) Liquid CO_2

254. The fruits and vegetables are known as
(*a*) Nutritive food
(*b*) Energy food
(*c*) Protective food
(*d*) None

255. Correlation coefficient lies between
(*a*) 0 to 1
(*b*) –1 to +1
(*c*) 1
(*d*) 0 to –1

256. In a small sample test, which method is used?
(*a*) t- test
(*b*) F- test
(*c*) Chi square test
(*d*) None

257. In mustard, which substance is present?
(*a*) Allyl propyl disulphide
(*b*) Diallyl disulphide
(*c*) Isothiocynate
(*d*) None

258. The browning in apple is due to
(*a*) Poly phenol oxidase
(*b*) Tyrosanee
(*c*) Hydrogenase
(*d*) None

259. The quick freezing is preferred over slow freezing due to
(*a*) Formation of large crystal
(*b*) Formation of small crystal
(*c*) To save time
(*d*) None

260. The FPO license is given by
(*a*) Govt. of India
(*b*) State Govt.
(*c*) Agricultural universities
(*d*) None

261. Square root of variance is referred to as
(*a*) Coefficient of variance
(*b*) Standard error
(*c*) Standard deviation
(*d*) Range

262. The standard error of a sample mean is given by
(a) σ/n (b) $\sigma/\sqrt{n}$
(c) σ^2/n (d) None

263. Food Corporation of India (FCI) was established in year
(a) 1950 (b) 1955
(c) 1960 (d) 1965

264. Milling is a general trade name normally applied for
(a) Grinding of wheat (b) Dehusking of paddy
(c) Dehusking and splitting of pulse (d) Extraction of juice or oil
(e) All of the above

265. Food grains are characterized by their
(a) Size (b) Shape
(c) Density (d) All of the above

266. The shape any individual grain particle can be expressed in the term of
(a) Roundness (b) Spherecity
(c) Equivalent diameter (d) Diameter

267. For spherical particles of diameter (D_p), the Spherecity is equal to
(a) Radius (b) Diameter
(c) Equivalent diameter (d) Roundness

268. The Spherecity of non-spherecity particles can be represented by
(a) $\Phi = 6V_p/ D_pS_p$ (b) $\Phi = 6D_p/V_pS_p$
(c) $\Phi = D_pS_p/6V_p$ (d) $\Phi = 6V_pD_pS_p$

269. The diameter of a sphere having the same ratio of surface to volume as that of the actual particles is called
(a) Spherecity (b) Roundness
(c) Equivalent diameter (d) Specific gravity

270. The equivalent diameter of granular material can be expressed as
(a) D_p = lbt (b) D_p = $(\text{lbt})^{1/3}$
(c) D_p = $(\text{l/bt})^{1/3}$ (d) D_p = $(\text{lbt})^3$

271. Performance of food grain mill is characterized by the
(a) Capacity
(b) Power requirement per unit of material reduced
(c) Size and shape of the product before and after reduction
(d) Range in size and shape of the end product
(e) All of the above

272. The size of agricultural products may be reduced by
(*a*) Compression or crushing
(*b*) Shearing
(*c*) Impact
(*d*) Cutting
(*e*) All of the above

273. When an external force applied on a material excess of its strength, the material fails because of its rupture in many direction is known as
(*a*) Impact
(*b*) Crushing
(*c*) Shearing
(*d*) Cutting

274. Food grains flour, grits and meal, ground feed for livestock are made by the
(*a*) Crushing
(*b*) Impact
(*c*) Cutting
(*d*) Shearing

275. One of the following size reduction procedure is applicable to extract oil from oil seed and juice from sugarcane is done by .
(*a*) Impact
(*b*) Cutting
(*c*) Crushing
(*d*) Shearing

276. When a material is subjected to sudden blow of force in excess of its strength, if fails is called
(*a*) Impact
(*b*) Shearing
(*c*) Crushing
(*d*) Cutting

277. A size reduction units consisting of a knife and bar is applied for the ____________ of materials.
(*a*) Shearing
(*b*) Impact
(*c*) Cutting
(*d*) All of the above

278. If size reduction is accomplished by forcing a sharp and thin knife through the materials is termed as
(*a*) Crushing
(*b*) Shearing
(*c*) Cutting
(*d*) Impact

279. In which process of size reduction, minimum deformation and rupture results and the new surface created is more or less undamaged.
(*a*) Impact
(*b*) Compression
(*c*) Crushing
(*d*) Cutting

280. The B-vitamins and minerals are added to ground meal which is known as
(*a*) Fortification
(*b*) Enrichment
(*c*) Enrobing
(*d*) Blending

281. The ratio of the overall surface area of the ground product to the overall surface area of the feed is expressed as
(*a*) Degree of grinding
(*b*) Grinding
(*c*) Grinding efficiency
(*d*) Milling efficiency

282. The product of coefficient of hulling '$E_{hulling}$' and coefficient of wholeness of kernel 'E_{wk}' is known as

(*a*) Degree of grinding (*b*) Coefficient of wholeness of kernel
(*c*) Grinding efficiency (*d*) Milling or hulling Coefficient

283. The overall efficiency of the food milling system (η) is expressed as

(*a*) $\eta = E_{hulling} \times E_{wk}$ (*b*) $\eta = E_{hulling} / E_{wk}$
(*c*) $\eta = E_{wk} / E_{hulling}$ (*d*) $\eta = E_{hulling} \times E_{wk} / 100$

284. The size reduction of agricultural materials in serrated roll crushers is done by the principle of

(*a*) Compression (*b*) Impact
(*c*) Shearing (*d*) All of the above

285. The size of particular material that can be caught by the smooth crushing rolls depends upon the coefficient of friction between the material and roll surface is estimated by the

(*a*) $d_p = 0.04g + R$ (*b*) $d_p = 0.04R + g$
(*c*) $d_p = 0.04\ g / \mathrm{R}$ (*d*) $d_p = 0.04R / g$

where R= radius of roll, and g = half of the width of gap between the rolls

286. The fineness of grinding of burr mill is controlled by

(*a*) Type of plates (*b*) Gap between plates
(*c*) Screen size (*d*) Both (*a*) and (*b*)

287. Fineness of grinding is controlled by

(*a*) Type of plates (*b*) Size of screen
(*c*) Gap between casing and hammer (*d*) Speed of rotor

288. The material in attrition mill is fed between the plates and reduced by

(*a*) Impact and compression (*b*) Crushing and shear
(*c*) Shear and cutting (*d*) Impact and shear

289. The hammer mill is used to reduce the size by

(*a*) Shear (*b*) Impact
(*c*) Cutting (*d*) Crushing

290. In a ball mill, size reduction of material is achieved by

(*a*) Shearing (*b*) Crushing
(*c*) Cutting (*d*) Impact

291. Purees, food pastes, pulp and other similar products are processed by

(*a*) Colloid mills (*b*) Concentric cylinder abrasive mill
(*c*) Ball mill (*d*) Reitz mill

292. Dissipated heat energy during size reduction (grinding) raise the temperature of
(*a*) Milled product
(*b*) The mill or machine
(*c*) The air
(*d*) All of the above

293. During grinding of grains in emery burr mill, the temperature rise is observed to be
(*a*) 10ºC
(*b*) 17ºC
(*c*) 20ºC
(*d*) 23ºC

294. The process of mechanical pressing of liquid out of liquid containing solids is called
(*a*) Disintegration
(*b*) Extraction
(*c*) Expression
(*d*) Extrusion

295. The approximately pressure applied by hydraulic press on oil sample in first stage is about
(*a*) 5MPa for 15-20 min
(*b*) 10MPa for 15-20 min
(*c*) 28MPa for 5-10 min
(*d*) 28MPa for 15-20 min

296. The process of separating a liquid from a liquid solid system with the use of a solvent is to be referred as
(*a*) Extraction
(*b*) Expression
(*c*) Leaching
(*d*) Extrusion

297. The process of cooking moistened starchy and or proteinous food material by a combination of heat, pressure and mechanical shear is called
(*a*) Thermal cooking
(*b*) Mechanical expression
(*c*) Extrusion cooking
(*d*) High pressure processing

298. The device used for extrusion cooking is termed as
(*a*) Pressure vessel
(*b*) Sterilizer
(*c*) HTST Cooker
(*d*) Extruder

299. An extruder is a ________ unit.
(*a*) Mechanical
(*b*) Thermodynamic
(*c*) Pneumatic
(*d*) Hydraulic

300. What is happened on food in extrusion cooking at high temperature in the presence of water?
(*a*) Promotes gelatinization of starch components
(*b*) Stretching of expandable components
(*c*) Cooking of food materials
(*d*) Destroyed the anti-nutritional factor
(*e*) All of the above

301. The transformation of product during extrusion is
(*a*) Irreversible
(*b*) Reversible
(*c*) Both
(*d*) None

302. The extrusion cooking is a ________________ process in which moist, soft grain is fed into the extruder where the desired temperature and pressure are obtained over the required period of residence time.

(*a*) UHT (*b*) LTLT
(*c*) HTST (*d*) All of the above

303. Heat for cooking of food materials in extrusion is achieved by

(*a*) Shear and friction (*b*) Crushing and shearing
(*c*) Pressure and friction (*d*) Compression and external heat source

304. Basic components of an extruder are

(*a*) Screw, die and barrel (*b*) Screw, feed hopper and breaker plate
(*c*) Die, feed hopper and breaker plate (*d*) Barrel, feed hopper and other parts

305. Which type food products can be prepared by extrusion

(*a*) Snacks (*b*) Functional and nutriceuticals foods
(*c*) Vegetable protein (*d*) Farinaceous food
(*e*) All of the above

306. Extrusion technology provides the advantage like

(*a*) Versatility in shape and type of product (*b*) Production of new food with lowest cost
(*c*) High quality product (*d*) Easy saleable
(*e*) All of the above

307. Which one of flow not occurs in extrusion processing?

(*a*) Drag flow (*b*) Heat flow
(*c*) Pressure flow (*d*) Leakage flow

308. Which type of flow results from the relative motion between screw and barrel in extruder?

(*a*) Heat flow (*b*) Pressure flow
(*c*) Leakage flow (*d*) Drag flow

309. The flow that occurs due to the presence of pressure gradient between the feed and the discharge screw of extruder is.

(*a*) Drag flow (*b*) Heat flow
(*c*) Pressure flow (*d*) Leakage flow

310. The flow occurs in the clearance between the screw flights and the barrel and within any slots in the barrel wall or surface.

(*a*) Drag flow (*b*) Heat flow
(*c*) Pressure flow (*d*) Leakage flow

311. Which types of flow reduces the output of an extruder.

(*a*) Leakage flow (*b*) Drag flow
(*c*) Heat flow (*d*) Pressure flow

312. An unit operation in which water or volatile solvents are removed to obtain concentrated liquid products from dilute liquid foods is called
(*a*) Molecular diffusion (*b*) Dehydration
(*c*) Evaporation (*d*) Distillation

313. The grain drying process in which all grains are fully exposed to the drying air under constant drying i.e. at constant air temperature and humidity is known as
(*a*) Deep layer drying (*b*) Thin layer drying
(*c*) Deep bed drying (*d*) Constant rate drying

314. The recommended grain bed thickness in thin layer drying is
(*a*) 20 cm (*b*) 40 cm
(*c*) 45 cm (*d*) 10 cm

315. Which one of the following used as the direct method of moisture content determination of grain?
(*a*) Air oven drying method (*b*) Brown duvel distillation method
(*c*) Vacuum oven drying method (*d*) All of the above

316. Resistance type moisture meter measures the electrical resistance of grain at the given
(*a*) Compaction (bulk density) and temperature (*b*) Temperature and moisture
(*c*) Temperature and porosity (*d*) Temperature and relative humidity

317. The electrical resistance of grain in resistance type moisture meter varies with
(*a*) Moisture (*b*) Temperature
(*c*) Degree of compaction (*d*) All

318. One of the following is a resistance type moisture meter.
(*a*) Universal moisture meter (USA) (*b*) Tag-Happenstall moisture meter (USA)
(*c*) Kett moisture meter (*d*) All of the above

319. One of the following is not a resistance type moisture meter.
(*a*) Motomco moisture meter (USA) (*b*) Universal moisture meter (USA)
(*c*) Tag-Happenstall moisture meter (USA) (*d*) Kett moisture meter

320. The equilibrium moisture content of a food (grain) is dependent upon
(*a*) Temperature (*b*) RH
(*c*) Variety and maturity of grain (*d*) All

321. A plot of equilibrium relative humidity and moisture content of particular food material at a particular temperature (usually, 25º) is known as
(*a*) Isotherm (*b*) Psychometric chart
(*c*) Isobar (*d*) Isochoric

322. Generally EMC is determined by two methods *i.e.*
(*a*) Static and dynamic method (*b*) Aerodynamic and hydrodynamic method
(*c*) Adsorption and desorption method (*d*) Mixing and non mixing method

323. The moisture contained by a material which exerts equilibrium vapour pressure less than that of the pure liquid at the same temperature is known as
(*a*) Unbound water (*b*) Bound water
(*c*) Free water (*d*) Surface water

324. The moisture contained by a material which exerts equilibrium vapour pressure equal to that of the pure liquid at the same temperature is known as
(*a*) Unbound water (*b*) Bound water
(*c*) Free water (*d*) Surface water

325. The moisture contained by a material in excess of the equilibrium moisture is termed as
(*a*) Unbound water (*b*) Bound water
(*c*) Free water (*d*) Surface water

326. The plot of moisture content verses drying time or drying rate versus drying time or drying rate versus moisture content are known as
(*a*) Drying curve (*b*) Isotherm
(*c*) Hysteresis (*d*) EMC curve

327. The moisture content of material at which the drying rate of a material changes from constant rate to falling rate is called
(*a*) Equilibrium Moisture Content (EMC) (*b*) Critical Moisture Content (CMC)
(*c*) Moisture Content (db) (*d*) Moisture Content (wb)

328. The rate of drying of food material during constant drying rate period is dependent upon
(*a*) Difference between the temperature of air and temperature of the wetted surface at constant air velocity and relative humidity
(*b*) Difference in humidity between air stream and wet surface at constant air velocity and temperature
(*c*) Air velocity at constant air temperature and humidity
(*d*) All of the above

329. The average critical moisture content (CMC) for a given type of food material depends upon
(*a*) Surface moisture content (*b*) Bed thickness of the food material
(*c*) Rate of drying (*d*) All of them

330. Liquid movement due to moisture concentration differences during drying is called
(*a*) Capillary flow (*b*) Surface diffusion
(*c*) Liquid diffusion surface (*d*) Vapour diffusion

331. Liquid movement in material due to surface forces during drying is done by
(*a*) Surface diffusion
(*b*) Capillary flow
(*c*) Liquid diffusion surface
(*d*) Hydrodynamic flow

332. Liquid movement due to moisture diffusion in material in the pores during drying of material takes place by
(*a*) Thermal diffusion
(*b*) Hydrodynamic flow
(*c*) Vapour diffusion
(*d*) Surface diffusion

333. Vapour movement due to differences in vapour pressure under drying process is caused by
(*a*) Vapour diffusion
(*b*) Thermal diffusion
(*c*) Surface diffusion
(*d*) None of them

334. Vapour movement due to temperature differences during drying process by the principles of
(*a*) Vapour diffusion
(*b*) Thermal diffusion
(*c*) Surface diffusion
(*d*) Hydrodynamic flow

335. Liquid and vapour movement due to total pressure difference is caused by
(*a*) Hydrodynamic flow
(*b*) Capillary flow
(*c*) Surface diffusion
(*d*) Vapour diffusion

336. $\partial M/\partial \theta = \partial \nabla^2 K_{11} M$ related to the theory of grain drying is given by
(*a*) Henderson
(*b*) Laplace
(*c*) Luikov
(*d*) None of them

337. $M - M_e / M_0 - M_e$ is known as
(*a*) Moisture ratio
(*b*) Critical moisture content
(*c*) Equilibrium moisture content
(*d*) Drying ratio

338. In $\partial C/\partial \theta = D_v(\partial^2 C/\partial x^2)$, the D_v is expressed as
(*a*) Drying constant
(*b*) Diffusivity
(*c*) Desorption
(*d*) None

339. From the theory of grain drying, the drying constant, $K = D_v . \pi^2/r^2$ is used for
(*a*) Sphere
(*b*) Rectangular
(*c*) Cylindrical
(*d*) Square

340. In the theory of grain drying, $B = 6/\pi^2$. B is represented by
(*a*) Shape factor
(*b*) Diffusivity
(*c*) Blanching
(*d*) Boiling

341. Half time period of drying ($\theta_{1/2}$) is equal to
(*a*) ln2/K
(*b*) K/ln2
(*c*) K.ln2
(*d*) ln4/K

342. The rate of drying increase with the rise of air temperature but EMC
(*a*) Increase (*b*) Decrease
(*c*) Remain same (*d*) None

343. When the humidity of air increases, the drying rate slightly
(*a*) Increase (*b*) Decrease
(*c*) Remain constant (*d*) None

344. The volume of drying zone under deep bed drying varies with
(*a*) Temperature and RH of entering air (*b*) The moisture content of grain
(*c*) Velocity of air movement (*d*) All of these

345. The time period taken by the drying front to reach the top of the bin is called
(*a*) Maximum dry rate period (*b*) Mass flow rate
(*c*) Decreasing rate period (*d*) Shape factor

346. The total drying time for grain in the bin is the sum of time required for the
(*a*) Maximum drying rate period (*b*) Decreasing drying rate period
(*c*) Both above (*d*) None

347. Deep bed drying problems are solved by
(*a*) Hukill's analysis (*b*) Dimensional analysis
(*c*) Hysteresis effect (*d*) Henderson equation

348. In general, an air temperature is recommended for deep bed drying about
(*a*) 40°C (15°C rise) (*b*) 45°C (15°C rise)
(*c*) 30°C (10°C rise) (*d*) 40°C (10°C rise)

349. The ratio of latent heat of evaporation credit to the heat energy of the fuel charged is called
(*a*) Mechanical efficiency (*b*) Drying efficiency
(*c*) Thermal efficiency (*d*) Drying performance

350. The ratio of temperature decrease due to cooling of the air during drying and the temperature increase due to heating of air is defined as
(*a*) Heat utilization factor (*b*) Thermal efficiency
(*c*) Coefficient of performance (*d*) Effective heat efficiency

351. Heat utilization factor for drier is expressed as
(*a*) Heat utilized/heat supplied (*b*) Heat supplied/heat utilized
(*c*) Heat supplied-heat utilized (*d*) None of them

352. The relationship between heat utilization factor (HUF) and coefficient of performance (COP) is as follows
(*a*) HUF = COP – 1 (*b*) HUF = 1 – COP
(*c*) HUF = 1/COP (*d*) HUF = 1 + COP

353. According to the mode of heat transfer, drying methods can be divided into
   (*a*) Conduction drying (*b*) Convective drying
   (*c*) Radiation drying (*d*) All of them

354. When the heat is transferred to the wet solid for drying mainly by conduction through a solid surface (usually metallic). This phenomenon is known as
   (*a*) Dielectric drying (*b*) Sorption drying
   (*c*) Contact drying (*d*) Vacuum drying

355. One of the following is a continuous conduction drier.
   (*a*) Cylinder dryer (*b*) Drum dryer
   (*c*) Steam tube rotary dryer (*d*) All of the above

356. Which one is the example of batch conduction dryer?
   (*a*) Vacuum dryer (*b*) Freeze dryer
   (*c*) Agitated pan dryer (*d*) All of the above

357. One of the following drying is most popular for the grain.
   (*a*) Conduction drying (*b*) Convective drying
   (*c*) Radiation drying (*d*) All of them

358. One of the following is not the example of conduction drying.
   (*a*) Continuous tray dryer (*b*) Freeze dryer
   (*c*) Drum dryer (*d*) Agitated pan dryer

359. When the hot gas (drying agent) is supplied at a velocity higher than the terminal velocity of the wet solid, the drying of the wet solid occurs in a suspended or fluidized state. This phenomenon is known as
   (*a*) Heated air drying (*b*) Osmotic drying
   (*c*) Dielectric drying (*d*) Fluidized bed drying or pneumatic drying

360. One of the following is not the example of radiation drying.
   (*a*) Sun drying (*b*) Infrared drying
   (*c*) Dielectric drying (*d*) None

361. Microwave oven as dryer is best example of
   (*a*) Radiation drying (*b*) Dielectric drying
   (*c*) Pneumatic drying (*d*) Vacuum drying

362. An atomizer in spray drying controls the
   (*a*) Droplet size (*b*) Size distribution
   (*c*) Trajectory (*d*) Speed
   (*e*) All of the above

363. Commercially sterilized food in hermetically sealed container is called
(*a*) Sterilized food (*b*) Canned food
(*c*) Irradiated food (*d*) Protective food

364. The potential to cause harm to the consumer and /or the product and can be biological or physical is called
(*a*) Critical limit (*b*) Critical control point
(*c*) HACCP (*d*) Hazard

365. A value that defines the difference between acceptable and unacceptable is defined as
(*a*) Critical limit (*b*) Critical control point
(*c*) HACCP (*d*) Hazard

366. A point, step or producer at which control can be applied and a food safety hazard prevented, eliminated, or reduced to acceptable level.
(*a*) Critical limit (*b*) Critical control point
(*c*) HACCP (*d*) Hazard

367. To conduct a planned sequence of observations or measurements to assess the control of CCP.
(*a*) Corrective actions (*b*) Monitor
(*c*) Hazard (*d*) Critical limit

368. Those actions to be taken when the monitoring of a CCP induced a loss of control.
(*a*) Corrective actions (*b*) Monitor
(*c*) Hazard (*d*) Critical limit

369. The presence of any objectionable material on the surface of a container or in a food is called
(*a*) Contamination (*b*) Adulteration
(*c*) Blending (*d*) Hazard

370. The treatment of a canned food in its original container recovered in a salvage operation followed by a scheduled process.
(*a*) Thermal processing (*b*) Micro wave processing
(*c*) Exhaust (*d*) Reprocessing

371. Any appropriate process or procedure by which food is recovered from a suspect lot of canned food and by which its safety and fitness for consumption is ensured.
(*a*) Salvage (*b*) Reprocessing
(*c*) Salvor (*d*) Scheduled process

372. The person responsible for carrying out the salvage operations including any or all of the on site operations.
(*a*) Salvage (*b*) Reprocessing
(*c*) Salvor (*d*) Scheduled process

373. The thermal process chosen by the processor for a given product and container size to achieve at least commercial sterility

(a) Salvage (b) Reprocessing

(c) Salvor (d) Scheduled process

374. Formation of a hard impermeable skin on some foods during drying, which reduces the rate of drying and produces a food with a dry surface and a moist interior.

(a) Cavitation (b) Centrifugation

(c) Bone dry condition (d) Case hardening

375. Production of bubbles in foods by ultrasound and their rapid expansion/contraction.

(a) Cavitation (b) Centrifugation

(c) Fogging (d) Case hardening

376. The separation of immiscible liquid or solid from liquid by the application of centrifugal force.

(a) Cavitation (b) Centrifugation

(c) Emulsification (d) Clinching

377. A chemical method used to measure the polluting potential of material in water.

(a) BOD (b) COD

(c) OD (d) DO

378. Abrupt increase in respiration rate in some fruits during ripening is called

(a) Climacteric (b) Non climacteric

(c) Both (d) None

379. A branch of engineering specializing in technical operations at very low temperature, about –160°C to –50°C

(a) Refrigeration (b) Chilling

(c) Freezing (d) Cryogenic

380. A refrigerant that absorbs latent heat and changes phase from solid or liquid to gas e.g. subliming or evaporating CO_2 or liquid N_2.

(a) Refrigerant (b) Cryogen

(c) Chilling agent (d) Sublimation

381. Diffusion coefficient has a dimension of ____________

(a) MLT^{-2} (b) ML^2T^{-1}

(c) $M^oL^3T^{-1}$ (d) None of them

ANSWER KEY (FOOD PROCESSING AND ENGINEERING)

1.D 2. D 3.C 4.D 5.D 6.C 7.C 8.D 9.C 10.A
11. C 12.B 13.D 14.A 15.A 16.B 17.A 18.B 19.A 20.C
21.B 22.C 23.B 24.C 25.A 26.A 27.C 28.D 29.C 30.A
31.B 32.C 33.B 34.C 35.D 36.A 37.C 38.B 39.B 40.B
41.C 42.A 43.C 44.C 45.C 46.C 47.D 48.A 49.D 50.A
51.D 52.A 53.B 54.C 55.D 56.D 57.B 58.D 59.D 60.B
61.C 62.B 63.B 64.B 65.A 66.C 67.A 68.B 69.D 70.C
71.C 72.C 73.B 74.D 75.B 76.A 77.B 78.B 79.C 80.C
81.A 82.B 83.B 84.C 85.A 86.D 87.C 88.C 89.C 90.B
91.C 92.B 93.D 94.C 95.A 96.B 97.A 98.C 99.C 100.D
101.A 102.B 103.B 104.D 105.A 106.C 107.D 108.A 109.D 110.D
111.A 112.D 113.B 114.C 115.B 116. C 117.A 118.A 119.B 120.B
121.D 122.D 123.C 124.A 125.D 126.C 127.C 128.D 129.D 130.B
131.C 132.C 133.A 134.D 135.C 136.A 137.A 138.B 139.D 140.C
141.A 142.C 143.D 144.D 145.A 146.D 147.C 148.A 149.A 150.B
151.A 152.A 153.D 154.D 155.A 156.B 157.A 158.C 159.B 160.C
161.B 162.B 163.A 164.B 165.A 166.B 167.B 168.C 169.C 170.D
171.A 172.B 173.C 174.C 175.B 176.C 177.D 178.A 179.B 180.C
181.C 182.C 183.A 184.B 185.A 186.A 187.B 188.A 189.B 190.A
191.B 192.C 193.A 194.A 195.B 196.C 197.B 198.C 199.A 200.D
201.A 202.B 203.B 204.C 205.D 206.A 207.D 208.A 209.B 210.A
211.C 212.D 213.C 214.D 215.A 216.E 217.E 218.D 219.A 220.A
221.A 222.D 223.A 224.E 225.B 226.D 227.A 228.C 229.E 230.C
231.C 232.D 233.E 234.C 235.D 236.A 237.D 238.E 239.C 240.B
241.A 242.D 243.B 244.E 245.B 246.A 247.C 248.A 249.A 250.C
251.B 252.A 253.A 254.C 255.B 256.A 257.C 258.A 259.A 260.A
261.C 262.B 263.D 264.E 265.D 266.B 267.A 268.A 269.C 270.B
271.E 272.E 273.B 274.A 275.C 276.A 277.A 278.C 279.D 280.B
281.A 282.D 283.A 284.D 285.B 286.D 287.B 288.B 289.B 290.D
291.A 292.D 293.C 294.C 295.A 296.A 297.C 298.D 299.B 300.E
301.A 302.C 303.A 304.A 305.E 306.E 307.B 308.D 309.C 310.D
311.A 312.C 313.B 314.A 315.D 316.A 317.D 318.D 319.A 320.D
321.A 322.A 323.B 324.A 325.C 326.A 327.B 328.D 329.D 330.C
331.B 332.D 333.A 334.B 335.A 336.C 337.A 338.B 339.A 340.A
341.A 342.B 343.B 344.D 345.A 346.C 347.A 348.A 349.C 350.A
351.A 352.B 353.D 354.C 355.D 356.D 357.B 358.A 359.D 360.C
361.B 362.E 363.B 364.D 365.A 366.B 367.B 368.A 369.A 370.D
371.A 372.C 373.D 374.D 375.A 376.B 377.B 378.A 379.D 380.B
381. D

Engineering Properties and Quality of Biomaterials

1. One of the following is not a physical property of food materials.
   (*a*) Porosity (*b*) Specific gravity
   (*c*) Specific heat (*d*) Frontal area
2. Mechanical damage to seeds and grains which occurs during
   (*a*) Harvesting (*b*) Threshing
   (*c*) Handling (*d*) All of the above
3. Mechanical damage to seeds and grains seriously affect the
   (*a*) Viability and germination power (*b*) Growth vigour
   (*c*) Insects and fungi attack (*d*) Quality of the final products
   (*e*) All of them
4. Which physical properties are necessary to evaluate aerodynamic properties of food in case of terminal velocity?
   (*a*) Density size (*b*) Shape
   (*c*) Drag coefficient (*d*) All of the above
5. The thermal processing may include
   (*a*) Heating (*b*) Cooling
   (*c*) Drying (*d*) Freezing
   (*e*) All of them
6. Heat or cooling of agricultural products may be accomplished by the method of
   (*a*) Convection (*b*) Conduction
   (*c*) Radiation (*d*) All of them
7. The water activity for high moisture foods such as fruits, vegetables, meat etc. is more than
   (*a*) 0.75 (*b*) 0.85
   (*c*) 0.97 (*d*) 0.95

8. The water activity of intermediate moisture foods varies from
   (*a*) 0.20-0.36 (*b*) 0.40-0.50
   (*c*) 0.60-0.90 (*d*) 0.90-0.95

9. The water activity of non-fat dry milk should be below
   (*a*) 0.35 (*b*) 0.40
   (*c*) 0.50 (*d*) 0.60

10. The plot of equilibrium moisture content of a food material versus relative humidity of the environment at a given temperature is referred to as
    (*a*) Sorption-desorption isotherms (*b*) Drying curve
    (*c*) Psychometric chart (*d*) Freezing curve

11. Water activity is important in the design of processing operations as well as product development like
    (*a*) Food dehydration (*b*) Control of chemical reaction
    (*c*) Moisture diffusion between membranes
    (*d*) Moisture transfer between food components in processing of intermediate moisture foods and during storage
    (*e*) All of the above

12. Those properties having to do the behavior of the material under applied force are known as
    (*a*) Frictional properties (*b*) Mechanical properties
    (*c*) Rheological properties (*d*) All of the above

13. A science devoted to the study of deformation and flow is called
    (*a*) Rheology (*b*) Dielectrics
    (*c*) Optics (*d*) Mechanics

14. A damage to plant tissue by external forces causing physical change in texture and/or eventual chemical alteration of colour, flavour and texture is known as
    (*a*) Abrasion (*b*) Bruising
    (*c*) Distortion (*d*) Crack

15. A change in shape of intact fruits or vegetables which is not characteristics for the variety is referred as
    (*a*) Abrasion (*b*) Bruising
    (*c*) Distortion (*d*) Puncture

16. A cleavage without complete separation of the parts of the foods is called
    (*a*) Crack (*b*) Cut
    (*c*) Shatter cracks (*d*) Skin break

17. Penetration or division by the sharp edge of an object is termed as
    (*a*) Puncture (*b*) Cut
    (*c*) Shatter cracks (*d*) Cracks

18. A small hole or wound on the surface of the fruits made by a pointed object or stem of other fruits is denoted by
    (*a*) Split (*b*) Crack
    (*c*) Puncture (*d*) Cut

19. One or more tortuous cracks radiating from a point of impact is known as
    (*a*) Shatter cracks (*b*) Skin break
    (*c*) Swell cracking (*d*) Split

20. Fracture of the periderm or a crack limited to the skin of fruits and vegetables is called
    (*a*) Cut (*b*) Puncture
    (*c*) Skin break (*d*) Stem end tearing

21. The separation of periderm from the plant part by scraping, rubbing etc. is known as
    (*a*) Skinning (*b*) Feathering
    (*c*) Skin break (*d*) Swell cracking

22. Division or separation into parts of foods is called
    (*a*) Cut (*b*) Split
    (*c*) Puncture (*d*) Crack

23. Skin break caused by separation of stem from the fruit is also called
    (*a*) Skin break (*b*) Skinning
    (*c*) Feathering (*d*) Stem end tearing

24. Cracking due to uptake of water by osmotic pressure is denoted as
    (*a*) Swell cracking (*b*) Stem end tearing
    (*c*) Shatter cracks (*d*) Distortion

25. The shape which is formed when an ellipse rotates about its major axis like a lemon is known as
    (*a*) Prolate spheroid (*b*) Oblate spheroid
    (*c*) Right circular cone (*d*) None

26. The shape which is formed when an ellipse rotates about its minor axis like a grape fruits is called
    (*a*) Prolate spheroid (*b*) Elliptical
    (*c*) Oblate spheroid (*d*) Right circular cone

27. The volume of prolate spheroid (V) is expressed as if 'a' and 'b' is major and minor axis, respectively as follows
    (*a*) $4/3\ (\pi ab^2)$
    (*b*) $4/3\ (\pi ab)^2$
    (*c*) $4/3\ (\pi a)\ b^2$
    (*d*) $4/3\ (\pi ab)^3$

28. The volume of oblate spheroid (V) is expressed as if 'a' and 'b' is major and minor axis, respectively as follows
    (*a*) $4/3\ (\pi ab)^2$
    (*b*) $4/3\ (\pi b)a^2$
    (*c*) $4/3\ (\pi a^2 b)$
    (*d*) $4/3\ (\pi ab)^3$

29. A fast and accurate method for determining volume and weight density of food material is accomplished by
    (*a*) Platform balance
    (*b*) Specific gravity balance
    (*c*) Specific gravity gradient tube
    (*d*) Air compression pycnometer

30. Air compression pycnometer is used to measure the
    (*a*) Weight
    (*b*) Volume
    (*c*) Specific gravity
    (*d*) Temperature

31. Which solvent is used to measure the specific gravity of seed and grains in pycnometer?
    (*a*) Benzene
    (*b*) n-hexane
    (*c*) Toluene
    (*d*) Water

32. Which method is not used to measure the specific gravity of fruit and vegetables?
    (*a*) Pycnometer method
    (*b*) Platform balance
    (*c*) Specific gravity gradient tube
    (*d*) Specific gravity balance

33. An empirical equation to determine the surface area of food is as follows
    (*a*) $S = K/W^m$
    (*b*) $S = KW^m$
    (*c*) $S = KW^{1/m}$
    (*d*) None

34. The surface of the pores in a porous media exposed to fluid flow either per unit volume or per unit weight of the solid particles is referred as
    (*a*) Specific surface
    (*b*) Surface area
    (*c*) Porosity
    (*d*) Specific gravity

35. For measurement of specific surface of non uniform pore space is accomplished by Carma-Kozeny equation is given as
    (*a*) $K = CP^3/S^2$
    (*b*) $K = PC^3/S^2$
    (*c*) $K = S^2/CP^3$
    (*d*) $K = CS^2/P^3$
    where K = permeability, S = specific surface area, P = porosity, C = dimensionless constant

36. Modified Kozeny equation for non uniform pore space is follows

(*a*) $K = P^3/5S^2(1 - P^2)$
(*b*) $K = P^2/5S^2(1 - P)^2$
(*c*) $K = P^3/5S^2(1 - P)^2$
(*d*) None

37. The energy necessary to cause a change in the particle size is directly proportional to the size of the particles raised to a power 'n' is given as

(*a*) Bond's law
(*b*) Kick's law
(*c*) Rittinger's law
(*d*) Hooke's law

38. Stress is directly proportional to strain is known as

(*a*) Bond's law
(*b*) Kick's law
(*c*) Rittinger's law
(*d*) Hooke's law

39. When the Hooke solid is under tensile or compressive stress, the modulus of elasticity or Young's modulus is given as

(*a*) $E = \sigma/\varepsilon$
(*b*) $G = \tau/\nu$
(*c*) $K = -p/\varepsilon_v$
(*d*) None

40. When the Hooke solid is subjected to distortion by shear stress, the shear modulus or modulus of rigidity is given as

(*a*) $E = \sigma/\varepsilon$
(*b*) $K = -p/\varepsilon_v$
(*c*) $G = \tau/\nu$
(*d*) None

41. When the Hooke solid is under hydrostatic pressure, the bulk modulus or incompressibility is given by

(*a*) $E = \sigma/\varepsilon$
(*b*) $G = \tau/\nu$
(*c*) $K = -p/\varepsilon_v$
(*d*) None

42. The classical ideal material is represented as

(*a*) Hooken body
(*b*) St.Venant body
(*c*) Newtonian liquid
(*d*) All of them

43. Hooken body acts as

(*a*) Ideal elastic behavior
(*b*) Ideal plastic behavior
(*c*) Ideal viscous behavior
(*d*) None

44. Ideal plastic behavior is represented by

(*a*) Newtonian liquid
(*b*) St. Venant body
(*c*) Hooken body
(*d*) All of the above

45. Ideal viscous behavior is observed in

(*a*) Newtonian liquid
(*b*) St. Venant body
(*c*) Hooken body
(*d*) All of the above

46. Fluidity is the reciprocal of
(*a*) Density
(*b*) Porosity
(*c*) Viscosity
(*d*) Specific gravity

47. Mechanical models consisting of spring and dashpot are used to explain and interpret the rheological behavior of
(*a*) Linear visco-elastic material
(*b*) Non linear visco-elastic material
(*c*) Elastic material only
(*d*) Plastic material only

48. Mechanical model for determination of visco-elastic behavior contains
(*a*) Spring only
(*b*) Dashpot only
(*c*) Both above
(*d*) None

49. On the basis of experiments, agricultural products are
(*a*) Visco-plastic
(*b*) Elastic or plastic
(*c*) Newtonian liquid
(*d*) Visco-elastic

50. The two basic mechanical elements used in mechanical models are spring which obey __________ and a dashpot with a __________.
(*a*) Hooke' law , Newtonian liquid
(*b*) Newtonian liquid, Hooke's law
(*c*) Hooke's law, St. Venant body
(*d*) Hooke's law, bond's law

51. Mechanical properties such as hardness, cohesiveness, viscosity, elasticity, adhesiveness, gumminess etc. dealing with
(*a*) Strength
(*b*) Flow characteristics
(*c*) Both above
(*d*) None

52. Geometric properties such as graininess, flakiness, lumpiness etc. denoted to
(*a*) Shape, size and orientation
(*b*) Arrangement of the constituents
(*c*) Appearance of food
(*d*) All of the above

53. The frictional force acting between surfaces of contact at rest with respect to each is called
(*a*) Internal friction
(*b*) Static friction
(*c*) Kinetic friction
(*d*) All of them

54. The frictional force existing between the surfaces in relative motion is called
(*a*) Internal friction
(*b*) Static friction
(*c*) Kinetic friction
(*d*) All of them

55. The angle of repose is generally __________ than angle of internal friction for grain of approximately at the same moisture content.
(*a*) Smaller
(*b*) Higher
(*c*) No change
(*d*) None

56. Artificial ageing brought about usually by some (dry or wet) heat treatment of the grain is called
(*a*) Conditioning (*b*) Steaming
(*c*) Curing (*d*) Parboiling

57. Rice with whole milled kernels having length breadth ratio between 2.0 and 3.0 is known as
(*a*) Parboiled rice (*b*) Shelled rice
(*c*) Cargo rice (*d*) Bold rice

58. Kelvin model for equilibrium moisture content (EM(*C*) is based on
(*a*) Multilayer molecular adsorption (*b*) Capillary condensation theory
(*c*) Potential field theory (*d*) All of the above

59. Which EMC model works on multilayer molecular adsorption theory?
(*a*) Henderson equation (*b*) BET equation
(*c*) Langmuir equation (*d*) Chung –Pfost equation

60. Harkin Jura equation related to EMC model is given as
(*a*) $V = f - g \ln [1 - (P_v/P_{vs})]$ (*b*) $\ln(P/P_v) = d - (e/V^2)$
(*c*) $V = V_m (bP_v/1 + bP_v)$ (*d*) $\ln (P_v/P_{vs}) = - (2\sigma V \cos (a)/rR_oT_{abs}$

61. The Chung – Pfost equation fits grain EMC data is over
(*a*) 0-20 % RH (*b*) 20-40% RH
(*c*) 15-30% RH (*d*) 20-90% RH

62. Henderson equation is very much popular and based on the
(*a*) Potential field theory (*b*) Capillary condensation theory
(*c*) Gibb's adsorption equation (*d*) Multilayer molecular adsorption theory

63. BET equation for EMC model is satisfactory, if the relative humidity is below than
(*a*) 20% (*b*) 30%
(*c*) 50% (*d*) 70%

64. The major forces acting in cyclone separator are
(*a*) Centripetal force (*b*) Centrifugal force
(*c*) Drag force (*d*) both (*b*) and (*c*)

65. The equation $dC/dt = KC$, represents a ________ order kinetic reaction.
(*a*) Zero (*b*) First
(*c*) Second (*d*) Third

66. Amount of water taken by molecular or physical action by the flour is called
(*a*) Absorption (*b*) Adsorption
(*c*) Steeping (*d*) Desorption

67. The production of granule from powder particles is known as
(*a*) Agglomeration
(*b*) Emulsification
(*c*) Palletization
(*d*) Blending

68. Heat sterilization of food before filling into pre sterilized container is called
(*a*) UHT process
(*b*) Aseptic processing
(*c*) HTST sterilization
(*d*) Extrusion cooking

69. A device to form fine droplets of food (e.g. in spray drier) is known as
(*a*) Fountain
(*b*) Sprinkler
(*c*) Nozzle
(*d*) Atomizer

70. Heating foods, especially vegetables, to below 100ºC for a short time, to both inactivate enzyme which would cause a loss of quality during storage and to remove air and soften the food.
(*a*) Pasteurization
(*b*) Sterilization
(*c*) Blanching
(*d*) Cooking

71. The crispiness of dry cereal based foods such as crackers, fried snacks and potato chip is a function of
(*a*) Water activity
(*b*) Temperature
(*c*) Moisture content
(*d*) None

72. The crispiness of dry based foods decreases as water activity
(*a*) Decreased
(*b*) Increased
(*c*) No change
(*d*) None

73. Which drier is more efficient to viscous and semi-viscous foods?
(*a*) Drum drier
(*b*) Spray drier
(*c*) FBD
(*d*) Tray drier

74. The ratio of the capacitance of a food to the capacitance of air or vacuum under the same condition is known as
(*a*) Dielectric constant
(*b*) Heat transfer coefficient
(*c*) Electrical conductivity
(*d*) Gas constant

75. The capacity of a material to conduct electricity is called
(*a*) Dielectric constant
(*b*) Current
(*c*) Electrical conductivity
(*d*) Resonance

76. The separation of electrolytes into anions and cations by the application of a direct electrical current and the use of ion-selective membranes.
(*a*) Electrophoresis
(*b*) Electro-dialysis
(*c*) Reverse osmosis
(*d*) Entrainment

77. A devise that qualitatively or quantitatively measures the dose of irradiation received by a food.
    (*a*) Dosimeter (*b*) Tintometer
    (*c*) Jelmeter (*d*) Salometer

78. The time required to lower the temperature of a food from an initial value to a predetermined temperature at the thermal centre.
    (*a*) Freezing time (*b*) Effective chilling time
    (*c*) Thawing time (*d*) Effective freezing time

79. Oil droplets that are carried over in steam produced by vigorously frying foods, leading to loss of oil, or loss of concentrated droplets of product with vapour evaporation by boiling.
    (*a*) Enrobing (*b*) Entrainment
    (*c*) Fouling (*d*) Expression

80. A process that involves the combination of several unit operations including mixing, cooking, kneading, shearing, shaping and forming to produce the high quality foods.
    (*a*) Extrusion (*b*) Expression
    (*c*) Expulsion (*d*) Milling

81. Automatic control of a process using information from sensors to adjust processing conditions.
    (*a*) Feed forward control (*b*) Feed back control
    (*c*) Both above (*d*) None

82. Comparison of processing conditions with a model system, used in automatic process control.
    (*a*) Feed forward control (*b*) Feed back control
    (*c*) Both above (*d*) None

ANSWER KEY (ENGINEERING PROPERTIES AND QUALITY OF BIOMATERIALS)

1.C	2. D	3.E	4.D	5.E	6.D	7.C	8.C	9.D	10.A
11. E	12.B	13.A	14.B	15.C	16.A	17.B	18.C	19.A	20.C
21.A	22.B	23.D	24.A	25.A	26.C	27.A	28.C	29.C	30.B
31.C	32.A	33.B	34.A	35.A	36.C	37.A	38.D	39.A	40.C
41.C	42.D	43.A	44.B	45.A	46.C	47.A	48.C	49.D	50.A
51.C	52.D	53.B	54.C	55.B	56.C	57.D	58.B	59.B	60.B
61.D	62.C	63.C	64.D	65.B	66.A	67.A	68.B	69.D	70.C
71.A	72.B	73.A	74.A	75.C	76.B	77.A	78.D	79.B	80.A
81.B	82.A								

Food Packaging, Material Handling and Transportation

1. Protects or conserves much of the energy expended during the production and processing of the product (food product) is known as
(*a*) Processing (*b*) Packaging
(*c*) Permeability (*d*) Practice

2. Food Packaging surrounds, enhances and ____________ the goods.
(*a*) Protects (*b*) Spoilage
(*c*) Transport (*d*) Process

3. Packaging is a ________ system of preparing goods for transport, distribution, storage, retailing and end use.
(*a*) Permanent (*b*) Temporary
(*c*) Coordinated (*d*) Modern

4. Packaging means of ensuring safe delivering to ultimate consumer in sound condition at
(*a*) Maximum cost (*b*) Minimum cost
(*c*) Optimum cost (*d*) Zero cost

5. Packaging is a techno-economic function aimed at minimizing cost of delivering while
(*a*) Minimum sales (*b*) Maximizing sales
(*c*) Net sale (*d*) None

6. The _______________ is one which is in direct contact with the contained product.
(*a*) Primary package (*b*) Secondary package
(*c*) Tertiary package (*d*) None

7. The ____________ contains a number of primary packages.
(*a*) Primary package (*b*) Secondary package
(*c*) Tertiary package (*d*) None

8. The package contain a number of secondary packages is called
(*a*) Primary package (*b*) Secondary package
(*c*) Tertiary package (*d*) None

9. To control post harvest losses, chemicals are heated to aid circulation is known as
(*a*) Evaporative cooling (*b*) Thermal fogging
(*c*) Aeration (*d*) Ventilation

10. Most of the food products require percentage relative humidity for storage should be about
(*a*) 60-65 (*b*) 50-55
(*c*) 70-75 (*d*) 80-90

11. Toddy is prepared from
(*a*) Grape (*b*) Apple
(*c*) Coconut (*d*) Cashew

12. Vodka is prepared from
(*a*) Grape (*b*) Potato
(*c*) Barley (*d*) Sugarcane

13. Hypobaric storage is also known as
(*a*) CA storage (*b*) MA storage
(*c*) Low pressure storage (*d*) Cold storage

14. Which microorganism is involved in vinegar production?
(*a*) *Lactobacillus cerevisiae* (*b*) *Saccharomyces cerevisiae*
(*c*) *Acetobacter aceti* (*d*) *Pediococcus cerevisiae*

15. Which of the following is known as 'wine yeast'?
(*a*) *Pediococcus cerevisiae* (*b*) *Saccharomyces cerevisiae*
(*c*) *Candida utilis* (*d*) *Torulopsis saphaerica*

16. Fruits juices are dried by
(*a*) Kiln drier (*b*) Cabinet drier
(*c*) Drum drier (*d*) Tunnel drier

17. Citric acid is commercially produced by
(*a*) *Aspergillus niger* (*b*) *Bacillus cereus*
(*c*) *Saccharomyces rouxii* (*d*) *Saccharomyces cerevisiae*

18. The ability of packaging to protect foods from mechanical damage is measured by
(*a*) Tensile strength (*b*) Young's modulus of elasticity
(*c*) Yield strength (*d*) All of them

19. Cold process is also known as
(*a*) Pre cooling (*b*) Freezing
(*c*) Cold storage (*d*) Ultra filtration

20. In Pickles, white cottony surface is due to
(*a*) Rhyzopus (*b*) Lactobacillus
(*c*) Mycoderma (*d*) Mucor

21. Which group of microorganism converts sugar into alcohol?
(*a*) Bacteria (*b*) Moulds
(*c*) Yeast (*d*) Above All

22. Which refrigerant is commonly used in cold storage in our country?
(*a*) Ozone (*b*) Nitrogen
(*c*) Ammonia (*d*) Carbon dioxide

23. Which is the cheapest preservation method of fruits and vegetables?
(*a*) Canning (*b*) Freezing
(*c*) Drying (*d*) Fermentation

24. Which metal foil is used for food packaging?
(*a*) Iron (*b*) Zinc
(*c*) Aluminium (*d*) Copper

25. Dehydrated foods have relative humidity
(*a*) High (*b*) Medium
(*c*) Low (*d*) Equilibrium

26. Dehydrated foods are
(*a*) Hydraulic (*b*) Hypobaric
(*c*) Hygroscopic (*d*) None of these

27. ERH data are also important when we consider storage of ______ product.
(*a*) Wet (*b*) Dried
(*c*) Fresh (*d*) Semi solid

28. The physical factors affecting heat and mass transfer during drying are
(*a*) Temperature and humidity (*b*) Air velocity
(*c*) Physical properties of foods (*d*) Above all

29. Case hardening is particularly common with foods that contain dissolved sugar and other solutes in concentration.
(*a*) Low (*b*) High
(*c*) Medium (*d*) All of the above

30. Porosity may be developed by creating ______ within a product during drying.
    (*a*) Steam pressure (*b*) Temperature
    (*c*) Air pressure (*d*) None

31. Browning reaction may be caused by enzymatic oxidation during food dehydration.
    (*a*) Poly phenols (*b*) Other susceptible compounds
    (*c*) Both (*a*) and (*b*) (*d*) None

32. Packaging is an essential part of ______ and ______ foods.
    (*a*) Processing and distribution (*b*) Heating and cooling
    (*c*) Manufacturing and marketing (*d*) All of them

33. The most important general requirement of food packaging are
    (*a*) Be non toxic (*b*) Protect against contamination
    (*c*) Acts as barrier (*d*) All of these

34. Food packaging can be classified into ______ type container.
    (*a*) Primary (*b*) Secondary
    (*c*) Tertiary (*d*) All of these

35. Deterioration in the quality of green tea during storage accelerate
    (*a*) Reduction in ascorbic acid content
    (*b*) Change in colour from bright green to olive green
    (*c*) Taste lacking
    (*d*) All of these

36. The quality of fruit juice depends essentially on the species and ______ of the fresh fruits.
    (*a*) Maturity (*b*) Size
    (*c*) Quantity (*d*) All of these

37. The deteriorative reaction in fruit juices are
    (*a*) Microbiological spoilage (*b*) Degradation of flavour component
    (*c*) Degradation of nutrients (*d*) All of these

38. Pasteurization of liquid foods is originally used as a means of controlling
    (*a*) Micro flora (*b*) Temperature
    (*c*) Relative humidity (*d*) Insect and pest

39. The filling of commercially sterile product into sterile container under aseptic conditions and sealing of the container so that reinjection is prevented is called
    (*a*) Aseptic packaging (*b*) Active packaging
    (*c*) Tetra packaging (*d*) Sterilization

40. The important phase through which fruits and vegetables pass are
(*a*) Growth
(*b*) Maturation
(*c*) Senescence
(*d*) All of these

41. Ethylene is a natural plant hormone and plays a central role in the initiation of
(*a*) Ripening
(*b*) Growth
(*c*) Senescence
(*d*) All of the above

42. A numbers of operation that can be executed either by hand (manual) or by mechanical mean or device to convey material and to reduce the human drudgery is express as
(*a*) Bulk storage
(*b*) Material separation
(*c*) Aeration
(*d*) Material handling

43. Selection of material handling system should be taken according to
(*a*) Conveying material
(*b*) Stability of conveyor against all normal and climatic condition
(*c*) Capacity and speed of conveyor
(*d*) Dead load of conveyor
(*e*) All of the above

44. The design of belt conveyor system is based on
(*a*) Available space
(*b*) Horizontal conveying length and conveying lift
(*c*) Characteristics of the material to be conveyed
(*d*) Capacity requirement
(*e*) All of the above

45. For transportation of grains, belt speed in belt conveyor should not be more than
(*a*) 2.5 m/s
(*b*) 3.5 m/s
(*c*) 4.5 m/s
(*d*) 5.0 m/s

46. Generally, for grain conveying by belt conveyor belt speed is recommended as
(*a*) 1.5-1.75 m/s
(*b*) 2.0-2.25 m/s
(*c*) 2.5-2.8 m/s
(*d*) 3.0-3.5 m/s

47. The selection of belt width in belt conveyor depends upon the
(*a*) Capacity requirement
(*b*) Speed of operation
(*c*) Angle of inclination of belt conveyor
(*d*) Trough angle and depth
(*e*) All

48. The capacity of belt conveyor can be expressed as
(*a*) Capacity (m^3/hr) = (area of cress section, m^2) × (belt of speed, m/min) × 60
(*b*) Capacity (m^3/hr) = (area of cress section, m^2) × (belt of speed, m/min) × 90
(*c*) Capacity (m^3/hr) = (length of belt, m) × (belt of speed, m/min) × 60
(*d*) Capacity (m^3/hr) = (mass of material, kg) × (belt of speed, m/min) × 60

49. A trough angle (B) in belt conveyor is best suited for paddy and other grain should be
(*a*) 10° (*b*) 15°
(*c*) 20° (*d*) 22°

50. The pitch of idler in belt conveyor is determined by
(*a*) the idler rating or the carrying capacity of each idler
(*b*) sag of belt between idlers
(*c*) belt tension and belt speed
(*d*) all of these

51. Recent innovations in belt conveyor system for grain handling are
(*a*) Use of higher belt speed (upto 4.5 m/s)
(*b*) Use of steeper troughing angle (upto 45°)
(*c*) Use of inclined belt to elevate grain, thereby eliminating the need for bucket elevator
(*d*) All of the above

52. High capacity belt conveyor has conveying capacity of
(*a*) 100 tonnes/hr of grain (*b*) 2000 tonnes/hr of grain
(*c*) 4500 tonnes/hr of grain (*d*) 1000 tonnes/hr of grain

53. High capacity belt conveyor has the belt width and troughing angle is as
(*a*) 1500 mm and 45° (*b*) 1000 mm and 45°
(*c*) 1500 mm and 20° (*d*) 1000 mm and 45°

54. Belt conveyor of high capacity can be conveyed the grain upto the distance of
(*a*) 50 m (*b*) 100 m
(*c*) 120 m (*d*) 200 m

55. Belt speed of high capacity belt conveyer is around
(*a*) 2.5 m/s (*b*) 2.8 m/s
(*c*) 3.5 m/s (*d*) 4.5 m/s

56. Inclined belt conveyor has maximum feed on angle of 10° but conveying angle should be
(*a*) 10° (*b*) 12°
(*c*) 14° (*d*) 15°

57. The vertical lift of the bucket elevator may range between
(*a*) 10-15 m (*b*) 20-40 m
(*c*) Few meter to 50 m (*d*) More than 100 m

58. Capacities of bucket elevator may vary from
(*a*) 10-20 t/h (*b*) 2-1000 t/h
(*c*) 200-500 t/h (*d*) 1000-3000t/h

59. Which one of the following bucket elevator is used for grain handling?
    (*a*) Centrifugal discharge bucket elevator (*b*) Positive discharge bucket elevator
    (*c*) Marine leg elevator (*d*) High speed elevator

60. The gram of water vapour transmitted from 1 m^2 of film area in 24 hours is known as
    (*a*) Water vapour permeance (*b*) Water vapour transmission rate (WVTR)
    (*c*) Water vapour permeability (*d*) Moisture transfer coefficient

61. The gram of water vapour transmitted through 1 m^2 of film area in 24 hours when a vapour pressure difference of 1 mm Hg is called
    (*a*) Water vapour permeance (*b*) Water vapour transmission rate (WVTR)
    (*c*) Water vapour permeability (*d*) Moisture transfer coefficient

62. A gram of water vapour permeated through 1 m^2 of film area in 24 hours when the vapour pressure difference is 1 mm Hg and 1 cm of film thickness is termed as
    (*a*) Water vapour permeance (*b*) Water vapour transmission rate (WVTR)
    (*c*) Water vapour permeability (*d*) Moisture transfer coefficient

63. Which chemical compound is used as sterilizing medium in aseptic packaging of food?
    (*a*) SO_2 (*b*) H_2S
    (*c*) H_2O_2 (*d*) H_2O

64. Clostridium botulinum exo-toxin can be destroyed by heating at
    (*a*) 80°C for 6 min (*b*) 50°C for 30 min
    (*c*) 70°C for 10 min (*d*) None of these

ANSWER KEY

1.B	2. A	3.C	4.B	5.B	6.A	7.B	8.C	9.B	10.D
11. C	12.B	13.C	14.C	15.B	16.C	17.A	18.D	19.D	20.A
21.A	22.C	23.C	24.C	25.D	26.C	27.B	28.D	29.B	30.A
31.C	32.A	33.D	34.D	35.D	36.A	37.D	38.A	39.A	40.D
41.A	42.D	43.E	44.E	45.B	46.C	47.E	48.A	49.C	50.D
51.D	52.B	53.A	54.D	55.C	56.B	57.C	58.B	59.A	60.B
61.A	62.C	63.C	64.						

Fluid Mechanics

1. A perfect gas
   (*a*) Has zero viscosity
   (*b*) Has constant viscosity
   (*c*) Satisfies the relation PV= m RT
   (*d*) Is incompressible
   (*e*) None of the above.

2. An ideal fluid is
   (*a*) Similar to a perfect gas
   (*b*) Frictionless and incompressible
   (*c*) One which obey Newton's Law of viscosity
   (*d*) One which satisfies continuity equation
   (*e*) One which flows through pipes with least friction.

3. An ideal flow of any fluid must fulfill the following
   (*a*) Newton's Law of viscosity
   (*b*) Newton's Laws of motion
   (*c*) Boundary layer theory
   (*d*) Pascal theory
   (*e*) Continuity equation.

4. For an irotational flow, the equation $\delta^2\Phi / \delta^2 x + \delta^2\Phi / \delta^2 y = 0$ is known as
   (*a*) Bernoulli equation
   (*b*) Reynolds equation
   (*c*) Cauchy Riemann's equation
   (*d*) Euler's equation
   (*e*) Laplace equation.

5. The velocity of a fluid particle at the center of the pipe section is
   (*a*) Maximum
   (*b*) Minimum
   (*c*) Average
   (*d*) r.m.s.
   (*e*) logarithmic average.

6. The stress-strain relation of the Newtonian fluid is
   (*a*) Linear
   (*b*) Parabolic
   (*c*) Hyperbolic
   (*d*) Involutic
   (*e*) There is no such relation.

7. The unit of kinematic viscosity are
   (*a*) kg/m^2-sec
   (*b*) kg sec/ m^2
   (*c*) m/kg-sec
   (*d*) m^2 /sec
   (*e*) m kg/ sec.

8. The unit of dynamic viscosity are
(*a*) Newton – sec/m^2
(*b*) Newton/m^2 –sec
(*c*) kg-sec/m
(*d*) m^2 kg/sec
(*e*) m^2 /sec.

9. The unit of surface tension is
(*a*) Same as units of force
(*b*) Energy/unit area
(*c*) Force/unit area
(*d*) It is dimensionless
(*e*) None of the above.

10. Density in terms of viscosity is
(*a*) kinematic viscosity/dynamic viscosity
(*b*) dynamic viscosity/ kinematic viscosity
(*c*) kinematic viscosity × dynamic viscosity
(*d*) Any of the above
(*e*) None of the above.

11. Newton's Law of viscosity relates to
(*a*) Shear stress and rate of angular deformation in a fluid
(*b*) Yield shear stress, rate of angular deformation and viscosity
(*c*) Pressure, velocity and viscosity
(*d*) Shear stress, temperature, viscosity and velocity
(*e*) Pressure, viscosity and rate of angular deformation.

12. SI unit of viscosity
(*a*) 10 times poise
(*b*) 9.81 times poise
(*c*) 1/9.81 times poise
(*d*) Equal to poise
(*e*) None of the above.

13. The Reynolds number for pipe flow is given by
(*a*) VD/ν
(*b*) $VD\mu/\delta$
(*c*) $\rho VD/\mu$
(*d*) VD/μ
(*e*) None of the above.

14. The Reynolds number may be defined as the ratio of
(*a*) Viscous forces to inertial forces
(*b*) Elastic forces to pressure forces
(*c*) Viscous forces to gravity forces
(*d*) Gravity forces to inertial forces
(*e*) None of the above.

15. Select the quantity from the following that is not a dimensionless parameter
(*a*) Pressure coefficient
(*b*) Froude Number.
(*c*) Darcy weisbatch friction factor
(*d*) Kinematic viscosity
(*e*) Weber number.

16. The Weber number is the ratio of
    (*a*) Inertial forces to surface tension forces
    (*b*) Inertial forces to viscous forces
    (*c*) Elastic forces to pressure forces
    (*d*) Viscous forces to gravity
    (*e*) Elastic forces to gravity.

17. Froude number is useful in calculation for
    (*a*) Water hammer
    (*b*) Flow through pipes
    (*c*) Hydraulic jump
    (*d*) Surface tension forces
    (*e*) Compressible flow problem.

18. The bulk modulus of elasticity
    (*a*) Is independent of temperature
    (*b*) Increases with the pressure
    (*c*) Is independent of pressure and viscosity
    (*d*) Is larger when the fluid is more compressible
    (*e*) None of the above.

19. One poise is equivalent to
    (*a*) 360 kg/m-hr
    (*b*) 1 dyne sec/cm^2
    (*c*) 1/98.1 kgf sec/m^2
    (*d*) 10^{-1} kg/ m-sec
    (*e*) Any of the above.

20. A barometer is used to measure
    (*a*) Very low pressure
    (*b*) Very high pressure
    (*c*) Pressure difference between two points
    (*d*) Pressure of fluid in a pipe line
    (*e*) Atmospheric pressure.

21. One atm pressure is equivalent to
    (*a*) 1.01315×10^5 N/m^2
    (*b*) 700 mm Hg
    (*c*) 1.0133×10^5 kg/ m-sec^2
    (*d*) 1.0133×10^6 gm/cm-sec^2
    (*e*) Any of the above.

22. An unit of kinematic viscosity
    (*a*) 10^{-5} m^2/sec
    (*b*) 38.75 ft^2/hr
    (*c*) 3.6 m^2/hr
    (*d*) 1 cm^2/sec
    (*e*) Any of the above.

23. A simple pitot tube is used to measure
    (*a*) The pressure in static fluid
    (*b*) The viscosity in a flowing stream
    (*c*) The total pressure
    (*d*) The dynamic pressure
    (*e*) The undisturbed fluid pressure.

24. One mm Hg pressure is equivalent to
    (*a*) 1.333×10^2 N/m
    (*b*) 1.333×10^{-3} bar
    (*c*) 1.3158×10^{-3} atm.
    (*d*) 1.332×10^2 kg/m-sec^2
    (*e*) Any of the above.

25. One N/m^2 pressure is equivalent to
(*a*) 1 Pascal (*b*) 10^{-5} bar
(*c*) 10^{-2} kg/m-sec^{-2} (*d*) 7.5 × 10^{-3} mm Hg.
(*e*) Any of the above.

26. A Pitot tube is used for the measurement of
(*a*) Pressure (*b*) Velocity
(*c*) Viscosity (*d*) Surface tension
(*e*) Flow rate.

27. A Rotameter is a device used to measure
(*a*) Velocity of fluid in pipe (*b*) Velocity of flow
(*c*) Vortex flow (*d*) Flow of fluid
(*e*) Density of fluids.

28. A hydrometer is used to determine
(*a*) Relative humidity (*b*) Buoyancy force
(*c*) Specific gravity of fluid (*d*) Viscosity of liquids
(*e*) Surface tension.

29. The flow of water in a pipe of diameter 3000 mm can be measured by
(*a*) Venturimeter (*b*) Rotameter
(*c*) Nozzle (*d*) Pitot tube
(*e*) Orifice plate.

30. A fluid is substance that
(*a*) Always expands until it fills any container
(*b*) Has the same shear stress at a point regardless of its motion
(*c*) Cannot remain at rest under action of any shear force
(*d*) Cannot be subjected to shear forces
(*e*) Is practically incompressible.

31. Mercury is used in barometers generally because
(*a*) It is the heavy liquid available (*b*) It is perfect liquid
(*c*) It is volume dose not change with pressure
(*d*) The height barometer required will be less
(*e*) Its expansion with temperature is negligible

32. An ideal fluid is
(*a*) Very viscous
(*b*) One which obey Newton's law of viscosity
(*c*) A useful assumption in problems in conduit flow
(*d*) Frictionless and incompressible
(*e*) None of these

33. The equation of continuity of flow is applicable when
    (*a*) The flow is one dimensional
    (*b*) The flow is steady
    (*c*) The flow is compressive
    (*d*) The velocity is uniform over the cross-section
    (*e*) All the above conditions together.

34. One dimensional flow is
    (*a*) Restricted to flow in a straight line
    (*b*) Flow which neglect changes in a transverse direction
    (*c*) Steady uniform flow
    (*d*) Uniform flow
    (*e*) None of the above.

35. The continuity equation
    (*a*) Requires that Newton's second law of motion be satisfied at every point in the fluid
    (*b*) Expresses the relation between energy and work
    (*c*) States that the velocity at a boundary must be zero relative to the boundary for a real fluid
    (*d*) Relates the momentum per unit volume for two points on a streamline
    (*e*) Relates mass rate of flow along a stream tube.

36. The equation of continuity of fluid is applicable only when
    (*a*) The flow is steady
    (*b*) The flow is one dimensional
    (*c*) The velocity is uniform over the cross-sections
    (*d*) The flow is compressive
    (*e*) All the above.

37. The equation of continuity of flow is
    (*a*) Momentum
    (*b*) Mass
    (*c*) Energy
    (*d*) All above
    (*e*) None of the above.

38. The continuity equation in idea fluid flow
    (*a*) States that the net rate of in-flow into any small volume must be zero
    (*b*) Applies to irotational flow only
    (*c*) Implies the existence of a velocity potential
    (*d*) States that the energy remains constant along a stream line
    (*e*) States that the energy is constant everywhere in the fluid.

39. The ratio $\eta=\tau / (du/dy)$ for turbulent flow is
    (*a*) A physical property of the fluid
    (*b*) Dependent upon the flow and density
    (*c*) The viscosity divided by the density
    (*d*) A function of temperature and pressure of fluid
    (*e*) Independent of the nature of flow.

40. A fluid in which resistance to deformation is independent of the shear stress is known as
(*a*) Bingham plastic fluid
(*b*) Pseudo plastic fluid
(*c*) Dilatant fluid
(*d*) Newtonian fluid
(*e*) None of the above.

41. Steady flow is motion in which
(*a*) Velocity is independent of time
(*b*) Velocity is zero
(*c*) Velocity varies with head
(*d*) Rate of change of velocity is unity
(*e*) Velocity is same at every point at a given instant of time.

42. In a centrifugal pump casing the flow of water leaving the impeller is
(*a*) Rectilinear flow
(*b*) Centrifugal flow
(*c*) Radial flow
(*d*) Forced vortex motion
(*e*) Free vortex motion.

43. 10^5 N/m^2 pressure is equivalent to
(*a*) 1 bar
(*b*) 750.06 mm Hg
(*c*) 0.98682 atm
(*d*) 401.85 m H_2O
(*e*) Any of the above.

44. A piezometer is not used for pressure measurement in pipes when
(*a*) The pressure is very low
(*b*) The pressure drop in the pipe is low
(*c*) Velocity of fluid is high
(*d*) Velocity of fluid is low
(*e*) The fluid in the pipe is a gas.

45. In a Venturimeter
(*a*) Length of divergent cone is equal to the length of convergent cone
(*b*) Length of divergent cone is half that of convergent cone
(*c*) Divergent cone is kept shorter than the of convergent cone
(*d*) Divergent cone is kept longer than the of convergent cone
(*e*) None of the above.

46. The flow in pipe is laminar if the Reynolds number is
(*a*) Less than 2000
(*b*) Between 2000and 4000
(*c*) Between 4000 and 6000
(*d*) Equal to 6300
(*e*) Equal to 10000.

47. The mass per unit volume of a liquid at a standard temperature and pressure is called
(*a*) Specific weight
(*b*) Mass density
(*c*) Specific gravity
(*d*) None of the above

48. The weight per unit volume of a liquid at standard temperature and pressure is called
(*a*) Specific weight
(*b*) Mass density
(*c*) Specific gravity
(*d*) None of the above

49. The specific weight of water in MKS units is taken as
(*a*) 1 gm/cm^3
(*b*) 0.001 kg/cm^3
(*c*) 1000 kg/m^3
(*d*) 1000 liters/m^3
(*e*) Any one of the above

50. The specific weight of water in SI units is taken as
(*a*) 9.81 kN/ m^3
(*b*) 9.81×10^3 N/m^3
(*c*) 9.81×10^{-6} N/mm^3
(*d*) Any one of the above
(*e*) None of the above

51. The ratio of specific weight of a liquid to the specific weight of pure water at standard temperature is called
(*a*) Density of water
(*b*) Specific gravity of liquid
(*c*) Compressibility of liquid
(*d*) Surface tension of liquid

52. The specific gravity of water is taken
(*a*) 0.001
(*b*) 0.01
(*c*) 0.1
(*d*) 1.0

53. The density of a liquid in gm/cm^3 is numerically equal to specific gravity.
(*a*) True
(*b*) False

54. The variation in the volume of a liquid with the variation of pressure is called
(*a*) Surface tension
(*b*) Compressibility
(*c*) Capillary
(*d*) Viscosity

55. The property of a liquid which enables it to resist tensile stress is called its surface tension.
(*a*) Agree
(*b*) Disagree

56. The viscosity of a liquid is due to cohesion of its particles.
(*a*) Correct
(*b*) Incorrect

57. The viscosity of a liquid _______________ its rate of flow through a hole in a vessel.
(*a*) Effect
(*b*) Does not effect

58. The pressure measured with the help of a pressure gauge is called
(*a*) Atmospheric pressure
(*b*) Gauge pressure
(*c*) Absolute pressure
(*d*) Mean pressure

59. The Atmospheric pressure at sea level is
(*a*) 1.03 kg/cm^2
(*b*) 10.3 m of water
(*c*) 760 mm of mercury
(*d*) All of the above

60. When a pressure intensity at a point is less than the local atmospheric pressure, then the difference of these two pressures is called vacuum pressure
   (*a*) Gauge pressure (*b*) Absolute pressure
   (*c*) Positive gauge pressure (*d*) Vacuum pressure

61. The absolute pressure is equal to
   (*a*) Gauge pressure + atmospheric pressure (*b*) Gauge pressure – atmospheric pressure
   (*c*) Atmospheric pressure – gauge pressure (*d*) Gauge pressure – vacuum pressure

62. The pressure of a liquid measured with the help of piezometer tube is
   (*a*) Atmospheric pressure (*b*) Gauge pressure
   (*c*) Absolute pressure (*d*) Vacuum pressure

63. The pressure measured with the help of a piezometer tube is in
   (*a*) kg/cm^2 (*b*) kg/m^2
   (*c*) N/mm^2 (*d*) N/m^2
   (*e*) Head of liquid

64. A piezometer tube is used only for measuring
   (*a*) Low pressure (*b*) High pressure
   (*c*) Moderate pressure (*d*) Vacuum pressure

65. The liquid used in manometers should have
   (*a*) Low density (*b*) High density
   (*c*) Low surface tension (*d*) High surface tension

ANSWER KEY (FLUID MECHANICS)

1.C	2. B	3.E	4.E	5.A	6.A	7.D	8.A	9.B	10.B
11.A	12.A	13.C	14.E	15.D	16.A	17.C	18.C	19.C	20.E
21.E	22.D	23.C	24.E	25.E	26.B	27.D	28.C	29.D	30.B
31.D	32.D	33.E	34.B	35.E	36.E	37.B	38.E	39.B	40.D
41.A	42.E	43.E	44.E	45.B	46.A	47.B	48.A	49.E	50.D
51.B	52.D	53.A	54.B	55.A	56.A	57.A	58.B	59.D	60.C
61.A	62.B	63.E	64.C	65.D					

Thermodynamics

1. A definite area or a space where some thermodynamic process takes place is known as
   (*a*) Thermodynamic cycle
   (*b*) Thermodynamic process
   (*c*) Thermodynamic system
   (*d*) Thermodynamic law

2. One atmospheric pressure is equal to
   (*a*) 1 kg/cm^2
   (*b*) 1.033 kg/cm^2
   (*c*) 1.1 kg/cm^2
   (*d*) 1.133 kg/cm^2
   (*e*) 14.7 kg/cm^2

3. When neither mass nor energy is allowed to cross the boundary of a system, it is called
   (*a*) Closed system
   (*b*) Open system
   (*c*) Isolated system
   (*d*) None of these

4. For water, boiling and melting point will coincide at
   (*a*) 0°K
   (*b*) 100°K
   (*c*) 200°K
   (*d*) 273.16°K
   (*e*) 216.73°K

5. Which of the following is the extensive property of a thermodynamic system?
   (*a*) Pressure
   (*b*) Volume
   (*c*) Temperature
   (*d*) Density

6. An isolated system
   (*a*) Is a specified region where transfers of energy and/or mass take place
   (*b*) Is a region of constant mass and only energy is allowed to cross the boundaries
   (*c*) Can not transfer either energy or mass to or from the surroundings
   (*d*) Is one in which mass within the system is not necessarily constant
   (*e*) None of the above

7. Which of the following is an intensive property of a thermodynamic system?
   (*a*) Volume
   (*b*) Temperature
   (*c*) Mass
   (*d*) Energy

8. The fact that the product of pressure and volume of a fixed amount of gas is of approximately constant value is known as
(*a*) Charle's law
(*b*) Gay-Lussac's law
(*c*) Dalton's law
(*d*) Lewis law
(*e*) None of these

9. Which of the following is not a thermodynamic property?
(*a*) Pressure
(*b*) Temperature
(*c*) Heat
(*d*) Specific volume

10. The statement that energy can be neither created nor destroyed but only converted from one form to another is known as
(*a*) Kinetic theory of gases
(*b*) Avogadro's hypothesis
(*c*) Zeroth law of thermodynamics
(*d*) First law of thermodynamics
(*e*) Second law of thermodynamics

11. When a process or processes are performed on a system in such a way that the final state is identical with the initial state, it is known as
(*a*) Thermodynamic cycle
(*b*) Thermodynamic property
(*c*) Thermodynamic process
(*d*) Zeroth law of thermodynamics

12. The sequence of processes that eventually returns the working substances to its original state is known as
(*a*) Event
(*b*) Process
(*c*) Thermodynamic property
(*d*) Thermodynamic cycle
(*e*) Zeroth law of thermodynamics

13. Atmospheric pressure is equal to
(*a*) 1.013 bar
(*b*) 101.3 kN/m^2
(*c*) 760 mm of Hg
(*d*) All of these

14. If all the variables of a steam are independent of time it is said to be in
(*a*) Steady flow
(*b*) Unsteady flow
(*c*) Uniform flow
(*d*) Closed flow
(*e*) Constant flow

15. First law of thermodynamics deals with
(*a*) Conservation of heat
(*b*) Conservation of momentum
(*c*) Conservation of mass
(*d*) Conservation of energy

16. A control volume refers to
(*a*) A fixed region in a space
(*b*) A specified mass
(*c*) An isolated system
(*d*) A reversible process only
(*e*) A closed system

17. Second law of thermodynamics defines
(*a*) Heat (*b*) Work
(*c*) Entropy (*d*) Internal energy

18. Internal energy of a perfect gas depends on
(*a*) Temperature, specific heats and pressure
(*b*) Temperature, specific heats and enthalpy
(*c*) Temperature, specific heats and entropy
(*d*) Temperature, specific heats
(*e*) Temperature only

19. Kelvin-Plank's law deals with
(*a*) Conservation of work (*b*) Conservation of heat
(*c*) Conservation of mass (*d*) Conservation of heat into work

20. The water vapour of water at 30°C in Pascal
(*a*) 0.44 (*b*) 7.18
(*c*) 223 (*d*) 431.5
(*e*) 4315

21. If the temperature remains constant, the volume of a given mass of a gas is inversely proportional to the pressure. This is known as
(*a*) Charles' law (*b*) Boyle's law
(*c*) Joule's law (*d*) Gay Lussac's law

22. The Reversible Polytropic process is
(*a*) Adiabatic frictionless *b*) P/ρ^k = constant
c) $P\rho^k$ = constant *d*) P/ρ^n = constant
(*e*) None of the above

23. The state of a substance whose evaporation from its liquid state is complete, is known as
(*a*) Steam (*b*) Vapour
(*c*) Air (*d*) Perfect gas

24. A reversible polytropic process is given by
(*a*) $T_1/T_2 = (\rho_1/\rho_2)^{n-1}$ (*b*) $T_1/T_2 = (\rho_2/\rho_1)^n$
(*c*) $T_1/T_2 = (P_1/P_2)^{n-1}$ (*d*) $T_1/T_2 = (\rho_1/\rho_2)^{(n-1)/n}$
(*e*) None of the above

25. The characteristics equation of a gas is
(*a*) $pv = constant$ (*b*) $pv = mR$
(*c*) $pv = mRT$ (*d*) $pv = RT^m$
where, p, v, T and m = pressure, volume, absolute temperature and mass of the gas respectively and R = gas constant.

26. In a reversible polytropic process
(*a*) Some heat transfer occurs (*b*) The entropy remains constant
(*c*) The enthalpy remains constant (*d*) The internal energy remains constant
(*e*) The temperature remains constant

27. The value of gas constant (R) is
(*a*) 287 J/kg K (*b*) 28.7 J/kg K
(*c*) 2.87 J/kg K (*d*) 0.287 J/kg K

28. The value of universal gas constant (R_u) is
(*a*) 8.314 J/kg K (*b*) 83.14 J/kg K
(*c*) 831.4 J/kg K (*d*) 8314 J/kg K

29. An isentropic process is always
(*a*) Irreversible and adiabatic (*b*) Reversible and isothermal
(*c*) Frictionless and irreversible (*d*) Reversible and adiabatic
(*e*) None of the above

30. The gas constant (R) is equal to the ____________ of two specific heats.
(*a*) Sum (*b*) Difference
(*c*) Product (*d*) Ratio

31. Specific heat at constant volume is defined as
(*a*) $K.C_p$ (*b*) $(\partial u/\partial T)_p$
(*c*) $(\partial T/\partial u)v$ (*d*) $(\partial u/\partial T)_v$
(*e*) None of the above

32. The specific heat at constant pressure is __________ that of the specific heat at constant volume.
(*a*) Equal to (*b*) Less than
(*c*) More than (*d*) None

33. Specific heat at constant pressure is defined as
(*a*) $K.C_p$ (*b*) $(\partial h/\partial T)_p$
(*c*) $(\partial T/\partial u)v$ (*d*) $(\partial u/\partial T)_v$
(*e*) None of the above

34. The ratio of specific heat at constant pressure (c_p) and specific heat at constant volume (c_v) is
(*a*) Equal to one (*b*) Less than one
(*c*) More than one (*d*) None of these

35. The value of c_p/c_v for air is
(*a*) 1.0 (*b*) 1.4
(*c*) 1.8 (*d*) 2.3

36. When the gas is heated at constant pressure, then the heat supplied
    (*a*) Raises the temperature of gas
    (*b*) Increases the internal energy of the gas
    (*c*) Does some external work during expansion
    (*d*) Both (*a*) and (*b*)
    (*e*) Both (*b*) and (*c*)

37. An ordinary mercury thermometer can be used for temperature about
    (*a*) 100°C (*b*) 200°C
    (*c*) 300°C (*d*) 400°C
    (*e*) 500°C

38. When a gas is heated at constant volume
    (*a*) Its temperature will increase
    (*b*) Its pressure will increase
    (*c*) Both temperature and pressure will increase
    (*d*) Neither temperature nor pressure will increase

39. The heating of a gas at constant pressure is governed by
    (*a*) Boyle's law (*b*) Charles' law
    (*c*) Gay Lussac's law (*d*) Joule's law

40. A process, in which the gas is heated or expanded in such a way that the product of its pressure and volume remains constant, is called
    (*a*) Isothermal process (*b*) Isobaric process
    (*c*) Adiabatic process (*d*) Polytropic process

41. Isentropic flow is
    (*a*) Irreversible adiabatic flow (*b*) Perfect gas flow
    (*c*) Real fluid flow (*d*) Reversible adiabatic flow
    (*e*) Frictionless reversible flow

42. The hyperbolic process is governed by
    (*a*) Boyle's law (*b*) Charles' law
    (*c*) Gay Lussac's law (*d*) Joule's law

43. The heating of gas at constant __________ is governed by Boyle's law.
    (*a*) Volume (*b*) Pressure
    (*c*) Temperature (*d*) None of these

44. Triple point temperature and pressure for water are
    (*a*) 100°C and 1 ata (*b*) 0°C and 1 ata
    (*c*) 0°C and 0.1 ata (*d*) 0.1°C and 0.06028 ata
    (*e*) 0.01°C and 0.006028 ata

45. In an isothermal process
(*a*) Internal energy increases
(*b*) Internal energy decreases
(*c*) There is no change in internal energy
(*d*) Internal energy first decreases and then increases

46. The expansion ratio (r) is the ratio of
(*a*) v_1/v_2 (*b*) v_2/v_1
(*c*) $(v_1+v_2)/v_1$ (*d*) $(v_1+v_2)/v_2$
where, v_1= volume at beginning of expansion, and v_2 = volume at the end of expansion

47. When the expansion or compression of the gas takes place according to the law $pv^n = C$, then the process is known as
(*a*) Isothermal process (*b*) Isobaric process
(*c*) Adiabatic process (*d*) Polytropic process

48. An adiabatic process is one is which
(*a*) No heat enters or leaves the gas
(*b*) The temperature of the gas changes
(*c*) The change in internal energy is equal to the work done
(*d*) All of the above

49. The general law of expansion or compression is $pv^n = C$. The process is said to be hyperbolic, if n is equal to
(*a*) 0 (*b*) 1
(*c*) γ (*d*) 8

50. The value of $n = 0$ in the general law $pv^n = C$, then the process is called
(*a*) Isochoric process (*b*) Isobaric process
(*c*) Isothermal process (*d*) Isentropic process

51. In the polytropic process equation pv^n = constant, if the value of n is infinitely large, the process is termed as
(*a*) Constant volume (*b*) Constant pressure
(*c*) Constant temperature (*d*) Adiabatic
(*e*) Isothermal

52. In the polytropic process equation pv^n = constant, if the value of $n = 0$, the process is termed as
(*a*) Constant volume (*b*) Constant pressure
(*c*) Constant temperature (*d*) Adiabatic
(*e*) Isothermal

53. In the polytropic process equation pv^n = constant, if the value of n =1, the process is termed as
(*a*) Constant volume
(*b*) Constant pressure
(*c*) Constant temperature
(*d*) Adiabatic
(*e*) Isothermal

54. The work done in free expansion process is
(*a*) Zero
(*b*) Minimum
(*c*) Maximum
(*d*) Positive

55. In a steady flow process,
(*a*) The mass flow rate is constant
(*b*) The heat transfer rate is constant
(*c*) The work transfer rate is constant
(*d*) All of the above

56. The work done in steady flow process is given by
(*a*) $\int_1^2 p\,dv$
(*b*) $-\int_1^2 p\,dv$
(*c*) $\int_1^2 v\,dp$
(*d*) $-\int_1^2 v\,dp$

57. The throttling process is a
(*a*) Non-flow process
(*b*) Steady flow process
(*c*) Non steady flow process
(*d*) All of these

58. The heat absorbed or rejected by the working substance is given by
a) $\delta Q = TdS$
b) $\delta Q = T/dS$
c) $\delta Q = dS/T$
where, δQ = heat absorbed or rejected, dS = increase or decrease of entropy, and T = absolute temperature

59. The property of a working substance which increases or decreases as the heat is supplied or removed in a reversible manner, is known as
(*a*) Enthalpy
(*b*) Internal energy
(*c*) Entropy
(*d*) External energy

60. The entropy may be expressed as a function of
(*a*) Pressure and temperature
(*b*) Temperature and volume
(*c*) Heat and work
(*d*) All of these

61. The entropy of water at 0°C is assumed to be
(*a*) 1
(*b*) 0
(*c*) –1
(*d*) 10

62. A system consists of more than one phase is known as
(*a*) Open system (*b*) Closed system
(*c*) Isolated system (*d*) Non uniform system
(*e*) Heterogeneous system

63. The change of entropy when heat is absorbed by the gas is
(*a*) Positive (*b*) Negative
(*c*) Positive or negative (*d*) None

64. Carnot cycle consists of
(*a*) Two constant volume and two reversible adiabatic processes
(*b*) Two isothermal and two reversible adiabatic processes
(*c*) Two constant pressure and two reversible adiabatic processes
(*d*) One constant volume, one constant pressure and two reversible adiabatic processes

65. Whenever a system undergoes either a change in state or an energy or mass transfer at a steady state, it is said to be undergo
(*a*) A change of state (*b*) A process
(*c*) A steady state transfer (*d*) An equilibrium
(*e*) A thermodynamic change

66. The efficiency of Carnot cycle may be increased by
(*a*) Increasing highest temperature (*b*) Decreasing highest temperature
(*c*) Increasing lowest temperature (*d*) Decreasing highest temperature

67. The processes or systems that do not involve heat are called
(*a*) Isothermal processes (*b*) Equilibrium processes
(*c*) Thermal processes (*d*) Steady processes
(*e*) Adiabatic processes

68. Second law of thermodynamics defines
(*a*) Heat (*b*) Work
(*c*) Enthalpy (*d*) Entropy
(*e*) Internal energy

69. A cycle consisting of two constant volume and two isothermal process is known as
(*a*) Carnot cycle (*b*) Joule cycle
(*c*) Diesel cycle (*d*) Stirling cycle

70. The efficiency of Ericsson cycle is __________ Carnot cycle.
(*a*) Greater than (*b*) Less than
(*c*) Equal to (*d*) none of these

71. Otto cycle is also known as
(*a*) Constant pressure cycle
(*b*) Constant volume cycle
(*c*) Constant temperature cycle
(*d*) Constant entropy cycle

72. For the same compression ratio, the efficiency of Diesel cycle is ___________ Otto cycle.
(*a*) Greater than
(*b*) Less than
(*c*) Equal to
(*d*) none of these

73. The efficiency of Diesel cycle approaches to Otto cycle efficiency when
(*a*) Cut-off is increased
(*b*) Cut-off is decreased
(*c*) Cut-off is zero
(*d*) Cut-off is constant

74. The air standard efficiency ratio of an Otto cycle is given by
(*a*) $1-(1/r^{\gamma-1})$
(*b*) $1+(1/r^{\gamma-1})$
(*c*) $1-r^{\gamma-1}$
(*d*) $1+r^{\gamma-1}$

75. The efficiency of the dual combustion cycle, for the same compression ratio is ____ diesel cycle.
(*a*) Greater than
(*b*) Less than
(*c*) Equal to
(*d*) none of these

76. For the same maximum pressure and temperature of Otto, Diesel and dual combustion air standard cycles,
(*a*) The compression ratio will be same
(*b*) The heat supplied to be cycles will be same
(*c*) The air standard efficiency will have the same value
(*d*) The heat rejected by the engine will be same

77. Water at atmospheric conditions
(*a*) Behaves as an ideal gas
(*b*) Is mostly liquid
(*c*) Is far above its critical state
(*d*) Is far below its critical state

78. The locus of saturated liquid line and saturated vapour line meets at
(*a*) Boiling point
(*b*) Critical point
(*c*) Ice point
(*d*) Triple point

79. For steam
(*a*) The critical temperature is 221.2°C and critical pressure is 374.15 bar
(*b*) The critical temperature is 374.15 °C and critical pressure is 221.2 bar
(*c*) The critical temperature is 221.2°C and critical pressure is 221.2 bar
(*d*) The critical temperature is 374.15 °C and critical pressure is 374.15 bar

80. The latent heat of steam at atmospheric pressure is
(*a*) 1535 kJ/kg
(*b*) 1875 kJ/kg
(*c*) 2257 kJ/kg
(*d*) 2685 kJ/kg

81. With the increase in pressure
(*a*) The boiling point of water decreases and enthalpy of evaporation increases
(*b*) The boiling point of water decreases and enthalpy of evaporation decreases
(*c*) Both the boiling point of water and enthalpy of evaporation decreases
(*d*) Both the boiling point of water and enthalpy of evaporation increases

82. The dryness fraction of steam is equal to
(*a*) $\frac{m_g}{m_g+m_f}$ (*b*) $\frac{m_f}{m_g+m_f}$
(*c*) $\frac{m_g}{m_f}$ (*d*) $\frac{m_f}{m_g}$
where, m_g = mass of dry steam, and m_f = mass of water in suspension

83. The specific volume of water when heated from 0°C
(*a*) Increases steadily (*b*) Decreases steadily
(*c*) First decreases and then increases (*d*) First increases and then decreases

84. The enthalpy of dry saturated steam __________ with the increase in pressure.
(*a*) Decreases (*b*) Increases
(*c*) Remain constant (*d*) None of them

85. The amount of heat absorbed to evaporate 1 kg of water from its saturation temperature without change of temperature is called
(*a*) Sensible heat of water (*b*) Latent heat of vaporization
(*c*) Enthalpy of steam (*d*) Entropy of steam

86. The internal energy (u) of a dry steam is given by
(*a*) $u = h_g + 100\ pv_g$ (*b*) $u = h_g - 100\ pv_g$
(*c*) $u = h_g / 100\ pv_g$ (*d*) $u = 100\ pv_g / h_g$
where, h_g = enthalpy of 1 kg of dry steam in kJ/kg; p = pressure of steam in bar, and v_g = specific volume of 1 kg of dry steam in m^3/kg.

87. The entropy of water at 0°C is
(*a*) 0 (*b*) 1
(*c*) 2.3 (*d*) 4.2

88. When water is heated from freezing point ot the boiling point, then the increase in entropy is given by
(*a*) $c_w \log(273)$ (*b*) $c_w \log(T)$
(*c*) $c_w \log(T/273)$ (*d*) $c_w \log(273/T)$
where, c_w = specific heat of water, and T = boiling point in °K.

89. The isentropic process, on the Mollier diagram, is represented by a
(*a*) Horizontal line (*b*) Vertical line
(*c*) Inclined line (*d*) Curve

90. The throttling process on h-s diagram will be a
(a) Horizontal line (b) Vertical line
(c) Inclined line (d) Curve

91. In a constant volume process, heat transferred is equal to
(a) Work done (b) Change in internal energy
(c) Change in enthalpy (d) None of these

92. The process in which heat transferred is equal to the change of enthalpy is known as
(a) Constant pressure process (b) Constant volume process
(c) Constant temperature process (d) Constant entropy process

93. The heating of wet steam at a constant temperature till it become dry saturated is similar to that of heat at a
(a) Constant volume (b) Constant pressure
(c) Constant entropy (d) None of these

94. The superheating of steam at ________ may be regarded as hyperbolic.
(a) Constant volume (b) Constant pressure
(c) Constant entropy (d) Constant temperature

95. In an isentropic process
(a) Work done is zero (b) Change in internal energy is zero
(c) Change in entropy is zero (d) Change in enthalpy is zero

96. The non-flow process, in which the work done is equal to the change in internal energy is known as
(a) Isentropic process (b) Isothermal process
(c) Isobaric process (d) Isochoric process

97. In a steady flow reversible adiabatic process, work done is equal to
(a) Change in internal energy (b) Change in entropy
(c) Change in enthalpy (d) Heat transferred

98. The isentropic process on h-s diagram will be a
(a) Horizontal line (b) Vertical line
(c) Inclined line (d) Curve

99. In a throttling process, ________ remains constant.
(a) Enthalpy (b) Temperature
(c) Entropy (d) All of the above

100. The water tubes in a simple vertical boiler are
(a) Horizontal (b) Vertical
(c) Inclined (d) None

101. Lancashire boiler is a
(*a*) stationary fire tube boiler
(*b*) internally fired boiler
(*c*) horizontal boiler
(*d*) natural circulation boiler
(*e*) all of the above

102. The diameter of internal flue tubes of a Lancashire boiler is about _____ that of its shell.
(*a*) One-fourth
(*b*) One-third
(*c*) Two-fifth
(*d*) One-half

103. Which of the following is a water tube boiler?
(*a*) Lancashire boiler
(*b*) Babcock and Wilcox boiler
(*c*) Locomotive boiler
(*d*) Cochran boiler

104. In fire tube boilers
(*a*) Water passes through the tubes which are surrounded by flames and hot gases
(*b*) The flames and hot gases pass through the tube which are surrounded by water
(*c*) Forced circulation takes place
(*d*) None of the above

105. Which of the following boiler is best suited to meet the fluctuating demand of steam?
(*a*) Locomotive boiler
(*b*) Lancashire boiler
(*c*) Cornish boiler
(*d*) Babcock and Wilcox boiler

106. Water tube boilers produce steam at a __________ pressure than that of fire tube boilers.
(*a*) Lower
(*b*) higher
(*c*) Equal
(*d*) None

107. The locomotive boiler has
(*a*) 137 fire tubes and 44 superheated tubes
(*b*) 147 fire tubes and 34 superheated tubes
(*c*) 157 fire tubes and 24 superheated tubes
(*d*) 167 fire tubes and 14 superheated tubes

108. La-Mont boiler is a ________ pressure water tube steam boiler working on forced circulation.
(*a*) Low
(*b*) High
(*c*) Equal
(*d*) None

109. A device attached to the steam chest for preventing explosions due to excessive internal pressure of steam is called
(*a*) Safety valve
(*b*) Water level indicator
(*c*) Pressure gauge
(*d*) Fusible plug

110. A safety valve mainly used with locomotive and marine boiler is
(*a*) Lever safety valve
(*b*) High pressure and low water safety valve
(*c*) Dead weight safety valve
(*d*) Spring loaded safety valve

111. A device used in boiler to control the flow of steam from the boiler to the main pipe and to shut off the steam completely when required is known as

(*a*) Blow off cock (*b*) Fusible plug
(*c*) Stop valve (*d*) Economiser

112. A device used to put off fire in the furnace of the boiler when the level of water in the boiler falls to an unsafe limit is called

(*a*) Blow off cock (*b*) Fusible plug
(*c*) Super heater (*d*) Economiser

113. A device used to increase the temperature of saturated steam without raising its pressure is called

(*a*) Blow off cock (*b*) Fusible plug
(*c*) Super heater (*d*) Economiser

114. A devise used to heat feed water by utilizing the heat in the exhaust flue gases before leaving through chimney is known as

(*a*) Super heater (*b*) Economiser
(*c*) Blow off cock (*d*) Stop valve

115. Which of the following are boiler accessories?

(*a*) Safety valve (*b*) Pressure gauge
(*c*) Super heater (*d*) Economiser
(*e*) Both (*a*) and (*b*) (*f*) Both (*c*) and (*d*)

116. An economizer ____________ the steam raising capacity of boiler.

(*a*) Increases (*b*) Decreases
(*c*) Has no effect on

117. The pressure of feed water has to be raised before its entry into the boiler. The pressure is raised by a device known as

(*a*) Feed cock valve (*b*) Feed pump
(*c*) Pressure gauge (*d*) Injector

118. An air preheater

(*a*) Increases evaporation capacity of the boiler
(*b*) Increases the efficiency of the boiler
(*c*) Enables low grade fuel to be burnt (*d*) All of the above

119. A condenser where circulating water flows through tubes which are surrounded by steam is known as

(*a*) Surface condenser (*b*) Jet condenser
(*c*) Barometric condenser (*d*) Evaporative condenser

120. The ratio of actual vacuum to the ideal vacuum in a condenser is called
(*a*) Condenser efficiency (*b*) Vacuum efficiency
(*c*) Boiler efficiency (*d*) Nozzle efficiency

121. The actual vacuum in a condenser is equal to
(*a*) barometric pressure + actual pressure (*b*) barometric pressure – actual pressure
(*c*) gauge pressure + atmospheric pressure (*d*) gauge pressure – atmospheric pressure

122. In a four stroke engine, the working cycle is completed in
(*a*) One revolution of the crankshaft (*b*) Two revolution of the crankshaft
(*c*) Three revolution of the crankshaft (*d*) Four revolution of the crankshaft

123. A two stroke cycle engine gives _______ the number of power strokes as compared to the four stroke cycle engine at same engine speed.
(*a*) Half (*b*) Same
(*c*) Double (*d*) Four times

124. The thermal efficiency of a two stroke cycle engine is _____ a four stroke cycle engine.
(*a*) Equal to (*b*) Less than
(*c*) Greater than (*d*) Undefined

125. The theoretically correct mixture of air and petrol is
(*a*) 10:1 (*b*) 15:1
(*c*) 20:1 (*d*) 25:1

126. The thermodynamic cycle on which the petrol engine works is
(*a*) Otto cycle (*b*) Joule cycle
(*c*) Rankine cycle (*d*) Stirling cycle

127. A diesel engine has
(*a*) One valve (*b*) Two valves
(*c*) Three valves (*d*) Four valves

128. If petrol is used in diesel engine, then
(*a*) Low power will be produced (*b*) Efficiency will be low
(*c*) Higher knocking will occur (*d*) Black smoke will be produced

129. A petrol engine has compression ratio from
(*a*) 6 to 10 (*b*) 10 to 15
(*c*) 15 to 25 (*d*) 25 to 40

130. Super charging __________ the power developed by the engine.
(*a*) Has no effect on (*b*) Increases
(*c*) Decreases

131. A carburettor is used to supply
(*a*) Petrol, air and lubrication oil
(*b*) Air and diesel
(*c*) Petrol and lubrication oil
(*d*) Petrol and air

132. A spark plug gap is kept between
(*a*) 0.3 to 6.5 mm
(*b*) 0.2 to 0.8 mm
(*c*) 0.4 to 0.9 mm
(*d*) 0.6 to 1.0 mm

133. The knocking tendency in spark ignition engines may be decreased by
(*a*) Controlling the air fuel mixture
(*b*) Controlling the ignition timing
(*c*) Reducing the compression ratio
(*d*) All of these

134. Which of the following does not relate to a spark ignition engine?
(*a*) Ignition coil
(*b*) Spark plug
(*c*) Distributer
(*d*) Fuel injector

135. The thermal efficiency of petrol engine is about
(*a*) 15%
(*b*) 30%
(*c*) 50%
(*d*) 70%

136. Diffusion is
(*a*) Mixing of two portions of fluid
(*b*) Mixing of a gas in two containers at different pressures
(*c*) Mixing of two portions of a fluid at different temperatures
(*d*) Mixing of unlike fluids
(*e*) None of these

137. Food can be cooked quicker in a pressure cooker because
(*a*) Less heat is lost
(*b*) High pressure steam has high temperature
(*c*) Steam remains within the cooker
(*d*) It is made of a metal
(*e*) None of them

138. The difference between two specific heats of a gas is equal to universal gas constant when
(*a*) 1 gm of gas is heated
(*b*) 1 gm mole of gas is heated
(*c*) 1 mole of gas is heated
(*d*) Any amount of gas is heated

139. A cycle which consists of two adiabatic and two constant volume processes is known as
(*a*) Otto cycle
(*b*) Stirling cycle
(*c*) Diesel cycle
(*d*) Carnot cycle

140. Which of the following remains constant during an adiabatic process?
(*a*) Internal energy
(*b*) Enthalpy
(*c*) Entropy
(*d*) All of the above

ANSWER KEY (THERMODYNAMICS)

1. C	2. B	3. C	4. D	5. B	6. C	7. B	8. E	9. C	10. D
11. A	12. D	13. D	14. A	15. D	16. A	17. C	18. E	19. D	20. E
21. B	22. D	23. D	24. A	25. C	26. B	27. A	28. D	29. D	30. B
31. D	32. C	33. B	34. C	35. B	36. E	37. C	38. C	39. B	40. A
41. D	42. A	43. C	44. E	45. C	46. B	47. D	48. D	49. B	50. B
51. A	52. B	53. D	54. A	55. D	56. D	57. B	58. A	59. C	60. A
61. B	62. E	63. A	64. B	65. B	66. B	67. E	68. D	69. D	70. C
71. B	72. B	73. B	74. A	75. A	76. D	77. D	78. B	79. B	80. C
81. B	82. A	83. C	84. A	85. B	86. B	87. A	88. C	89. B	90. A
91. B	92. A	93. B	94. D	95. C	96. A	97. C	98. B	99. A	100. C
101. E	102. C	103. B	104. B	105. A	106. B	107. C	108. B	109. A	110. D
111. C	112. B	113. C	114. B	115. F	116. A	117. B	118. D	119. A	120. B
121. B	122. B	123. C	124. B	125. B	126. A	127. C	128. C	129. A	130. B
131. D	132. A	133. C	134. D	135. B	136. D	137. C	138. B	139. A	140. C

Heat Transfer

1. The heat transfer is constant when
   (*a*) Temperature remain constant with time
   (*b*) Temperature increase with time
   (*c*) Temperature decrease with time
   (*d*) None of the above.

2. Flow of heat by conduction is the result of
   (*a*) The transfer of vibrational energy from one molecule to another
   (*b*) The transfer of kinetic energy
   (*c*) The movement of free electrons
   (*d*) A combination of (*a*), (*b*) and (*c*).

3. Transfer of heat by conduction is generally
   (*a*) From one part of a body to another part of the same body
   (*b*) Between two bodies in physical constant
   (*c*) Neither of (*a*) nor of (*b*)
   (*d*) Either of (*a*) or (*b*)

4. When compared with non-metallic solids, the metallic solids usually have
   (*a*) Much higher thermal conductivity
   (*b*) Much lower thermal conductivity
   (*c*) Almost equal thermal conductivity

5. Among the liquid, water has a comparatively high value of thermal conductivity, which is due to
   (*a*) High viscosity
   (*b*) Partial ionization
   (*c*) High molality of the molecules

6. For gases, thermal conductivity values increases with
   (*a*) Molecular weight
   (*b*) Temperature
   (*c*) Density

7. Temperature distribution in case of heat flow by conduction for a cylindrical body with internal heat source is
   (*a*) Asymptotic in nature
   (*b*) Parabolic in nature
   (*c*) Hyperbolic in nature
   (*d*) Linear in nature

8. Rate of heat transfer in case of forced convection when compared with natural convection is
   (*a*) Higher (*b*) lower
   (*c*) Almost equal (*d*) Nothing in particular

9. Heat transfer from higher temperature to lower temperature takes place
   (*a*) According to First Law of Thermodynamics
   (*b*) According to Second Law of Thermodynamics
   (*c*) According to Zeroth Law of Thermodynamics
   (*d*) According to Fourier Law

10. Heat transfer by conduction follows
    (*a*) Stefan's Law (*b*) First Law of Thermodynamics
    (*c*) Fourier Law (*d*) Newton's law

11. The coefficient of thermal conductivity is defined as
    (*a*) Quantity of heat transfer of per unit time per unit area per one degree temperature drop per unit length
    (*b*) Quantity of heat transfer per unit time per unit area
    (*c*) Quantity of heat transfer of per one degree temperature drop per unit area
    (*d*) Quantity of heat transfer of per unit area one degree drop in temperature

12. Q = rate of heat transfer, A = area of flow normal to direction of flow, Δx = thickness of wall, k = thermal conductivity; then $kA/\Delta x$ is called
    (*a*) Thermal resistance of wall (*b*) Thermal conductance of wall
    (*c*) Heat flux density (*d*) Thermal loading wall

13. With usual notation, $\Delta x/kA$ is called
    (*a*) Thermal resistance of wall (*b*) Thermal conductance of wall
    (*c*) Heat flux density (*d*) Thermal loading wall

14. Interior wall is at 1000°C. The wall thickness is 0.6 m, it is 1.0 m wide, 1.5 m broad of material with thermal conductivity 0.4 W/m °K, the temperature of out side wall is 200 °C. The thermal resistance is
    (*a*) 1° K/W (*b*) 2° K/W
    (*c*) 1.8° K/W (*d*) 1.5° K/W

15. Interior wall is at 1000°C. The wall thickness is 0.6 m, it is 1.0 m wide, 1.5 m broad of material with thermal conductivity 0.4 W/m °K, the temperature of out side wall is 200 °C. The thermal conductance is
    (*a*) 1.0 W/ °K (*b*) 0.5 W/ °K
    (*c*) 1.5 W/ °K (*d*) 2.0W/ °K

16. The thermal conductivity is expressed as
    (*a*) kcal/hr-m^2 °C (*b*) kcal/hr-m °C
    (*c*) kcal/hr-°C (*d*) kcal/m^2-°C

17. The thermal conductivity is expressed as
(*a*) W/m² °K (*b*) W/m °K
(*c*) J/m² °K (*d*) Nm/m² °K

18. The radial heat transfer rate through hollow cylinder increases as the ratio of outer to inner radius
(*a*) increases (*b*) decreases
(*c*) constant (*d*) none of the above

19. The term $(\rho v l/\mu)$ is called
(*a*) The Reynolds number (*b*) Prandtl number
(*c*) Froude number (*d*) Nusselt number

20. The term (hl/k) is called
(*a*) The Reynolds number (*b*) Prandtl number
(*c*) Froude number (*d*) Nusselt number

21. The term $(C_p\mu/k)$ is called
(*a*) Reynolds number (*b*) Prandtl number
(*c*) Froude number (*d*) Nusselt number

22. For Laminar flow, Reynolds number will be
(*a*) Re > 2500 (*b*) Re < 2500
(*c*) Re < 2100 (*d*) Re >2100

23. Nusselt number is the function of
(*a*) Reynold number and Prandtl number (*b*) Reynold number and Grashof number
(*c*) Prandtl number and Grashof number (*d*) None of the above

24. Peclet number in heat transfer is a product of
(*a*) Reynold number and Nusselt number (*b*) Reynold number and Prandtl number
(*c*) Grashof number and Reynold number

25. With increase in concentration, thermal conductivity of liquid generally
(*a*) Increases (*b*) Decreases
(*c*) Remains almost unaffected

26. Stefan Boltzman Law is expressed as
(*a*) $E_b = \sigma T^4$ (*b*) $E_b = \sigma(\Delta T)^4$
(*c*) $E_b = \sigma(\Delta T)^{1.4}$ (*d*) $E_b = \sigma T^{1/4}$

27. The quantity of heat radiation is dependent on
(*a*) The shape of the body only (*b*) Area of the body only
(*c*) Temperature of the body only (*d*) On all (*a*), (*b*) and (*c*)

28. For turbulent flow, Ditters-Boelter correlation gives
    (*a*) $N_u = 0.023\ (\text{Re})^{0.4}(P_r)^{0.6}$
    (*b*) $N_u = 0.023\ (\text{Re})^{0.8}(P_r)^{0.6}$
    (*c*) $N_u = 0.23\ (\text{Re})^{0.2}(P_r)^{0.4}$
    (*d*) $N_u = 0.23\ (\text{Re})^{0.8}(P_r)^{0.4}$

29. Thermal conductivity of water
    (*a*) first increases with temperature then decreases with temperature
    (*b*) increases steadily with temperature
    (*c*) decreases with temperature
    (*d*) does not depend on temperature

30. Prandtl number of water varies from
    (*a*) 0.5-1.0
    (*b*) 1.0-5.0
    (*c*) 5.0-10.0
    (*d*) $10\text{-}10^2$

31. In case of convection heat transfer from a hot tube surface to a fluid flowing in it ,the bulk of the resistance to transfer is
    (*a*) Distributed uniformly throughout the fluid
    (*b*) Confined to a thin film near the surface
    (*c*) Confined to the central core of the fluid

32. *jH* factor of heat transfer is a function of
    (*a*) Reynolds number
    (*b*) Prandtl number
    (*c*) Froude number
    (*d*) Nusselt number

33. In forced convection, liquid viscosity effect (for viscous liquids) is taken into account in case of
    (*a*) Nusselt's equation
    (*b*) Sieder –Tate equation
    (*c*) Dittus – Boeltier equation

34. For convective transfer of heat in case of fluid flowing in tubes, the length to diameter ratio influences the heat transfer coefficient, when flow is
    (*a*) Highly turbulence
    (*b*) Laminar
    (*c*) Laminar as well as transition zone

35. The effect of baffle in heat exchanger is to
    (*a*) Hold the tubes in position
    (*b*) Increase turbulence
    (*c*) Clean the outer tube surface

36. For convective heat transfer, controlling film is one which has
    (*a*) Lower film resistance
    (*b*) Lower film coefficient
    (*c*) higher film coefficient

37. Convective heat transfer coefficient in case of turbulence flow in tubes varies with mass velocity as
(*a*) $h \propto G$ (*b*) $h \propto G^{0.5}$
(*c*) $h \propto G^{0.8}$

38. Natural convection is significant in case of
(*a*) Only non-viscous fluids
(*b*) Only viscous fluids
(*c*) Both viscous and non-viscous fluids

39. The value of overall heat transfer coefficient for a fluid-fluid heat transfer exchanger is controlled by the value of
(*a*) Film coefficient which is lower (*b*) Film coefficient which is higher
(*c*) The dirt factors for both films

40. The condensing film coefficient is
(*a*) Almost equal for both vertical and horizontal condensers
(*b*) Higher for horizontal condenser
(*c*) Higher for vertical condenser

41. Sub cooling of condensate is required when
(*a*) The condensate is corrosive at condensing temperature
(*b*) Better heat recovery is aimed at in condensers
(*c*) The condensate is volatile liquid to be sent for storage

42. Increase in the number of tube side passes, results in
(*a*) The decrease of shell side dirt factor
(*b*) The decrease of overall pressure drop
(*c*) A high rate of heat transfer

43. Condensation of vapour is carried out inside the tubes, when
(*a*) The condensate is corrosive
(*b*) A higher condensing film coefficient is desired
(*c*) A low pressure drop through the unit is desired

44. Average value of condensing film coefficient for steam on horizontal tubes varies between
(*a*) 500-1000 kcal/hr m^2 °C (*b*) 1000-5000 kcal/hr m^2 °C
(*c*) 5000-15000 kcal/hr m^2 °C

45. Presence of a non-condensable gas
(*a*) Increase the condensing film coefficient
(*b*) Decreases the condensing film coefficient
(*c*) Doesn't appreciably alter the value

46. The air is relatively transparent to electro-magnetic waves and absorbs the radiant energy
    (*a*) To a considerable amount
    (*b*) Completely
    (*c*) To a small extent

47. Wave length of radiation which are important to radiant heat transfer as found in ordinary industrial equipments
    (*a*) 0.75-400 μ
    (*b*) 10-500 μ
    (*c*) 100-1000 μ (where μ is micron)

48. Substance can emit or absorb radiant energy to varying extent provided their temperature is
    (*a*) Above 273 ºK
    (*b*) Above 298 ºK
    (*c*) Above 0 ºK

49. Which of the following surface has the maximum emissivity?
    (*a*) Smooth and black
    (*b*) Smooth and white
    (*c*) Rough and white
    (*d*) Rough and black

50. Multiple effect evaporators are adopted to achieve
    (*a*) Better quality of product
    (*b*) Higher steam quality
    (*c*) Higher throughout

51. For natural circulation evaporators, the entering velocity of liquid into the tubes is
    (*a*) 10-30 cm/sec
    (*b*) 30-90 cm/sec
    (*c*) 100-250 cm/sec

52. Concentration of viscous liquid is handled effectively in a
    (*a*) Short tube evaporator
    (*b*) Falling film evaporator
    (*c*) Forced circulation evaporator

53. Liquor with foaming tendency can be concentrated in a
    (*a*) Horizontal tube evaporator
    (*b*) Long tube evaporator
    (*c*) Agitated film evaporator

54. Heat sensitive material like orange juice is concentrated in a
    (*a*) Long tube evaporator
    (*b*) Basket evaporator
    (*c*) Falling film evaporator

55. In case of long tube forced circulation evaporator, the value of overall heat transfer increases, because
    (*a*) Increased liquor velocity in the tube
    (*b*) Increased in residence time
    (*c*) Neither of the above in (*a*) and (*b*)

56. The agitated film evaporator is particularly effective for
    (*a*) Foamy liquid
    (*b*) Heat sensitive liquid
    (*c*) Viscous and heat sensitive liquid

57. Capacity of an evaporator depends on
(*a*) Number of effects (*b*) Temperature difference
(*c*) Method of feeding

58. Which of the following statements is correct?
(*a*) Good radiators make good absorbers
(*b*) Good radiators make bad absorbers
(*c*) bad radiators make good absorbers

59. In multiple effect evaporation, if the feed is cold, then the method of feeding is of
(*a*) Mixed type (*b*) Forward type
(*c*) Backward type

60. With the increase in vacuum in the vapour space of an evaporator, the available temperature difference
(*a*) Increases (*b*) Decreases
(*c*) Remains constant

61. Intermittent tube cleaning can be done in case of
(*a*) Horizontal tube evaporator (*b*) Calendria type evaporator
(*c*) Basket type evaporator

62. Tubes used in short tube vertical evaporators have inside diameters in the range of
(*a*) 5-10 cm (*b*) 5-15 cm
(*c*) 10-25 cm

63. Tube length in case of long tube evaporator is
(*a*) 12-24 meter (*b*) 6-15 meter
(*c*) 3-10 meter

64. Number of effects in multiple effect evaporator system is limited by
(*a*) Final product concentration (*b*) Steam pressure
(*c*) Total boiling point rise

65. For solutions, Dühring's plot is used to calculate
(*a*) Boiling point rise (*b*) Concentration
(*c*) Enthalpy

66. With increase in vacuum in the effect, the value of hydrostatic head
(*a*) Increases (*b*) Decreases
(*c*) Remains almost unaffected

67. An important liquid property which determines the type of circulation (*viz.* natural or forced) in evaporators is
(*a*) Thermal conductivity (*b*) Viscosity
(*c*) Specific gravity

68. For conduction through a thick walled tube, the value of R_m (mean radius) used in the heat conduction equation is given by (where, R_1 = inner radius, R_2 = outer radius
(*a*) $(R_2 - R_1)/2$ (*b*) $(R_2 + R_1)/2$
(*c*) $(R_2 + R_1)/ \ln (R_2/R_1)$ (*d*) $(R_2 - R_1) / \ln (R_2/R_1)$.

69. Transmission of heat by molecular collision is
(*a*) Conduction (*b*) Convection
(*c*) Radiation (*d*) Scattering
(*e*) Transmission

70. Metals are good conductors of heat because
(*a*) They contain free electrons (*b*) Their atoms are relatively far apart
(*c*) Their atoms collide frequently (*d*) They have reflecting surfaces
(*e*) None of the above

71. Heat is shown to be closely related to
(*a*) A fluid (*b*) Energy
(*c*) Momentum (*d*) Temperature
(*e*) Wave form

72. Body *A* is kept in contact with body *B*. Heat will flow from *A* to *B*, if
(*a*) The heat of *A* is greater than that of *B*
(*b*) The temperature of *A* is greater than that of *B*
(*c*) The specific heat of *A* is greater than that of *B*
(*d*) The specific heat of *A* is lower than that of *B*
(*e*) None of the above

73. Cork is a good insulator because
(*a*) Its density is low (*b*) It is porous
(*c*) It can be powdered (*d*) It can be cast into rolls
(*e*) It is flexible

74. Which one of the following can not be unit of the thermal conductivity?
(*a*) Watt/m^2 °K (*b*) Kcal/m-hr °C
(*c*) BTU/ft-h-°F (*d*) Cal/cm-sec °K
(*e*) CHU/ft-hr-°C

75. Which one of the following will have least value of thermal conductivity?
(*a*) Copper (*b*) Silver
(*c*) Glass (*d*) Water
(*e*) Air

76. Which one of the following will have highest value of thermal conductivity?
(*a*) Steel
(*b*) Aluminium
(*c*) Brass
(*d*) Copper
(*e*) Lead

77. If thermal conductivity of a material varies as $K_o (1 + \alpha t)$, temperature at the centre of a 10 cm thick wall of material will be
(*a*) Average of the temperature at two ends of the wall
(*b*) More than the average of the temperature at two ends of the wall
(*c*) Less than the average of the temperature at two ends of the wall
(*d*) Depends on the thermal conductivity of the wall material
(*e*) Depends on the range of temperature

78. Heat conduction does not occur
(*a*) If a physical body is impermeable to any kind of rays
(*b*) If the parts of a body are not in motion relative to one another
(*c*) If the bodies are kept in vacuum
(*d*) If the bodies are immersed in water
(*e*) If the temperature difference between the bodies does not exist

79. Unsteady state of heat flow occurs in
(*a*) Flow of heat through furnace wall
(*b*) Flow of heat through insulated pipe with constant surface temperature
(*c*) Annealing of casting
(*d*) Flow of heat through refrigerator walls
(*e*) None of the above

80. Heat is transferred by conduction, convection and radiation in
(*a*) Insulated pipes carrying hot water
(*b*) Refrigerator freezer coils
(*c*) Melting of ice
(*d*) Boiler furnaces
(*e*) Condensation of steam in a condenser

81. Transient heat flow occurs in
(*a*) Electric conductors carrying current
(*b*) Insulated pipes carrying superheated steam
(*c*) Cooling of casting
(*d*) Heating and cooling of building due to sun
(*e*) Melting of ice

82. Essential requirements for the transfer of heat from one body to another is
(*a*) Both bodies must be solids
(*b*) Both bodies must be in contact
(*c*) Temperatures of the two bodies must be different
(*d*) One of the bodies must have internal source of heat generation
(*e*) All of the above

83. – 40 °C is equivalent to
(*a*) 172°R
(*b*) 273°K
(*c*) 500°R
(*d*) – 40°K
(*e*) – 40°F

84. The rate of heat flow from a 50 mm thick wall of material having thermal conductivity of 40 W/m°C for a temperature difference of 10°C will be
(*a*) 80 W/m^2-hr
(*b*) 800 W/m^2-hr
(*c*) 8000 W/m^2-hr
(*d*) 200 W/m^2-hr
(*e*) 2000 W/m^2-hr

85. The ratio of heat flow Q_A/Q_B from two walls of same thickness having thermal conductivity $K_A = 2K_B$ for the same temperature difference will be 1.0
(*a*) 1
(*b*) 0.5
(*c*) 2
(*d*) 0.25
(*e*) 4

86. All heat transfer processes
(*a*) Involve transfer of energy
(*b*) Involve temperature difference between the bodies
(*c*) Obey first law of thermodynamics
(*d*) Obey second law of thermodynamics
(*e*) Obey Newton's law

87. Steady state heat transfer occurs when
(*a*) The flow of heat is negligible
(*b*) The flow of heat is uniform
(*c*) The flow of heat is independent of time
(*d*) The flow of heat is uniformly increasing
(*e*) The flow of heat is uniformly decreasing

88. When K is the thermal conductivity, τ is the density and C_p is the specific heat of a substance, then thermal diffusivity
(*a*) $K / \tau C_p$
(*b*) $C_p / \tau K$
(*c*) $\tau C_p / K$
(*d*) $K \tau / C_p$
(*e*) $K C_p / \tau$

89. Thermal diffusivity is
(*a*) A mathematical formula only
(*b*) A thermal property of the material
(*c*) A configuration for heat conduction
(*d*) A dimensionless parameter
(*e*) A function of temperature

90. The unit of thermal diffusivity is
(*a*) m/hr
(*b*) m^2/hr
(*c*) m/hr°C
(*d*) m^2/ hr°C
(*e*) kcal/ m^2-hr

91. The value of thermal conductivity of water at room temperature is expressed to be
(*a*) 0.001 kcal/m-hr-°C
(*b*) 0.05 kcal/m-hr-°C
(*c*) 0.5 kcal/m-hr-°C
(*d*) 1.0 kcal/m-hr-°C
(*e*) 1.5 kcal/m-hr-°C

92. Thermal conductivity of air at room temperature is approximately
(*a*) 0.002 kcal/m-hr-°C
(*b*) 0.02 kcal/m-hr-°C
(*c*) 0.2 kcal/m-hr-°C
(*d*) 2.0 kcal/m-hr-°C
(*e*) 20.0 kcal/m-hr-°C

93. The value of Prandtl number for air is of the order of
(*a*) 1000
(*b*) 67
(*c*) 9.7
(*d*) 0.7
(*e*) 0.065

94. The numerical value of which property of air generally drops with rising temperature
(*a*) Density
(*b*) Dynamic viscosity
(*c*) Kinematic viscosity
(*d*) Thermal conductivity
(*e*) Thermal diffusivity

95. The surface temperatures of a flat wall are 650°C and 450°C respectively and the thermal conductivity of the wall material is 0.30kcal/m-hr-°C. The heat flux for a wall of 20 cm thickness will
(*a*) 300 kcal/m^2-hr
(*b*) 120 kcal/m^2-hr
(*c*) 60 kcal/m^2-hr
(*d*) 450 kcal/m^2-hr
(*e*) 600 kcal/m^2-hr

96. The rate of heat loss through a furnace wall is 1000 kcal/hr when inside temperature is 1000°C and outside wall temperature is 500°C. What should be the wall thickness in case the thermal conductivity of wall material is 0.5kcal/m-hr-°C.
(*a*) 12.5 cm
(*b*) 25 cm
(*c*) 40 cm
(*d*) 50 cm
(*e*) 60 cm

97. The density of water is maximum at
(*a*) 20 °C
(*b*) 4 °C
(*c*) 0 °C
(*d*) – 4 °C
(*e*) Absolute zero

98. A body cools from 80°C to 70°C in 10 minutes. The time taken by the body for further cooling to 60°C will be
(*a*) 5 minutes
(*b*) 7 minutes
(*c*) 10 minutes
(*d*) Less than 10 minutes
(*e*) More than 10 minutes

99. A thermistor is a
    (*a*) Thermo couple (*b*) Thermometer
    (*c*) Thermal resistance (*d*) Thermal insulation
    (*e*) Miniature resistance

100. The quantity of heat required to change the temperature of 1 gm of ice from –6°C to –5°C is known as
    (*a*) Latent heat of freezing (*b*) Specific heat
    (*c*) Heat of vaporization (*d*) Cold heat
    (*e*) Freezing heat

101. A substance above critical temperature exists as
    (*a*) Solid (*b*) Liquid
    (*c*) Dry vapour (*d*) Wet vapour
    (*e*) Gas

102. There are two spheres of copper the ratio of their radii being 1:3 both at the same temperature. The ratio of heat content will be
    (*a*) 1:1 (*b*) 1:3
    (*c*) 1:9 (*d*) 1:27
    (*e*) 1:81

103. When analogy is drawn between heat and electricity flow in circuits, the heat flow if thermal circuits is equated in the electrical circuit against
    (*a*) Voltage (*b*) Current
    (*c*) Resistance (*d*) Charge (coulombs)
    (*e*) 1/Resistance

104. Which one of the following can not be the unit of convective heat transfer coefficient
    (*a*) Watt/m-hr-°K (*b*) Cal/cm^2-sec-°K
    (*c*) BTU/ft^3 hr-°C (*d*) CTU/ft^2 hr-°C
    (*e*) kcal/ m^2 hr-°C

105. Heat is transferred from a insulated pipe to the surrounding still air mainly by
    (*a*) Conduction (*b*) Free convection
    (*c*) Forced convection (*d*) Radiation
    (*e*) All above together

106. Non-dimensional number generally involved in cases of heat transfer from horizontal cylinder by natural convection is
    (*a*) Reynold number (*b*) Nusslet number
    (*c*) Prandtl number (*d*) Grashoff number
    (*e*) (*b*), (*c*) and (*d*) Above

107. The nature of flow of a fluid inside a tube *i.e.* whether it is turbulent or laminar, is ascertained by
(*a*) Flow velocity
(*b*) Tube size
(*c*) Surface conditions
(*d*) Viscosity of fluid
(*e*) Reynolds number

108. A dimensionless heat transfer coefficient which gives a measure of the ratio of the heat transfer rate to the rate at which heat would be conducted within the fluid under a temperature gradient is known as
(*a*) Reynold number
(*b*) Grashoff number
(*c*) Nusselt number
(*d*) Paclet number
(*e*) Stanton number

109. .A dimensionless number which is the ratio of mass heat flow rate to the heat flow rate by conduction under a unit temperature gradient and through a thickness *L* is known as
(*a*) Peclet number
(*b*) Nusselt number
(*c*) Grashoff number
(*d*) Weber number
(*e*) Prandtl number

110. A non dimensionless number generally associated with natural convection heat transfer is
(*a*) Peclet number
(*b*) Nusselt number
(*c*) Prandtl number
(*d*) Grashoff number
(*e*) Weber number

111. A dimensionless number representing the ratio of inertia forces to viscous forces is known as
(*a*) Reynolds number
(*b*) Prandtl number
(*c*) Nusselt number
(*d*) Mach number
(*e*) Stanton number

112. A dimensionless group which represents the rate of the (product of inertial forces times buoyant forces) to viscous forces squared is known as
(*a*) Reynolds number
(*b*) Prandtl number
(*c*) Nusselt number
(*d*) Grashoff number
(*e*) Mach number

113. A dimensionless number which is the ratio of kinematic viscosity to thermal diffusivity is known as
(*a*) Prandtl number
(*b*) Reynolds number
(*c*) Nusselt number
(*d*) Peclet number
(*e*) Grashoff number

114. A non-dimensionless number which generally finds application in mass transfer problem is
(*a*) Stanton number
(*b*) Mach number
(*c*) Graetz number
(*d*) Grashoff number
(*e*) Weber number

115. Longest wave length
(*a*) Gamma rays
(*b*) X-rays
(*c*) UV rays
(*d*) Radio waves
(*e*) Infrared rays

116. The wave length of X-rays is
(*a*) 6 to 100000 micron
(*b*) 0.4 to 0.8 micron
(*c*) 0.1 to 0.4 micron
(*d*) 0.014 to 0.018 micron
(*e*) 0.8 to 400 micron

117. One micron is equal to
(*a*) 10^{-4} meters
(*b*) 10^{-5} meters
(*c*) 10^{-6} meters
(*d*) 10^{-8} meters
(*e*) 10^{-12} meters

118. If the absorptivity plus reflectivity of a body equals 1, then the body is known as
(*a*) Black body
(*b*) Grey body
(*c*) Opaque body
(*d*) White body
(*e*) None of the above

119. The relation between the emissive and absorptive power of a surface is developed under
(*a*) Plank's law
(*b*) Kirchhoff's law
(*c*) Electrical analogy
(*d*) Stefan-Boltzmann's law
(*e*) Law of radiations

120. For all substances the absorptivity must lie between
(*a*) 0 and 0.1
(*b*) –1 and +1
(*c*) – 0.5 and +0.5
(*d*) 0 and 1
(*e*) 0 and +0.5

121. Two bodies one hot and other cold are kept in vacuum. What will be happen to the temperature of the hot body after sometime?
(*a*) remains the same
(*b*) decreases due to convection
(*c*) increases due to convection
(*d*) increases due to radiation
(*e*) decreases due to radiation

122. A body cools from 90°C to 80°C in 5 minutes. Under the same external conditions to cool from 80°C to 70°C the body will be take
(*a*) 5 minutes
(*b*) 4 minutes
(*c*) 2.5 minutes
(*d*) More than 5 minutes
(*e*) Can not be predicted

123. The surface temperature of the sun is
(*a*) 2000 ºK
(*b*) 6000 ºK
(*c*) 20000 ºK
(*d*) Millions of degree
(*e*) Infinitely large

124. A body at high temperature TºK radiates heat at a rate proportional to
(*a*) T
(*b*) T^2
(*c*) T^4
(*d*) T^{-4}
(*e*) T^{-2}

125. A perfect black body is the one which
(*a*) absorbs all incident radiations
(*b*) reflects all incident radiations
(*c*) absorbs most of the incident radiations
(*d*) in coated with lamp black

126. A vapour containing liquid particles to the extent of 50 % by the volume is known as
(*a*) saturated liquid
(*b*) super saturated liquid
(*c*) wet vapour
(*d*) saturated vapour
(*e*) super saturated vapour

127. Hygrometer is used to measure
(*a*) convective heat transfer coefficient
(*b*) heat exchanged due to vapour
(*c*) density of liquids
(*d*) relative humidity
(*e*) none of the above

128. The mass of water vapour per kg of the pressure of saturated vapour at the air is known as
(*a*) Specific humidity
(*b*) Relative humidity
(*c*) Molar humidity
(*d*) Humidity ratio
(*e*) None of the above

129. The ratio of the actual vapour pressure to the pressure of saturated vapour at the prevailing dry bulb temperature is known as
(*a*) Specific humidity
(*b*) Vapour humidity
(*c*) Relative humidity
(*d*) Humidity ratio
(*e*) Humidity

130. In psychometric chart the abscissa represents
(*a*) Dry bulb temperature
(*b*) Wet bulb temperature
(*c*) Specific humidity
(*d*) Relative humidity
(*e*) Enthalpy

131. A liquid is boiling in a air tight vessel. Using an exhaust tube, the vapour is pumped out at a faster rate. What happen to the liquid?
(*a*) temperature goes down, boiling continues
(*b*) temperature goes up, boiling continues
(*c*) temperature goes down, boiling stops
(*d*) temperature goes up, boiling stops
(*e*) nothing changes

132. The conditions under which three phases of matter can coexist in equilibrium, is known as
(*a*) Absolute point
(*b*) Basic point
(*c*) Common point
(*d*) Standard point
(*e*) Triple point

133. The triple point temperature for water is
(*a*) 0 ºK
(*b*) 0 ºC
(*c*) 0.1 ºC
(*d*) 0.01 ºC
(*e*) 0.001 ºC

134. Absolute zero is
(*a*) 0 ºC
(*b*) 0 ºF
(*c*) 0 ºR
(*d*) 0 ºK
(*e*) 273 ºC

135. The temperature reading taken using Centigrade and Fahrenheit thermometers are plotted on a graph- ºC along *X*-axis, ºF along *Y*-axis. The shape of graph would be
(*a*) Straight line passing through the origin
(*b*) Straight line cutting an intercept on the positive *Y*-axis
(*c*) Straight line cutting an intercept on the positive *X*-axis
(*d*) Straight line cutting an intercept on the negative *Y*-axis
(*e*) Straight line cutting an intercept on the negative *X*-axis

136. An isobar is a graph
(*a*) Connecting points of equal temperature
(*b*) Showing pressure volume changes at constant temperature
(*c*) Connecting points of equal pressure
(*d*) Showing product of pressure and volume against pressure
(*e*) Constant temperature

137. The phenomenon of boiling the milk in an open container when milk spills over the vessel is termed as
(*a*) Film boiling
(*b*) Pool boiling
(*c*) Nuclear boiling
(*d*) Sub cooled boiling
(*e*) Interface evaporation

138. A 40 cm thick wall has the surface temperature of 50ºC and 300ºC. if the thermal conductivity of the wall is given by $k = 0.005\,t - 5 \times 10^{-6}\,t^2$, the heat loss in kcal/m^2 hr will be
(*a*) 109
(*b*) 218
(*c*) 436
(*d*) 872
(*e*) 1744

139. A heat exchanger with heat transfer surface A and overall heat transfer coefficient U handles two fluids of heat capacities C_1 and C_2 such that $C_1 > C_2$. The NTU of the heat exchanger will be

(*a*) NTU = $1 - e^{-AU/(C_1+C_2)}$
(*b*) NTU = $1 - e^{-AU/(C_1C_2)}$
(*c*) NTU = $1 - e^{AU/C_1}$
(*d*) NTU = AU/C_2
(*e*) NTU = AU/C_1

140. In case of natural convection, the Nusselt number is function of

(*a*) Reynolds number
(*b*) Reynolds number and Prandtl number
(*c*) Grashoff's number and Prandtl number
(*d*) Weber number and Mach number
(*e*) None of the above

141. 'Fouling factor'

(*a*) is dimensionless
(*b*) measures the heat transfer efficiency
(*c*) is virtually a factor of safety in heat exchanger design
(*d*) is used when a liquid exchanges heat with a gas
(*e*) is used only in case of Newtonian fluids.

142. Heat sensitive liquids are concentrated

(*a*) by using insulated vessels
(*b*) by using small quantities
(*c*) by using deoxidizers
(*d*) by using vacuum
(*e*) by using low intensity heating

143. The purpose of using multiple pass heat exchangers is to

(*a*) reduce fluid flow friction losses
(*b*) reduce pressure drop
(*c*) increase pressure drop
(*d*) increase the rate of heat transfer
(*e*) neutralize viscous friction

144. Floating heads are provided in heat exchangers to

(*a*) increase the pressure drop
(*b*) reduce the pressure drop
(*c*) facilitate maintenance
(*d*) regulate the flow
(*e*) avoid deformation of tubes due to thermal expansion

145. The first stage of crystal formation is

(*a*) Nucleation
(*b*) Separation
(*c*) Foaming
(*d*) Vortexing
(*e*) None of the above

146. In sugar mills cane juice is evaporated in

(*a*) Horizontal tube evaporators
(*b*) Long vertical tube evaporators
(*c*) Short vertical tube evaporators
(*d*) Zigzag tube evaporators
(*e*) Flash evaporators

147. Which evaporator is preferred for concentrating the fruit juice
(*a*) Long vertical tube evaporators
(*b*) Short vertical tube evaporators
(*c*) Agitated film evaporators
(*d*) Zigzag evaporators
(*e*) Falling film evaporators

148. The material preferred for the fabrication of fruit juice concentrator is
(*a*) Cast iron
(*b*) Mild steel sheet
(*c*) Stainless steel
(*d*) Copper
(*e*) Brass

149. Baffles are provided in heat exchangers to
(*a*) remove dirt
(*b*) increase heat transfer rate
(*c*) provide better mechanical strength
(*d*) reduce heat transfer
(*e*) reduce vibration

150. A 1-2 heat exchanger means
(*a*) single pass on shell side and double pass on tube side
(*b*) single pass on tube side and double pass on shell side
(*c*) single liquid cools two liquids at different temperatures
(*d*) two tubes of cold fluid pass through one tube of hot fluid
(*e*) two tubes of hot fluid pass through one tube of cold fluid

ANSWER KEY

1.A	2. D	3.D	4.A	5.B	6.B	7.B	8.A	9.B	10.C
11.A	12.B	13.A	14.A	15.B	16.B	17.B	18.B	19.A	20.D
21.B	22.C	23.C	24.B	25.B	26.A	27.C	28.B	29.A	30.C
31.B	32.A	33.B	34.C	35.B	36.B	37.C	38.A	39.A	40.B
41.C	42.C	43.A	44.C	45.B	46.C	47.A	48.C	49.D	50.B
51.B	52.C	53.B	54.C	55.A	56.C	57.B	58.A	59.C	60.A
61.C	62.A	63.C	64.C	65.A	66.A	67.B	68.D	69.B	70.C
71.D	72.B	73.D	74.A	75.E	76.D	77.B	78.E	79.C	80.D
81.D	82.C	83.E	84.C	85.C	86.B	87.E	88.A	89.B	90.B
91.C	92.B	93.D	94.A	95.A	96.B	97.B	98.E	99.E	100.B
101.E	102.D	103.B	104.A	105.B	106.E	107.E	108.C	109.A	110.D
111.A	112.D	113.A	114.A	115.A	116.A	117.C	118.C	119.B	120.D
121.E	122.D	123.C	124.C	125.A	126.C	127.D	128.A	129.C	130.A
131.A	132.E	133.D	134.A	135.B	136.C	137.A	138.C	139.D	140.C
141.C	142.D	143.D	144.E	145.A	146.B	147.E	148.C	149.B	150.A

Mass Transfer and Instrumentation

1. In a turbulence fluid the mass transfer is effected as a result of
   (*a*) Random motion of molecules (*b*) Transfer by currents
   (*c*) A combination of (*a*) and (*b*)

2. Separation of components of liquid by method of liquid extraction is preferred with
   (*a*) one of the component is the heat sensitive
   (*b*) one of the component has more affinity to word the solvent
   (*c*) extracting solvent is available at a low cost.

3. Solvent extracting is the name applied to
   (*a*) solid – liquid extraction (*b*) gas liquid transfer operation
   (*c*) liquid - liquid extraction

4. With increase in total pressure on a binary system, the boiling points of the pure component
   (*a*) increase (*b*) decrease
   (*c*) remain unchanged.

5. Ideal solutions obey
   (*a*) Dalton's law (*b*) Henry's law
   (*c*) Roult's law.

6. Separation becomes uneconomical when the value of relative volatility is
   (*a*) greater that 1.05 (*b*) less then 1.05
   (*c*) in between 1.50.

7. Perforated plate towers are unique for liquid-liquid extraction, because
   (*a*) they provide repeated coalescence and redispersion of the droop phase
   (*b*) better contact of the phase is achieved
   (*c*) they provide more contact area.

8. Steam distillation is applied to liquids which
   (*a*) form azeotropes
   (*b*) decompose if distilled directly at atmospheric pressure
   (*c*) are immiscible .

9. For a given L/V ratio, with decrease in feed temperature, the number of plates for Binary distillation.
   (*a*) Decreases (*b*) Increases
   (*c*) Remains constant

10. With increase in reflux rate, the number of plates in a binary distillation unit
    (*a*) Increases
    (*b*) Decreases
    (*c*) Remains constant

11. Use of packed towers in distillation is generally limited to
    (*a*) High pressure operations
    (*b*) Multi-component systems
    (*c*) Small sizes

12. Separating agents for extractive and azeotropic distillation are normally
    (*a*) Polar organic compounds
    (*b*) Non- Polar organic compounds
    (*c*) Inorganic electrolytes.

13. Most important factor for the choice of separating agent extractive distillation is
    (*a*) Toxicity (*b*) Availability
    (*c*) Selectivity

14. Psychrometry deals with the properties of
    (*a*) Air-water vapour mixture (*b*) Gas- vapour mixture
    (*c*) Air- vapour mixture

15. Lewis number is of significance for
    (*a*) Gas absorption (*b*) distillation
    (*c*) Humidification.

16. At a given temperature, the humid value is
    (*a*) Linear function of humidity (*b*) Inverse function of humidity
    (*c*) No- specific function of humidity

17. For air-water vapour mixture the value of Lewis number is
    (*a*) 10 (*b*) 1.0
    (*c*) 0.1.

18. Humidity can be determined by measurement of
    (*a*) Thermal conductivity of a fibre (*b*) Electrical resistance of a fibre
    (*c*) Mechanical strength of a fibre.

19. In case of mechanical draft cooling towers better performance is obtained in case of
(*a*) Induced draft (*b*) Forced draft
(*c*) Nothing in particular.

20. In case of induced draft, the fan is placed at
(*a*) The bottom of the tower (*b*) The top of the tower
(*c*) Any point in the tower.

21. Effectiveness of cooling tower is reduced by the re-circulation of the hot and humid discharged air in case of
(*a*) Forced draft tower (*b*) Induced- draft tower
(*c*) Natural- draft tower

22. Performance of a mechanical draft cooling tower improves with
(*a*) Decrease in air wet-bulb temperature
(*b*) Increase in air wet-bulb temperature
(*c*) Neither of the above.

23. Equilibrium moisture will be maximum for
(*a*) Inorganic solids (*b*) Rag paper
(*c*) Bamboo

24. It is easier to remove
(*a*) Equilibrium moisture (*b*) Unbound moisture
(*c*) Bound moisture

25. Equilibrium moisture of a solid is equal to
(*a*) Bound moisture (*b*) (Total moisture) – (Free moisture)
(*c*) (Total moisture) – (critical moisture).

26. Non-uniformity in moisture content of the finished product taken from various parts of the equipment is found in case of
(*a*) Tray drier (*b*) Rotary drier
(*c*) Freeze drier

27. Case-hardening during drying of high moisture cake
(*a*) Increases the drying rate (*b*) Decrease the drying rate
(*c*) Dose not have effect on drying rate

28. Disadvantages of batch tray driers are
(*a*) Non-uniform and low output
(*b*) High fabrication and maintenance
(*c*) Low heat economy and high labour cost.

29. Pharmaceutical product are dried by
    (*a*) Compartment tray drier (*b*) Vacuum shelf drier
    (*c*) Rotary drier.

30. In freeze drying of food stuffs, moisture is removed by
    (*a*) Evaporation (*b*) Dehydration
    (*c*) Sublimation

31. For the drying of hides, the suitable equipment is a
    (*a*) Tunnel drier (*b*) Tray drier
    (*c*) Rotary drier.

32. A turbo-type drier is a form of
    (*a*) Indirect drier (*b*) direct drier
    (*c*) semi-direct drier

33. Rotary drier is suitable for
    (*a*) Free flowing materials (*b*) Sticky materials
    (*c*) All types of materials

34. For rotary drier best performance is obtained when the hold up is in the range of
    (*a*) 0.1 – 0.3 (*b*) 0.05 – 0.25
    (*c*) 0.05 – 0.15

35. For the process of drying in a drum drier the controlling factor/s is/are
    (*a*) Heat transfer (*b*) Diffusion
    (*c*) Heat transfer and diffusion both.

36. A suitable device for the production of egg powder is
    (*a*) Drum drier (*b*) Spray drier
    (*c*) Vacuum self drier

37. A spouted bed drier can be used effectively for the drying of
    (*a*) Pastes (*b*) Detergent slurry
    (*c*) Peas

38. Use of an extracting solvent at its boiling temperature is called
    (*a*) Lixivation (*b*) Decoction
    (*c*) Elution.

39. Recovery of oil from vegetable seeds with organic solvent is an example of
    (*a*) Elution (*b*) Dissolution
    (*c*) Leaching

40. Dorr thickener is a device used in
    (*a*) Sedimentation (*b*) Leaching
    (*c*) Classification

41. Common Leaching solvent used in vegetable oil industry is
    (*a*) Hexane (*b*) Polyalcohols
    (*c*) Chlorinated hydrocarbons

42. Adsorption is an operation which generally in valves
    (*a*) Gas-solid systems (*b*) Liquid-solid systems
    (*c*) Fluid- solid systems

43. Vander walls adsorption is also called
    (*a*) activated adsorption (*b*) physical adsorption
    (*c*) Chemisororption.

44. Adsorbent used in the refining of sugar is
    (*a*) Silica gel (*b*) Bone char
    (*c*) Activated clay.

45. Cross-section of natural draft cooling tower is
    (*a*) Uniform throughout the height (*b*) Wider at the bottom
    (*c*) Wider at the top.

46. Natural draft towers have shell made up of generally
    (*a*) Mild steel (*b*) Any suitable metal
    (*c*) Ferro-concrete.

47. A natural draft cooling tower is filled with wooden grids, which cover
    (*a*) 10-15% of tower height (*b*) 40-50% of tower height
    (*c*) The entire tower height.

48. Draft eliminator is an important accessory for
    (*a*) Turbo-drier (*b*) Evaporator
    (*c*) Mechanical draft cooling tower.

49. Bubble- cup towers, though very effective for humidification, are not commonly used, because of
    (*a*) The difficulty in fabrication (*b*) High gas pressure drop
    (*c*) High loss of water.

50. Flights for rotary drier extend from the wall to a distance which is
    (*a*) 2-5%of shell inside diameter (*b*) 4-6% of shell inside diameter
    (*c*) 8-12% of shell inside diameter.

51. Recommended peripherical speed of a rotary drier is
(*a*) 5-10 meter/minute
(*b*) 10-30 meter/minute
(*c*) 15-50 meter/minute.

52. Rapid drying in case of spray driers can be achieved only when the particles in the feed are of diameter
(*a*) 10μ – 60 μ
(*b*) 100 μ – 200 μ
(*c*) 250 μ – 500 μ.

53. In liquid-liquid extraction the extraction per unit of total height for a short packed column is
(*a*) Less effective then that for a tall common
(*b*) Same as that for a tall common
(*c*) More effective then for a tall common.

54. If a binary system has no critical solution temperature it implies that
(*a*) On cooling solid phase will appear and on heating a vapour phase will appear
(*b*) The system consist of miscible liquids
(*c*) None of the above.

55. For a ternary system containing two pairs of partially miscible liquids, the no. of plait point (s) is
(*a*) One
(*b*) Two
(*c*) Three
(*d*) Zero.

56. A plait point is the point on the solubility curve where
(*a*) The solute concentrations in both the phases are equal
(*b*) The last tie line on the phase diagram
(*c*) The solute concentrations is maximum
(*d*) The tie line reduces to a point

57. Separation of a binary solution to its components by addition of a solvent is possible. If the distribution co-efficient is
(*a*) Greater than one
(*b*) Less than one
(*c*) Both (*a*) and (*b*)

58. Sieve tray towers are generally used for
(*a*) Low interfacial tension system
(*b*) high interfacial tension system
(*c*) None of the above

59. Mechanical agitations is required if the system has
(*a*) Low interfacial tension
(*b*) high interfacial tension
(*c*) Both (*a*) and (*b*)

60. Which of the following equipments uses centrifugal forces to separate the two phases?
(*a*) Rotating Disk Contactor
(*b*) Schiebel extractor
(*c*) Pedbielniak extractor
(*d*) Treybal extractor

61. Which type of extractor would be useful for handling antibiotics?
    (*a*) Rotating Disk Contactor (*b*) Podbielniak extractor
    (*c*) Sieve tray extractor (*d*) Pulsed column

62. Which type of extractor could be used for handling radioactive solutions?
    (*a*) Pulsed column (*b*) Packed column
    (*c*) Schiebel extractor (*d*) None of the above

63. In a Ballman Extractor the flow is
    (*a*) Co-current (*b*) Counter-current
    (*c*) Both

64. Solids to be leached should be small size if the controlling factor for the rate of extraction is
    (*a*) Phase change of the solute as it dissolves in the solvent
    (*b*) Diffusion of the solute through the solvent in the pores to the outside of the particle
    (*c*) None of the above

65. High degree of agitation is required if the controlling factor in case of leaching is
    (*a*) Phase change of the solute as it dissolves in the solvent
    (*b*) Diffusion of the solute through the solvent in the pores to the outside of the particle
    (*c*) None of the above

66. For leaching of sugar beet high temperature is desirable
    (*a*) True (*b*) False

67. Coarse solids can be leached in
    (*a*) Dorr classifier (*b*) Ballman extractor
    (*c*) Continuous rotary extractor (*d*) Pachuca tank

68. Fine solid can be leached in a
    (*a*) Pachuca tank (*b*) Dorr agitator
    (*c*) Dorr thickener (*d*) All of the above

69. For leaching vegetable seeds, extractor used is
    (*a*) Rotocel (*b*) Kennedy extractor
    (*c*) Ballman extractor (*d*) All of the above

70. The breakpoint time of an adsorption system decreases with
    (*a*) Increased bed height (*b*) Increased particle size of adsorbent
    (*c*) Decreased rate of flow of fluid
    (*d*) Decreased initial solute content of the feed

71. Freundlich equation is applicable to
    (*a*) Dilute solutions over small concentration range
    (*b*) Concentrated solutions
    (*c*) Both (*a*) and (*b*)

72. Higgins contactor is equipment used for
(*a*) Leaching (*b*) Solvent extractor
(*c*) Adsorption (*d*) Drying

73. The range of diffusivity of gases in cm^2/sec, at atmospheric pressure is
(*a*) Less than 1 (*b*) 1
(*c*) 5 to 1 (*d*) 10 to 5
(*e*) More than 10

74. The diffusivity in a binary gas mixture *T* by the relation
(*a*) $D \alpha T$ (*b*) $D \alpha \sqrt{T}$
(*c*) $D \alpha T^{2/3}$ (*d*) $D \alpha 1/\sqrt{D}$
(*e*) $D \alpha 1/T^2$

75. Which of the following provided maximum contact surface for a liquid vapour system?
(*a*) Bubbles cap tower (*b*) Packed tower
(*c*) Sieve plate column (*d*) Wetted wall column

76. Diffusion coefficient of water into air at 26°C is nearly
(*a*) 0.026 cm^2/s (*b*) 0.26 cm^2/s
(*c*) 2.6 cm^2/sec (*d*) 5.2 cm^2/s
(*e*) 2.6 cm^2/sec

77. When *P* is the vapour pressure, *p* the partial pressure and *X* the mole fraction in liquid phase, then according to Roult's law
(*a*) $p = P + X$ (*b*) $pX = P$
(*c*) $p = P/X$ (*d*) $p = P.X$
(*e*) $pPX= 1$

78. Schmidt number is given by
(*a*) $VD\delta / \mu$ (*b*) $D\delta / \mu$
(*c*) $\mu D / \delta$ (*d*) $\delta\mu / D$
(*e*) $\delta\mu / D^2$

79. $D_{BA} = D_{AB,}$ in case of
(*a*) Stagnet diffusion (*b*) Equimolar counter diffusion
(*c*) Either of (*a*) and (*b*) above (*d*) None of the above

80. The process of separation of solutes of different molecular sizes in a liquid solution by use of a suitable membrane is known as
(*a*) Hyper sorption (*b*) Distillation
(*c*) Adsorption (*d*) Leaching
(*e*) Dialysis

81. Which of the following is an example of absorption?
(*a*) extraction of sugarcane juice
(*b*) dechlorination of water
(*c*) removal of oil from cake
(*d*) removal of moisture from gasoline
(*e*) none of the above

82. The process of removal of oil from oil cakes is known as
(*a*) Leaching
(*b*) Decoction
(*c*) Osmosis
(*d*) Soret effect
(*e*) None of the above

83. Peclet number is
(*a*) Re/Pr
(*b*) Pr/Sc
(*c*) Pr/We
(*d*) Re/Sc
(*e*) Pr/Re.Sc

84. A wet solid is to be dried from 80% to 50% moisture (on wet basis). Calculate the moisture to be evaporated for 10 tonnes of dried products?
(*a*) 375 kg
(*b*) 3750 kg
(*c*) 7500 kg
(*d*) 13,750 kg
(*e*) 37,500 kg

85. The dimensions of specific heat are
(*a*) $LT\theta^2$
(*b*) $LT^{-1}\theta^{-1}$
(*c*) $LT^{-2}\theta$
(*d*) $LT^{-2}\theta^{-1}$

86. The dimensions of thermal conductivity are
(*a*) $MLT\theta^2$
(*b*) $MLT^2\theta^2$
(*c*) $MLT^3\theta^{-1}$
(*d*) $ML^{-1}T^{-2}\theta$

87. Grashoff number has significant role in heat transfer by
(*a*) Conduction
(*b*) Natural convection
(*c*) Forced convection
(*d*) Radiation

88. If a steam pressure of 10 kg/cm² is to be measured, the range of pressure gauge should be
(*a*) –10 kg/cm² to + 10 kg/cm²
(*b*) 0 to 10 kg/cm²
(*c*) 5 to 11 kg/cm²
(*d*) 0 to 20 kg/cm²
(*e*) 0 to 100 kg/cm²

89. If $(2n - 1)$ divisions on the main scale are equal to n divisions on the vernier scale and if S is the value of the smallest division, then least count would be
(*a*) $2n/S$
(*b*) $S/(2n - 1)$
(*c*) $S/2n$
(*d*) $2S/n$
(*e*) S/n

90. The elements that senses and convert the desired input to a more convenient and practicable form to be handled by the measurement system.
(*a*) Data presentation element (*b*) Signal conditioning element
(*c*) Transducer element (*d*) Auxiliary element

91. The element that are used for manipulating/processing the output of the transducer in a suitable form.
(*a*) Data presentation element (*b*) Signal conditioning element
(*c*) Transducer element (*d*) Auxiliary element

92. That elements implies for giving the information of the measurand or measured variable in the quantitative form.
(*a*) Data presentation element (*b*) Signal conditioning element
(*c*) Transducer element (*d*) Auxiliary element

93. The elements are those which may be incorporated in a particular system depending on the type of requirement, nature of measurement technique etc.
(*a*) Data presentation element (*b*) Signal conditioning element
(*c*) Transducer element (*d*) Auxiliary element

94. The activity in which increasing the amplitude of the signal without affecting its waveform is called
(*a*) Amplification (*b*) Attenuation
(*c*) Data presentation (*d*) Signal filtration

95. The reduction of the signal amplitude while retaining its original waveform is known as
(*a*) Amplification (*b*) Attenuation
(*c*) Data presentation (*d*) Signal filtration

96. The removal of unwanted noise signals that tend to obscure the transducer signal is referred as
(*a*) Amplification (*b*) Attenuation
(*c*) Data presentation (*d*) Signal filtration

97. The degree of perfection used in instruments, the methods and the observation, is known as
(*a*) Precision (*b*) Accuracy
(*c*) Efficiency (*d*) Least count
(*e*) Error

98. The accuracy depends upon
(*a*) Precision of instruments (*b*) Precision of method
(*c*) Good planning (*d*) All of the above
(*e*) None of above

99. A discrepancy is
(*a*) The difference between a measurement and true value of the quantity measured
(*b*) The difference between true value of the quantity and error
(*c*) The difference between measured true value and actual value
(*d*) The difference between two measured values of the same quantity
(*e*) None of the above

100. If a measuring tape is too long as compared to standard, the error will be known as
(*a*) Instrumental error
(*b*) Personnel error
(*c*) Natural error
(*d*) Manufacturing error
(*e*) Superficial error

101. Natural error in measurement may be due to
(*a*) Humidity
(*b*) Temperature
(*c*) Wind
(*d*) Gravity
(*e*) Any of the above

102. The error which form in the experience of the observer are known as
(*a*) Training error
(*b*) Handling error
(*c*) Personnel error
(*d*) Accidental error
(*e*) Mistakes

103. If an error under the same conditions will always be of the same size and sign, it is known as
(*a*) Uniform error
(*b*) Systematic error
(*c*) Cumulative error
(*d*) Either of (*a*) and (*b*) above
(*e*) Either of (*b*) and (*c*) above

104. The maximum allowable limit that a measurement may vary from the true value is called
(*a*) Permissible error
(*b*) Expected error
(*c*) Range of error
(*d*) Least error
(*e*) Safe error

105. The value of permissible error depends upon
(*a*) The scale
(*b*) The instrument available
(*c*) Class of work
(*d*) All of the above
(*e*) None of above

106. An error that under the same conditions will always be of the same size and sign is known as
(*a*) Mistakes
(*b*) Accidental error
(*c*) Cumulative error
(*d*) Systematic error
(*e*) Detectable error

107. An error that always follows the same definite mathematical or physical law is known as
(*a*) Systematic error
(*b*) Mathematical error
(*c*) Detectable error
(*d*) Physical error
(*e*) Cumulative error

108. The minimum change in the measured variable which produces an effective response of the instrument is known as
(*a*) Resolution sensitivity
(*b*) Hysteresis
(*c*) Accuracy
(*d*) Precision
(*e*) Deviation

Questions 109 to 111 refer to data given below:

A set of independent temperature measurements were taken and recorded as follows: 12.8°C, 12.2°C, 12.5°C, 13.1°C, 12.9°C and 12.4°C.

109. The average deviation is
(*a*) 0.283°C
(*b*) 0.299°C
(*c*) 0.315°C
(*d*) 0.414°C
(*e*) 0.444°C

110. The standard deviation is
(*a*) 0.283°C
(*b*) 0.300°C
(*c*) 0.355°C
(*d*) 0.399°C
(*e*) 0.415°C

111. The variance is
(*a*) 0.015°C
(*b*) 0.115°C
(*c*) 0.222°C
(*d*) 0.445°C
(*e*) 0.695°C

112. The reliability of an instruments means
(*a*) The maximum useful life of an instrument
(*b*) The service of an instrument between two repairs
(*c*) The range in which the characteristics of an instruments remain linear
(*d*) The degree to which repeatability continues to remains specified limits
(*e*) None of the above

113. The sensitivity accuracy of an instrument depends on
(*a*) Frequency response
(*b*) Amplitude distortion
(*c*) Temperature variations
(*d*) Hysteresis
(*e*) None of the above

114. A digital thermometer has 3½ digits display. The 1°C range can be read upto

(a) 1.000°C (b) 1.001°C

(c) 1.999°C (d) 0.999°C

(e) None of the above

115. A 0 - 100°C thermometer has accuracy of +2.5%. Its accuracy while reading 50 mA will be

(a) +1.25% (b) – 1.25%

(c) ± 2.5% (d) ± 5%

(e) ± 10%

116. For bridge circuit of stain gauge which source of power will be ideal?

(a) Low voltage dc source (b) High voltage dc source

(c) Low frequency ac source (d) High frequency ac source

117. An iron – constantan thermocouple can be used for the temperature

(a) 0°C to 600°C (b) –175°C to 900°C

(c) –250°C to 1500°C (d) –373°C to 1800°C

(e) 0°C to 2500°C

118. A copper - constantan thermocouple can be used for the range of temperature from

(a) 100°C to 250°C (b) –0°C to 250°C

(c) –50°C to 300°C (d) –175°C to 350°C

(e) –212°C to 1000°C

119. The radiant energy emitted by a hot body is proportional to (T = absolute temperature)

(a) T (b) T^2

(c) T^4 (d) $1/T^4$

(e) $1/T^2$

120. Mcleod gauge is used to measure

(a) Pressure (b) Vacuum

(c) Flow rate (d) Suspended matter

(e) pH value

121. The amount of moisture in air is measured by

(a) Single psychrometer (b) Orsat apparatus

(c) Mass spectrometer (d) Photo conductive cell

(e) Thermister

122. Which of the following can be used as sensing element for an instrument?

(a) Spring (b) Proving ring

(c) Burdon tube (d) Diaphragm

(e) Any one of the above

123. Which of the following can be used as thermal detector?
(*a*) Thermister (*b*) Pyrometer
(*c*) Thermocouple (*d*) Resistance thermometer
(*e*) Any one of above

124. Which of the following device is primarily used to measure pressure?
(*a*) Bourdon tube (*b*) Kundts tube
(*c*) Hygrometer (*d*) Rotameter
(*e*) All of the above

125. A pirani gauge works on the principles of change of
(*a*) Thermal conductivity of medium (*b*) Electrical resistivity
(*c*) Conductance (*d*) Capacitance
(*e*) Relative humidity

126. Which of the following meter is preferred for the measurement of pressure in steam condenser?
(*a*) Manometer (*b*) Kentometer
(*c*) Pitot tube (*d*) Mcleod gauge
(*e*) None of the above

127. Which of the following is an indirect method of pressure measurement?
(*a*) Mcleod gauge (*b*) Thermal conductivity gauge
(*c*) Ionization gauge (*d*) Radioactive vacuum gauge
(*e*) All of the above

128. Which of the following is an indirect method of pressure measuring device?
(*a*) Bourdon tube (*b*) Flat diaphragm
(*c*) Ionization gauge (*d*) Manometer
(*e*) Capsules

129. A transducer that converts the input physical phenomenon into an electrical output in the form of pulses is known as
(*a*) Secondary transducer (*b*) Analogy transducer
(*c*) Digital transducer (*d*) Tuned transistor
(*e*) None of the above

130. Which one of the following transducer is used to translate linear motion to electrical signals?
(*a*) LVDT (*b*) Thermister
(*c*) Bellows (*d*) Strain gauge
(*e*) Bourdon tube

131. A hot wire anemometer is used to measure
(*a*) Pressure of gases (*b*) Liquid discharges
(*c*) Very low pressures (*d*) Gas velocities
(*e*) Diameter of fine particles

132. A rotameter can be used to measure
   (*a*) Rotation (*b*) Flow
   (*c*) Viscosity (*d*) Specific gravity
   (*e*) Pressure of wind

133. Flow of water in a pipe of 250 cm diameter can be measured by
   (*a*) Orifice plate (*b*) Venturimeter
   (*c*) Rotameter (*d*) Pitot tube
   (*e*) Manometer

134. The advantage of piezo electric crystal transducer is
   (*a*) High frequency response (*b*) High output
   (*c*) Negligible phase shift (*d*) Small size
   (*e*) All of the above

135. A piezometer is used to measure
   (*a*) Very low pressure (*b*) Pressure differential
   (*c*) Atmosphere pressures (*d*) High pressures
   (*e*) Pressure above and below atmosphere pressure

136. Which of the following statement is wrong for thermocouple measuring instruments?
   (*a*) They read average values (*b*) They read r.m.s. values
   (*c*) They cannot take overheads
   (*d*) When calibrated on d.c. they cannot be used for ac signals
   (*e*) Their calibration does not change with time or temperature

137. The display of a digital numerical readout instrument is achieved by
   (*a*) Light emitting diodes (*b*) Light dependent resistors
   (*c*) Photo tube (*d*) Ionization of gases
   (*e*) Heated filaments

138. Which of the following device can be used to give an indication for temperature change?
   (*a*) Bourdon tube (*b*) Thermister
   (*c*) Thermocouple (*d*) Transistor
   (*e*) LED

139. Thermocouples are generally used for temperature measurement upto
   (*a*) 250°C (*b*) 500°C
   (*c*) 1000°C (*d*) 1600°C
   (*e*) 2600°C

140. The function of the reference electrode in a pH meter is to
   (*a*) Measure average pH value (*b*) Produce a constant voltage
   (*c*) Provide temperature compensation (*d*) Produce a constant current
   (*e*) None of the above

141. The pH value of púre water could be
(*a*) 0
(*b*) 1
(*c*) 7
(*d*) 10
(*e*) 14

142. Thermistor has
(*a*) Low and positive temperature coefficient
(*b*) Low and negative temperature coefficient
(*c*) High and negative temperature coefficient
(*d*) Zero temperature coefficient

143. A hydrometer can be used to measure
(*a*) Relative humidity of air
(*b*) Conductivity of gases
(*c*) Temperature coefficient of liquids
(*d*) Specific gravity of liquids
(*e*) Specific gravity of solids

144. Which of the following device can not be used to measure pressure?
(*a*) Manometer
(*b*) LVDT
(*c*) Strain gauge
(*d*) Pirani gauge
(*e*) Bolometer

145. Which of the following device can be used to measure temperatures of hot bodies' radiant energy in visible spectrum?
(*a*) Thermopile
(*b*) Piezometer
(*c*) Bolometer
(*d*) Optical pigometer
(*e*) Planimeter

146. A first order system when subjected to a step input has a temperature rise of 25°C after one hour and 37.5°C after two hours, starting from cold conditions. The final steady temperature rise will be
(*a*) 5°C
(*b*) 15°C
(*c*) 25°C
(*d*) 35°C
(*e*) 50°C

147. In the above problem the thermal time constant will be
(*a*) 1hr
(*b*) 1.44 hrs
(*c*) 2.05 hrs
(*d*) 2.55 hrs
(*e*) 3.36 hrs

148. Which of the following primary transducer does not convert velocity to pressure?
(*a*) Orifice
(*b*) Venturi
(*c*) Pitot tube
(*d*) Vanes
(*e*) All of the above

149. A pitot tube converts
(*a*) Pressure head into velocity head
(*b*) Velocity head into pressure head
(*c*) Pressure head into temperature rise
(*d*) Velocity head into temperature rise
(*e*) None of the above

150. Which of the following is not an element of electro pneumatic pressure transmitter ?
(*a*) LVDT
(*b*) Bellows
(*c*) Opamp
(*d*) Flapper nozzle mechanism
(*e*) All of the above

151. A pirani gauge is used to measure
(*a*) Pressure above 100 kg/cm^2
(*b*) Pressure upto 100 kg/cm^2
(*c*) Atmospheric pressure
(*d*) Very low pressure
(*e*) All pressures

152. A LVDT has
(*a*) One primary coil and two secondary coils
(*b*) Two primary coil and one secondary coils
(*c*) One primary coil and one secondary coils
(*d*) Two primary coil and two secondary coils

153. The 'dead time' of the instrument is
(*a*) The time required by an instrument for initial warming up
(*b*) The time required by an instrument to begin to respond to a change in the measurand
(*c*) The largest change of input quantity for which there is no output of the instrument
(*d*) None of the above

154. A thermometer is calibrated 150°C to 200°C. The accuracy is specified within ±0.25 percent. The maximum static error will be
(*a*) ± 0.25°C
(*b*) 0.5°C
(*c*) ± 0.125°C
(*d*) + 1°C
(*e*) – 1°C

155. A salinometer measures
(*a*) Temperature in cryogenic regions
(*b*) Intensity of illumination
(*c*) Sound level
(*d*) Density of brine
(*e*) Specific gravity of molasses

156. The advantage of digital instruments is
(*a*) No observation error
(*b*) Faster reading
(*c*) Output can be fed to a memory circuit
(*d*) Better accuracy
(*e*) All of the above

157. Those instruments are that present the physical variables of interest in the form of continuous or step less variations with respect to time.
(*a*) Digital type
(*b*) Analog type
(*c*) Null and deflection type
(*d*) Manually operated and automatic type

158. The instruments in which the physical variables are represented by digital quantities and which are discrete and vary in steps are
(*a*) Digital type
(*b*) Analog type
(*c*) Null and deflection type
(*d*) Manually operated and automatic type

159. The act or result of quantitative comparison between known standard and the output of the measuring system measuring the same quantity.
(*a*) Standard of measurement
(*b*) Calibration
(*c*) Digital computer
(*d*) Filter

160. The instrument with digital output is
(*a*) An aneroid barometer
(*b*) A milometer/odometer of the automobile
(*c*) Pendulum clock
(*d*) A micrometer screw gage

161. The closeness of the instrument output to the true value of the measured quantity (as per standards) is defined as
(*a*) Accuracy
(*b*) Precision
(*c*) Resolution
(*d*) Threshold

162. The ability of the instrument to reproduce a certain set of reading within a given accuracy.
(*a*) Resolution
(*b*) Accuracy
(*c*) Precision
(*d*) Threshold

163. Precision of an instrument is in fact dependent on the
(*a*) Accuracy
(*b*) Error
(*c*) Repeatability
(*d*) Resolution

164. The ability of the instrument to reproduce a group of measurement of the same measured quantity, made by the same observer using the same instruments, under the same condition.
(*a*) Accuracy
(*b*) Error
(*c*) Repeatability
(*d*) Resolution

165. Which statement is correct?
(*a*) Accuracy represents the degree of correctness of the measured value with respect to the true value

(*b*) Precision represents degree of repeatability of several independent measurement of the desired input at the same reference conditions.
(*c*) Both above
(*d*) None

166. The smallest increment in the measured value that can be detected with certainty by the instrument or the degree of fineness with which a measurement can be made.
(*a*) Threshold (*b*) Accuracy
(*c*) Precision (*d*) Resolution

167. The least count of any instrument is taken as the ________ of the instrument.
(*a*) Accuracy (*b*) Precision
(*c*) Resolution (*d*) Linearity

168. The smallest measurable input above the zero value is called
(*a*) Linearity (*b*) Threshold
(*c*) Accuracy (*d*) Error

169. The ratio of magnitude of response (output signal) to the magnitude of the quantity being measured (input signal) is called
(*a*) Static sensitivity (*b*) Linearity
(*c*) Dynamic sensitivity (*d*) Threshold

170. The largest change of the measurand to which the instrument does not respond is termed as
(*a*) Drift (*b*) Linearity
(*c*) Backlash (*d*) Dead Band

171. The maximum distance or angle through which any part of the mechanical system may be moved in one direction without causing motion of the next part.
(*a*) Drift (*b*) Linearity
(*c*) Backlash (*d*) Dead Band

172. The variation of output for a given input caused due to change in the sensitivity of the instrument to certain interfering inputs like temperature changes, component instabilities etc is defined as
(*a*) Drift (*b*) Linearity
(*c*) Backlash (*d*) Dead Band

173. Errors caused by the act of measurement on the physical system being tested is termed as
(*a*) Relative error (*b*) Human error
(*c*) Environmental error (*d*) Loading error

174. When the order of the governing equation of an instrument of a combination of instruments is high, it is convenient to plot the frequency response of the system by logarithmic plots known as
(*a*) Bode diagram (*b*) Frequency response
(*c*) Break point frequency (*d*) Dynamic response

175. An electrical output is produced due to an input of mechanical displacement or strain is called
(*a*) Electromechanical transducer (*b*) Inductive transducer
(*c*) Potentio-metric resistance type transducer (*d*) All of the above

176. An ______________ transducer may be employed to convert the displacement into electrical signal that can be released to the force, pressure or temperature input desired.
(*a*) Inductive transducer (*b*) Electromechanical transducer
(*c*) Optical electric transducer (*d*) None

177. Which strain gage transducer are widely used for measurement of several physical variables like strains, force, torque, pressure , vibration etc.
(*a*) Unbounded strain gage (*b*) Resistance strain gage bridges
(*c*) Bonded strain gage (*d*) None

178. An essentially a high input impedance voltage measuring device, capable of indicating voltage signals from the intermediate elements as function of time is known as
(*a*) Cathode ray oscilloscopes (CROs)
(*b*) Linear variable differential transducer (LVDT)
(*c*) Digital voltmeters (DVMs)
(*d*) Pulse duration modulation (PDM)

179. Which device is used for force measurement?
(*a*) Balance (*b*) Hydraulic load cells
(*c*) Pneumatic load cells (*d*) Elastic force device
(*e*) All of the above

180. Which device is used for the measurement of both static and dynamic force?
(*a*) Balance (*b*) Hydraulic load cells
(*c*) Pneumatic load cells (*d*) Elastic force device

181. An piezo-electric transducer employing a piezo-electrical crystal is a useful device for the measurement of
(*a*) Static force (*b*) Dynamic force
(*c*) Frictional force (*d*) Gravitational force

182. One torr is a pressure equivalent to
(*a*) 1 mm Hg (*b*) 10 mm Hg
(*c*) 780 Hg (*d*) None

183. Manometer and elastic element gage can be used to about
(*a*) 10 torr (*b*) 100 torr
(*c*) 1.0 torr (*d*) 0.1 torr

184. Micron is a unit of pressure is equivalent to ________ torr.
(*a*) 0.1 (*b*) 0.01
(*c*) 0.001 (*d*) 1.0

185. Which thermometer is based on the linear expansion of metals which increase in temperature?
(*a*) Liquid in glass thermometer (*b*) Pressure thermometer
(*c*) Electrical resistance thermometer (*d*) Solid rod thermometer

186. The thermometer is based on the principle of fluid expansion due to increase in the pressure in a given volume of the temperature measuring system.
(*a*) Liquid in glass thermometer (*b*) Pressure thermometer
(*c*) Electrical resistance thermometer (*d*) Solid rod thermometer

187. Total radiation pyrometer is used to measure the
(*a*) Radiation (*b*) Temperature
(*c*) Light intensity (*d*) Pressure

188. Selective radiation pyrometer is a temperature measuring device works on the principle of
(*a*) Plank's law (*b*) Newton's law
(*c*) Coulomb's law (*d*) Kirchhoff's law

189. A thermostatic cut out works on the principles of
(*a*) Thermal expansion of fluid (*b*) Variation of resistance with temperature
(*c*) Expansion due to air pressure (*d*) Thermal expansion of metals

190. According to the Stefan-Boltzmann's law, the amount of radiant energy per unit area is proportional to
(*a*) Absolute temperature (*b*) Square of absolute temperature
(*c*) Fourth of absolute temperature (*d*) Cube of absolute temperature

191. Bimetallic strips made of two different materials bend during a rise in temperature on account of
(*a*) Difference in coefficient of linear expansion (*b*) Difference in the elastic properties
(*c*) Difference in thermal conductivities (*d*) None

192. The principle of working of the constant volume thermometer is based on
(*a*) Boyle's law (*b*) Charle's law
(*c*) Gay-Lussac's law (*d*) Equation of state

193. Electrical resistance of a thermister
(*a*) Increase as the temperature increases (*b*) Decreases as the temperature increases
(*c*) Remain unaffected with the change in temperature
(*d*) Increase at low temperature and decreases at high temperatures.

194. The best use of iron-constantan thermocouple is in the temperature range of
(*a*) 63 to 1473°K (*b*) 223 to 2033°K
(*c*) 3 to 673°K (*d*) 3 to 1273°K

195. The best use of resistance thermometer is the range of
(*a*) Below 630°C (*b*) 0°C to 1400°C
(*c*) 630°C to 1200°C (*d*) 100°C to 400°C

196. A thermocouple arrangement is to be used to measure temperatures in the range of 700-800°C. Pointed out the pair that would be most suitable for this application
(*a*) Copper-Constantan (*b*) Iron-Constantan
(*c*) Chromel-Alumel (*d*) Platinum-Platinum Rhodium

197. The unit of Stefan-Boltzmann constant is
(*a*) W/cm^2K (*b*) W/cm^2-K^4
(*c*) W/cm^2-K^2 (*d*) Watt/cm-K^4

198. The Stefan-Boltzmann's law is applicable for heat transfer by
(*a*) Conduction and radiation combined (*b*) Convection and radiation combined
(*c*) Radiation alone (*d*) Convection alone

199. Optical pyrometer is used to measure
(*a*) Light intensity (*b*) Low temperature
(*c*) High temperature (*d*) Light intensity and high temperature

200. A piezo meter is used to measure
(*a*) Very low pressure (*b*) Pressure differential
(*c*) Atmospheric pressure (*d*) High pressure

ANSWER KEY (MASS TRANSFER AND INSTRUMENTATION)

1. C	2. B	3.C	4.A	5.C	6.B	7.A	8.B	9.A	10.B
11. C	12.A	13.C	14.B	15.C	16.A	17.B	18.B	19.A	20.B
21.A	22.B	23.C	24.B	25.B	26.A	27.B	28.C	29.B	30.C
31.A	32.B	33.A	34.C	35.A	36.B	37.C	38.B	39.C	40.A
41.A	42.C	43.B	44.B	45.B	46.C	47.A	48.C	49.B	50.C
51.B	52.A	53.C	54.A	55.D	56.D	57.C	58.A	59.B	60.C
61.B	62.A	63.C	64.B	65.C	66.B	67.A	68.D	69.D	70.B
71.A	72.C	73.A	74.C	75.D	76.B	77.D	78.B	79.B	80.E
81.D	82.A	83.A	84.E	85.D	86.C	87.B	88.D	89.E	90.C
91.B	92.A	93.D	94.A	95.B	96.D	97.A	98.D	99.D	100.A
101.E	102.E	103.E	104.A	105.D	106.A	107.A	108.A	109.A	110.D
111.B	112.D	113.C	114.C	115. D	116. D	117.B	118.D	119.C	120.B
121.A	122.E	123.E	124.A	125.A	126.B	127.E	128.C	129.C	130.A
131.D	132.B	133.D	134.E	135.A	136.D	137.A	138.C	139.D	140.B
141.C	142.C	143.D	144.E	145.D	146.E	147.B	148.D	149.B	150.D
151.D	152.C	153.B	154.C	155.D	156.E	157.B	158.A	159.B	160.D
161.A	162.C	163.C	164.C	165.C	166.D	167.C	168.B	169.A	170.D
171C	172.A	173.D	174.A	175.A	176.B	177.C	178.A	179.E	180.D
181.B	182.A	183.D	184.C	185.D	186.B	187.B	188.A	189.D	190.C
191.A	192.A	193.B	194.A	195.A	196.A	197.B	198.C	199.C	200.A

Refrigeration and Air Conditioning

1. In a mechanical refrigeration system the highest temperature of refrigerant occurs
   (*a*) Between condenser and evaporator
   (*b*) In evaporator
   (*c*) Before expansion valve
   (*d*) Between evaporator and compressor
   (*e*) Between compressor and evaporator.

2. As warm air cools, its
   (*a*) Relative humidity decreases
   (*b*) Relative humidity increases
   (*c*) Capacity to hold moisture increases
   (*d*) There is no effect moisture holding capacity
   (*e*) Dew point occurs.

3. When the air temperature reaches the point where relative humidity is 100% the air is saturated because it cannot hold any more moisture. This temperature is called
   (*a*) Freezing point
   (*b*) Partially Freezing point
   (*c*) Dew point
   (*d*) Dew point of water vapour
   (*e*) Dew point of air.

4. The chemical formula for feron-12 is
   (*a*) $CHClF_2$
   (*b*) CH_3Cl
   (*c*) CCl_2F_2
   (*d*) $CClF_3$
   (*e*) $CClF_2CF_3$

5. In a refrigeration cycle the heat is rejected by refrigerant at
   (*a*) Evaporator
   (*b*) Condenser
   (*c*) Expansion valve
   (*d*) Compressor
   (*e*) All of the above

6. The flow of the refrigerant in refrigeration cycle is controlled by
   (*a*) Compressor
   (*b*) Condenser
   (*c*) Evaporator
   (*d*) Expansion valve
   (*e*) Copper tubing.

7. Sub- cooling is
   (*a*) Cooling a part of liquid
   (*b*) Differential cooling of a liquid
   (*c*) Removing latent heat from a gas
   (*d*) Removing sensible heat froe a liquid
   (*e*) Cooling a liquid below 0 C.

8. A device designed to remove moisture from a refrigerant is called
   (*a*) Expansion valve
   (*b*) Dehumidifier
   (*c*) Drier
   (*d*) Solenoid
   (*e*) Humidistat

9. The ratio of sensible heat to total heat is known as
   (*a*) Dew point
   (*b*) Humidification ratio
   (*c*) Relative humidity
   (*d*) Specific humidity
   (*e*) Sensible heat factor

10. In a refrigeration system, if the compressor 'short cycles' the problem could be
    (*a*) Intermittent contact in electric control circuit
    (*b*) Dirty or iced evaporator
    (*c*) Lake of refrigerant
    (*d*) Over change of refrigerant
    (*e*) Any of the above

11. In Electrolux refrigerator
    (*a*) Ammonia is absorbed in water
    (*b*) Ammonia is absorbed in hydrogen
    (*c*) Hydrogen is absorbed in water
    (*d*) Ammonia evaporates in hydrogen
    (*e*) Hydrogen evaporates in ammonia.

12. The moisture in a refrigeration system is removed by
    (*a*) Drier
    (*b*) Dehumidifier
    (*c*) Evaporator
    (*d*) Coolers
    (*e*) None of the above

13. The most common type of vapour absorption system in use for industrial applications
    (*a*) Froen 12and water
    (*b*) Froen 22 and water
    (*c*) Ammonia and water
    (*d*) Brine and water
    (*e*) Hydrogen and water

14. In a vapour absorption as compared to vapour compression system, the compressor is replaced by
    (*a*) Evaporator
    (*b*) Receiver
    (*c*) Expansion valve
    (*d*) Condenser
    (*e*) Absorber and generator.

15. In a refrigerator system if compressor fails to operate, the probable reason could be
    (*a*) Power failure (*b*) Low voltage
    (*c*) Fuse blown (*d*) Any of the above

16. A refrigerant with lowest freezing point temperature
    (*a*) Froen11 (*b*) Froen12
    (*c*) Froen22 (*d*) Ammonia
    (*e*) Carbon dioxide.

17. A refrigerating system operates on the reversed carnot cycle. The higher temperature of the refrigerant in the system is 40° C and the lower is –15°C. the capacity is to be 10 tons. Neglected all losses, the coefficient of performance of the system would by
    (*a*) 2.67 (*b*) 3.75
    (*c*) 4.69 (*d*) 7.36
    (*e*) 9.41

18. A Bell-Colman cycle is
    (*a*) Reversed carnot cycle (*b*) Reversed joule cycle
    (*c*) Reversed Otto cycle (*d*) Reversed Rankine cycle
    (*e*) Reversed Sterling cycle

19. Cryogenics
    (*a*) is same as refrigeration (*b*) include heating and cooling of liquids
    (*c*) is a thermodynamic analysis of low temperature systems
    (*d*) is the field of engineering associated with development, production and use of equipment in the range -180°C to absolute zero
    (*e*) none of the above.

20. Air conditioning means
    (*a*) Cooling (*b*) Heating
    (*c*) Dehumidifying (*d*) Removal of air impurities
    (*e*) All of above

21. pH value of a brine solution is
    (*a*) 7.0 (*b*) More then 7
    (*c*) Less then 7 (*d*) 14
    (*e*) Could be any value.

22. Dehydration mean
    (*a*) Removing hydrogen from a mixture (*b*) Removing hydrogen from a hydrocarbon
    (*c*) Removing hydrogen from any matter (*d*) Removing steam from any matter
    (*e*) Removing water from any matter

23. A refrigeration cycle in usually a
    (*a*) Open cycle (*b*) Close cycle
    (*c*) Mixed cycle (*d*) Hybrid cycle

24. Water is used in a refrigerant in
    (*a*) Vapour absorption cycle (*b*) Steam jet refrigeration
    (*c*) Vapour adsorption cycle (*d*) Cascade system
    (*e*) None of the above

25. In a vapour compression cycle the vapour as it leaves the compressor is
    (*a*) in liquid form (*b*) wet vapour
    (*c*) dry vapour (*d*) dry and saturated vapour
    (*e*) none of the above

26. In a vapour compression cycle the refrigerant immediately after expansion valve is in
    (*a*) liquid form (*b*) wet vapour form
    (*c*) dry vapour form (*d*) dry and saturated vapour form
    (*e*) none of the above

27. The ability of an evaporator to absorb heat does not depends on
    (*a*) its size, surface condition and material (*b*) kinds of refrigerant used
    (*c*) temperature difference between the evaporator and the surrounding air
    (*d*) velocity of turbulence
    (*e*) compressor speed

28. A condenser removes from the refrigerant
    (*a*) Sensible heat only (*b*) Latent heat only
    (*c*) Sensible heat and Latent heat (*d*) Partly Sensible heat and Latent heat
    (*e*) None of the above

29. The critical temperature is the temperature
    (*a*) Below which gas is always liquefied
    (*b*) Above which gas will never be liquefied
    (*c*) Below which gas dose note obey gas laws
    (*d*) Above which gas may explode
    (*e*) None of the above

30. If a room has bad ammonia leaks
    (*a*) Before entering the room fans should be put on
    (*b*) Light should be switched off
    (*c*) One should wait for some time
    (*d*) Gas mask should be put on
    (*e*) No specific precautions are needed

31. Freon type (halocarbons) refrigerants are
(*a*) Poisonous
(*b*) Irritating
(*c*) Non-poisonous
(*d*) Coloured
(*e*) Have peculiar odour

32. In a psychrometric chart vertical lines indicate
(*a*) Dry bulb temperature
(*b*) Wet bulb temperature
(*c*) Dew point temperature
(*d*) Percentage humidity
(*e*) Relative humidity

33. In a psychrometric chart horizontal lines indicate
(*a*) Dry bulb temperature
(*b*) Wet bulb temperature
(*c*) Dew point temperature
(*d*) Percentage humidity
(*e*) Relative humidity

34. For better performance a desert cooler
(*a*) Should be placed inside the room
(*b*) Should be placed outside
(*c*) Should semi-mounted on window
(*d*) Should be placed at the roof level
(*e*) None of the above

35. In rainy season desert coolers become ineffective
(*a*) Because of low temperature
(*b*) Pleasant weather
(*c*) Rain serves the purpose of cooler
(*d*) Humidity of air increases
(*e*) None of the above

36. On a psychrometric chart sensible heating or cooling is represented by
(*a*) A horizontal line
(*b*) A vertical line
(*c*) Inclined line
(*d*) Curved line
(*e*) None of the above

37. Coefficient of Performance of a reversed Carnot cycle refrigerator working between high temperature T_2 and lower temperature T_1 :
(*a*) Will increase with increase in T_1 keeping T_2 fixed
(*b*) Will increase with increase in T_1 keeping T_2 fixed
(*c*) Will first increase with increase in T_1 and then decrease with increase in T_1 keeping T_2 fixed

38. Reversed Carnot cycle comprises
(*a*) Two isentropic processes and two adiabatic processes
(*b*) Two isentropic processes and two isothermal processes
(*c*) Two isentropic processes and two constant pressure processes
(*d*) Two isentropic processes and two constant volume processes.

39. In Reversed Carnot cycle working on perfect gas
    (*a*) Isothermal work of compression is equal to isothermal work of expansion
    (*b*) Isentropic work of compression is equal to isentropic work of expansion
    (*c*) Net work of the cycle is zero
    (*d*) Net heat transfer of cycle is zero

40. In Reversed Carnot cycle working on vapour
    (*a*) Isentropic work of compression is equal to isentropic work of expansion
    (*b*) Isothermal heat absorption is equal to isothermal heat rejection
    (*c*) There is no work done during isothermal processes
    (*d*) There is no work done during isentropic processes.

41. Reversed Carnot cycle assumes that all processes in the cycles are
    (*a*) Non flow only
    (*b*) Steady flow only
    (*c*) Non flow only or Steady flow only
    (*d*) Transient flow.

42. Two Carnot Refrigerators are employed, one for ice making and other for comfort cooling
    (*a*) The COP of refrigerators for ice making is higher than that for the other
    (*b*) The COP of refrigerators for ice making is lower than that for the other
    (*c*) The COP of refrigerators for ice making is same as that for the other
    (*d*) The COP of Carnot refrigerator will depend on refrigerant load.

43. A reverse Carnot cycle has a COP of 4. The ratio of higher temperature to lower will be
    (*a*) 1.5
    (*b*) 2.0
    (*c*) 1.25
    (*d*) 2.5

44. The moisture in a refrigeration system is removed by
    (*a*) driers
    (*b*) dehumidifiers
    (*c*) evaporators
    (*d*) coolers
    (*e*) none

45. A Carnot refrigerator works with a temperature range of –5°C to 40°C on (*i*) R-12 (*ii*) R-717 (*iii*) air
    (*a*) COP with R-12 is greater than COP with R-717
    (*b*) COP with R-12 is greater than COP with air
    (*c*) COP with R-12 is equal to COP with R-717
    (*d*) COP with R-717 is greater than COP with air.

46. The COP of a refrigerator working on reversed Carnot cycle with T_1 being lower temperature and T_2 being higher temperature is
    (*a*) $(T_2 - T_1)/ T_1$
    (*b*) $T_1 / (T_2 - T_1)$
    (*c*) $T_2 / (T_2 - T_1)$
    (*d*) $(T_2 - T_1) / T_2$.

47. The COP of a Carnot refrigerator can be expressed : (r = Volume compression ratio for isentropic compression)

(*a*) $r^{\gamma-1}$ (*b*) $1/(1-r^{\gamma-1})$

(*c*) $1/(r^{\gamma-1}-1)$ (*d*) $1/(1+r^{\gamma-1})$.

48. The COP of Carnot Refrigerator is 3 and it produces 1 ton of refrigeration. The work that will be done is equal to

(*a*) 210 kJ/min (*b*) 70 kJ/min

(*c*) 500 kJ/min (*d*) 200 kJ/min

49. The ratio of high temperature to low temperature for Reversed Carnot refrigerator is 1.25. The COP will be

(*a*) 2 (*b*) 3

(*c*) 4 (*d*) 5

50. A Carnot Refrigerator operates between 300.3°K and 273°K. The fraction of cooling effect required as work input

(*a*) 20 % (*b*) 10 %

(*c*) 50 % (*d*) Not possible to find with the data

51. The relationship between $COP_{heat\ pump}$ and $COP_{refrigerator}$ for the same range of temperature operation is

(*a*) $COP_{heat\ pump} = COP_{refrigerator}$ (*b*) $COP_{heat\ pump} - COP_{refrigerator} = 1$

(*c*) $COP_{heat\ pump} = 2COP_{refrigerator}$ (*d*) None

52. Carnot refrigerator absorbs heat at –13°C and requires 1kW for each 6.5 kW of heat absorbed, the COP and temperature of heat rejections respectively are (0=273°K)

(*a*) COP = 6.5, t = 27°C (*b*) COP = 7.5, t = 27°C

(*c*) COP = 6.5, t = 30°C (*d*) COP = 7.5, t = 27°C

53. The dense air system as compared to open air system for same range of temperatures using Bell-Colemann system gives

(*a*) High power/ton of refrigeration (*b*) Low power/ton of refrigeration

(*c*) Same power/ton of refrigeration (*d*) Unpredictable results.

54. In Bell-Colemann refrigerator for the same temperature range

(*a*) COP of dense air system is higher than COP of open air system

(*b*) COP of dense air system is lower than COP of open air system

(*c*) COP of dense air system is equal to COP of open air system

(*d*) COP of dense air system may be higher or lower than COP of open air system depending upon pressure ratio.

55. COP of Bell-Colemann refrigerator for pressure ratio = r_p is expressed as

(*a*) $1/1 - r_p^{(\gamma-1)/\gamma}$ (*b*) $1/r_p^{(\gamma-1)/\gamma} - 1$

(*c*) $1/r_p^{1/\gamma} - 1$ (*d*) $1/r_p^{\gamma/(\gamma-1)} - 1$

56. In vapour compression refrigeration system the heat is absorbed from the environments to be cooled by
(*a*) evaporator (*b*) condenser
(*c*) compressor (*d*) throttle valve

57. In simple saturated vapour compression cycle plant, heat removed in condenser = 10500 kJ. The work input to the compressor = 2100 kJ. Refrigeration capacity of this plant is
(*a*) 20 ton (*b*) 60 ton
(*c*) 40 ton (*d*) not possible to calculate

58. Heat is rejected by the refrigerant during vapor compression refrigeration cycle in
(*a*) condenser (*b*) evaporator
(*c*) throttle valve (*d*) compressor

59. In a simple saturated vapour compression cycle, the refrigerant is in superheated condition
(*a*) before entering compressor (*b*) before entering condenser
(*c*) before entering throttle valve (*d*) before entering evaporator

60. In a simple saturated vapour compression refrigeration cycle the following results were obtained. Heat rejected in condenser = 140 kJ/kg. Compressor work = 28 kJ/kg. The COP refrigeration is
(*a*) 5 (*b*) 4
(*c*) 6 (*d*) Not possible to find with given data.

61. The effect of liquid sub-cooling in vapour compression refrigeration system is
(*a*) to reduce the rate of flow of refrigerant per ton of refrigeration
(*b*) to reduce the volume of vapour handled by compression per ton of refrigeration
(*c*) to reduce power per ton refrigeration
(*d*) All of the above.

62. The effect of superheating the vapour before suction to compression in vapour compression refrigeration system
(*a*) increase the work of compression
(*b*) increase the heat rejection in condenser
(*c*) may increase COP or decrease COP depending upon the refrigerant used
(*d*) All of the above.

63. A domestic refrigerator capacity may be
(*a*) 1 ton (*b*) 0.1 ton
(*c*) 5 ton (*d*) 10 ton

64. A domestic window type air conditioner capacity may be approximately
(*a*) 1 ton (*b*) 0.1 ton
(*c*) 5 ton (*d*) 10 ton

65. COP of domestic air conditioner as compared to that of domestic refrigerator will be
(*a*) lower (*b*) higher
(*c*) same (*d*) un-predictable

66. In a domestic vapour compression refrigerator the refrigerant used is
(*a*) CO_2 (*b*) Freon-12
(*c*) Ammonia (*d*) All of above

67. The order in which main components of vapour compression refrigeration system are used is
(*a*) Compressor-evaporator-condenser-throttle valve
(*b*) Compressor-condenser -evaporator-throttle valve
(*c*) Compressor-condenser -throttle valve- evaporator
(*d*) Compressor-throttle valve -evaporator-condenser

68. Sub-cooling is a process of cooling the refrigerant in vapour compression refrigeration system
(*a*) After compression (*b*) Before compression
(*c*) Before throttling (*d*) None of the above

69. In vapour compression refrigeration cycle the following data is available, heat rejected in condenser = 65 kW, work done in compressor = 10 kW, then, COP (refrigerator) is
(*a*) 6.5 (*b*) 5.5
(*c*) 4.5 (*d*) Not possible to find out with data

70. Ammonia is used with
(*a*) Reciprocating refrigerant compressors (*b*) Centrifugal refrigerant compressors
(*c*) Axial flow refrigerant compressors (*d*) All of the above

71. Freon-11 is used with
(*a*) Reciprocating refrigerant compressors for central plants
(*b*) Centrifugal refrigerant compressors for large capacity central plants
(*c*) Window type air conditioning units
(*d*) All of the above

72. Freon-12 is used with
(*a*) Reciprocating refrigerant compressors in small units like domestic refrigerator
(*b*) Reciprocating refrigerant compressors in large units for central air conditioning plants
(*c*) Centrifugal refrigerant compressors for large units for central air conditioning plants
(*d*) All of the above

73. Freon-22 is used with
(*a*) Reciprocating refrigerant compressors in large units such a package units or central air conditioning plants

(*b*) Centrifugal refrigerant compressors in large units and central air conditioning plants
(*c*) Only small domestic refrigerators or water coolers
(*d*) All of the above

74. Normal boiling point temperature of R-13 is
(*a*) Greater than NBP temperature of R-12
(*b*) Greater than NBP temperature of R-22
(*c*) Greater than NBP temperature of R-11
(*d*) Less than NBP of all above namely R-12, R-22 and R-11

75. The refrigerant which gives evaporator pressure
(*a*) As high as possible is desirable
(*b*) As low as possible is desirable
(*c*) Near about atmosphere pressure but below atmosphere pressure is desirable
(*d*) Near about atmosphere pressure but above atmosphere pressure is desirable.

76. Refrigerant R-12 as compared to R-717 is
(*a*) Costly
(*b*) Cheap
(*c*) Same cost
(*d*) None of the above.

77. Operating pressures of the refrigerant at the evaporator in the ascending order for any application are
(*a*) R-11, R-12, R-717, R-22
(*b*) R-22, R-717, R-12, R-11
(*c*) R-11, R-12, R-22, R-717
(*d*) R-717, R-22, R-11, R-12

78. For high COP
(*a*) Critical temperature in general should be low
(*b*) Critical temperature in general should be high
(*c*) Critical temperature is insignificant
(*d*) Critical pressure is important.

79. High boiling point refrigerants are perfectly suited for
(*a*) Reciprocating compressors
(*b*) Centrifugal compressors
(*c*) Small screw type compressors
(*d*) For all the above.

80. Low boiling point refrigerants are perfectly suited for
(*a*) Reciprocating compressors
(*b*) Large centrifugal compressors
(*c*) Large axial flow rotary compressors
(*d*) All the above.

81. The following refrigerant is most miscible with oil
(*a*) R-717
(*b*) R-11
(*c*) R-22
(*d*) R-12

82. For vapour compression refrigeration system using R-717 (NH_3), the material of pipes are
(*a*) Iron and steel
(*b*) Copper
(*c*) Brass
(*d*) Aluminium

83. It is desirable for refrigerant to have
    (*a*) High thermal conductivity for high heat transfer co-efficient
    (*b*) Low thermal conductivity for low heat transfer co-efficient
    (*c*) Any thermal conductivity as it does not affect performance
    (*d*) All the above depending upon components of plant

84. A good refrigerant should be
    (*a*) Low specific heat ratio (γ)   (*b*) High latent heat
    (*c*) High thermal conductivity   (*d*) All of the above

85. The refrigerant R-744 stands for
    (*a*) Ammonia   (*b*) Carbon dioxide
    (*c*) Sulphur dioxide   (*d*) Methyl chloride

86. The refrigerant R-717 stands for
    (*a*) Ammonia   (*b*) Carbon dioxide
    (*c*) Sulphur dioxide   (*d*) Methyl chloride

87. The refrigerant R-764 stands for
    (*a*) Ammonia   (*b*) Carbon dioxide
    (*c*) Sulphur dioxide   (*d*) Methyl chloride

88. The refrigerant R-22 stands for
    (*a*) $CFCl_2$ (*b*) CF_2Cl_2
    (*c*) $CHClF_2$ (*d*) CHF_3Cl

89. The refrigerant R-290 stands for
    (*a*) C_2H_4 (*b*) C_3H_8
    (*c*) C_4H_{10} (*d*) $(CH_2)_2CH$

90. Freezing point temperature for R-717 (NH_3) is
    (*a*) –94°C   (*b*) –77.8°C
    (*c*) –157.5°C   (*d*) –160°C

91. Compound compression is adopted for refrigeration system
    (*a*) To improve volumetric efficiency
    (*b*) To reduce mechanical work input per unit mass
    (*c*) To take care of very large pressure ratios
    (*d*) To achieve all the above

92. The function of compressor in vapour compression refrigeration system is
    (*a*) To continuously circulate the refrigerant through the system
    (*b*) To compress the vapour form low pressure to high pressure
    (*c*) To compress the vapour from temperature to high temperature
    (*d*) To do all of the above

93. The reciprocating refrigerant compressors are very suitable for
   (*a*) Small displacements and high condensing pressures
   (*b*) Large displacements and high condensing pressures
   (*c*) small displacements and low condensing pressures
   (*d*) Large displacements and low condensing pressures

94. The refrigerant R-11 is very suitably for
   (*a*) Large reciprocating refrigerant compressors
   (*b*) Large centrifugal refrigerant compressors
   (*c*) Small rotary positive displacement refrigerant compressors
   (*d*) Small reciprocating refrigerant compressors

95. Centrifugal refrigerant compressors are employed for the following refrigerants
   (*a*) R-113 (*b*) R-22
   (*c*) R-717 (*d*) R-21

96. A thrust bearing is generally necessary in
   (*a*) Reciprocating refrigerant compressors
   (*b*) Centrifugal refrigerant compressors
   (*c*) Rotary vane type refrigerant compressors
   (*d*) None of the above

97. In refrigeration system having water cooled condenser, in a hot humid day
   (*a*) The head pressure will be lower
   (*b*) The head pressure will be higher
   (*c*) The head pressure of water cooled condenser is not affected by high humidity
   (*d*) The head pressure of water cooled condenser is not affected by dry bulb temperature

98. The heat rejection ratio is defined as
   (*a*) Heat absorbed at evaporator/heat rejected at condenser
   (*b*) Heat rejected at condenser/heat absorbed at evaporator
   (*c*) Heat rejected at condenser/work done on the compressor
   (*d*) Heat absorbed at evaporator/work done on compressor

99. For a fixed condensing temperature, the heat rejection ratio
   (*a*) Increases with decrease in suction temperature
   (*b*) Decreases with decrease in suction temperature
   (*c*) Remains unaffected with decreases in suction temperature
   (*d*) Is not predictable with decreases in suction temperature.

100. In shell and tube condenser
   (*a*) Water flows in the shell and refrigerant flows in the tubes
   (*b*) Water flows in the tube and refrigerant flows in the shell
   (*c*) Any one of (*a*) and (*b*) may be there (*d*) None of the above

101. In shell and coil condenser
(*a*) Water flows in the coil and refrigerant in shell
(*b*) Water flows in the shell and refrigerant in coil
(*c*) Any one of (*a*) and (*b*) may be there
(*d*) None of the above

102. For air cooled refrigeration condensers the typical values of the temperature difference between refrigerant and air is
(*a*) 3-4°C (*b*) 8-10°C
(*c*) 14-17°C (*d*) 20-30°C

103. For water cooled refrigeration condensers the typical values of the temperature difference between refrigerant and entering water is
(*a*) 3-4°C (*b*) 8-10°C
(*c*) 14-17°C (*d*) 20-30°C

104. Evaporative condenser used
(*a*) Only air as cooling medium (*b*) Water and air both as cooling media
(*c*) Only water as the cooling medium (*d*) None of the above

105. Natural convection air cooled condensers are used for
(*a*) Domestic refrigerator (*b*) Domestic air conditioners
(*c*) Both (*a*) and (*b*) (*d*) None of the above

106. For one ton refrigeration plant, if the heat rejected at the condenser is 280 kJ, heat rejection ratio is
(*a*) 2.333 (*b*) 1.333
(*c*) 1.2 (*d*) 2.2

107. For one ton refrigeration plant, if the COP of the plant is 5, the heat rejection ratio of the condenser is
(*a*) 2.2 (*b*) 1.2
(*c*) 2.333 (*d*) 1.333

108. A water cooled condenser is fitted with
(*a*) Water pressure regulator connection
(*b*) Liquid refrigerant shut off valve connection
(*c*) Purge cock connection
(*d*) Between evaporator and compressor
(*e*) All of the above

109. A window air conditioners uses
(*a*) Natural convection air cooled condenser (*b*) Shell and coil type condenser
(*c*) Forced convection air cooled condenser (*d*) Double pipe condenser

110. In double pipe condenser
(*a*) Water flows through inner pipe and refrigerant through annular outer pipe
(*b*) Refrigerant flows through inner pipe and Water through annular outer pipe
(*c*) Both (*a*) and (*b*) are used
(*d*) None of the above

111. The ambient temperature as recorded by ordinary thermometer is called
(*a*) Wet bulb temperature
(*b*) Dew point temperature
(*c*) Dry bulb temperature
(*d*) Saturation temperature

112. Standard barometric pressure is
(*a*) 1.0332 kgf/cm^2
(*b*) 760 mm Hg
(*c*) 1013.25 m bar
(*d*) All of the above

113. In a mechanical refrigeration system the highest temperature of refrigerator occurs
(*a*) Between condenser and evaporator
(*b*) In evaporator
(*c*) Before expansion valve
(*d*) between evaporator and compressor
(*e*) between compressor and condenser

114. Which of the following refrigerant is highly toxic?
(*a*) Freon 12
(*b*) Freon 22
(*c*) Ammonia
(*d*) Carbon dioxide
(*e*) Sulphur dioxide.

115. The ratio of specific humidity of moist air to the specific humidity of the saturated air at the same temperature, is known as
(*a*) Absolute humidity
(*b*) Degree of saturation
(*c*) Relative humidity
(*d*) Any of the above
(*e*) None of the above.

116. The ratio of water vapour to the mass of dry air in a given mixture of air-water vapour mixture is known as
(*a*) Humidity ratio
(*b*) Relative humidity
(*c*) Degree of saturation
(*d*) Any of the above
(*e*) None of the above.

117. The rate of evaporation of water will be low when
(*a*) Relative humidity is zero
(*b*) Relative humidity is les then 50%
(*c*) Relative humidity is high
(*d*) Rate of evaporation does not depend on relative humidity.

118. The ratio of sensible heat to total heat is known as
(*a*) Dew point
(*b*) Humidification ratio
(*c*) Relative humidity
(*d*) Specific humidity
(*e*) Sensible heat factor.

119. When moisture is added to air at constant dry bulb temperature, the process is known as
(a) Humidification (b) Sensible cooling
(c) Wetting (d) Chilling
(e) None of the above.

120. When moisture is removed from air at constant dry bulb temperature, the process is known as
(a) Aeration (b) Deionization
(c) Dehumidification (d) Drying
(e) Sensible cooling.

121. The difference between wet and dry bulb temperature
(a) Increases at atmospheric temperature (b) Increases as air gets wet
(c) Increases as air gets drier (d) Remains constant
(e) None of the above.

Questions 122 to 125 refer to Fig. 1

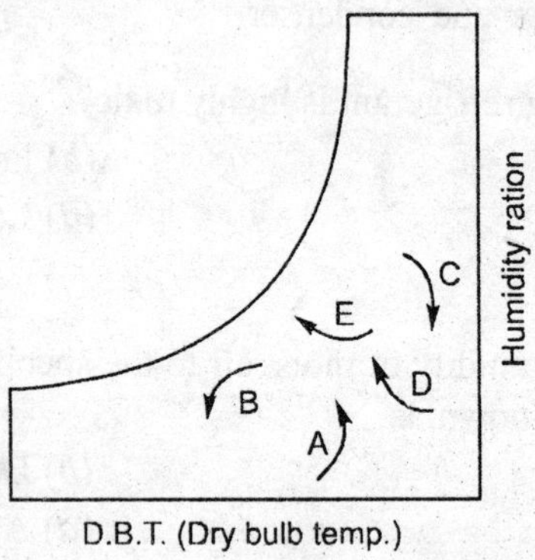

Fig. 1

122. Curve *A* represents
(a) Cooling process (b) Cooling with dehumidification
(c) Heating with humidification (d) Heating with dehumidification
(e) None of the above.

123. Sensible cooling process is represented by
(a) Curve *B* (b) Curve *C*
(c) Curve *D* (d) Curve *E*
(e) None of the above

124. Curve *B* represents
(a) Cooling and dehumidification (b) Sensible heating
(c) Adiabatic mixing of to streams (d) Sensible cooling
(e) None of the above.

125. Curve C represents
   (*a*) Moisture addition by streams
   (*b*) Moisture addition by water
   (*c*) Heating with electric heater
   (*d*) Chemical dehumidification
   (*e*) None of the above.

Questions 126 to 128 refer to Fig. 2

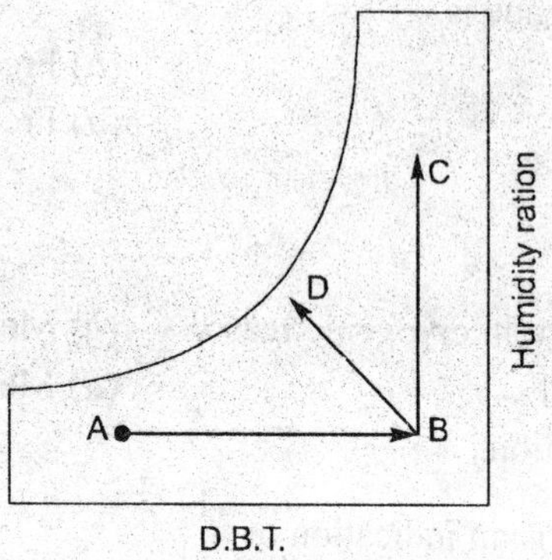

Fig. 2

126. The process *AB* represents
   (*a*) Sensible cooling
   (*b*) Sensible heating
   (*c*) Dehumidification
   (*d*) Humidification
   (*e*) None of the above

127. The process *BC* represents
   (*a*) Sensible cooling
   (*b*) Sensible heating
   (*c*) Dehumidification
   (*d*) Humidification
   (*e*) None of the above.

128. The process *BD* represents
   (*a*) Heating by an electric heater
   (*b*) Chemical dehumidification
   (*c*) Adiabatic cooling
   (*d*) Adiabatic heating
   (*e*) Humidification by water spray

129. If wet bulb temperature is same as dew point, the relative humidity of air will be
   (*a*) 0
   (*b*) 25
   (*c*) 50
   (*d*) 67
   (*e*) 100

130. Dew point is
   (*a*) Percentage of moisture contained in a sample of air
   (*b*) Partial pressure of water vapour in air starts
   (*c*) The temperature of water vapour in air starts
   (*d*) The temperature at which condensation of water vapour in air starts
   (*e*) The temperature at which ice starts melting in presence of air without external aid.

131. For the measurement of temperature in cryogenic range the thermometers used are
(*a*) Pyrometers
(*b*) Radiation pyrometers
(*c*) Mercury thermometers
(*d*) Gas thermometers
(*e*) Electro digital instrument.

132. Dichloromono fluoroethane is
(*a*) Freon
(*b*) Freon 12
(*c*) Freon 21
(*d*) Freon 22
(*e*) None of the above

133. A salinometer
(*a*) Measure temperature in cryogenic reason
(*b*) Measure intensity of illumination
(*c*) Measure sound level
(*d*) Measure density of brine
(*e*) Measure solar radiation.

134. Dew point temperature is an indication of
(*a*) Moisture content of air
(*b*) Dryness of air
(*c*) Lowest temperature where air and water vapour starts separating
(*d*) Condition air
(*e*) Any of the above

135. During dehumidification process
(*a*) Specific humidity increases
(*b*) Specific humidity decreases
(*c*) Relative humidity increases
(*d*) Relative humidity decreases
(*e*) Relative humidity dose not change.

136. Under increased atmospheric pressure (as compared to the standard conditions) the steam will condense at
(*a*) More then 100° C
(*b*) 100° C
(*c*) Less then 100° C
(*d*) Has nothing to do with pressure
(*e*) Depends on other conditions.

137. Dalton's law of partial pressure state that
(*a*) All gases exert partial pressures only
(*b*) All gases or mixture of gases exert partial pressures
(*c*) All gases or mixture of gases in a container will exert their own pressure on the container entirely in depends on the others
(*d*) Total pressure exert by a mixture of gases or vapour is equal to the pressure exerted by individual gases or vapour
(*e*) None of the above

138. Anhydrous substances is
   (*a*) Free of hydrogen
   (*b*) Free of hydrogen and other chemicals containing hydrogen
   (*c*) Free of water especially water of crystallization
   (*d*) Free of humidity
   (*e*) Dry substances

139. Dehumidification means
   (*a*) Removing humidity from a space
   (*b*) Reducing the quantity of water vapour in a space
   (*c*) Reducing of moisture content of a substance
   (*d*) Any of the above
   (*e*) None of the above

140. The type of compressor used in refrigeration system is
   (*a*) Reciprocating
   (*b*) Centrifugal
   (*c*) Rotary sliding vane
   (*d*) Rotary screw compressor
   (*e*) All above.

141. The rate of evaporation of water will be low when
   (*a*) Relative humidity is zero
   (*b*) Relative humidity is less than 50%
   (*c*) Relative humidity is high
   (*d*) Rate evaporation does not depend on relative humidity

142. The ratio of sensible heat to total heat is known as
   (*a*) Dew point
   (*b*) Humidification
   (*c*) Relative humidity
   (*d*) Specific humidity
   (*e*) Sensible heat factor

143. When moisture is removed from air at constant dry bulb temperature, the process is known as
   (*a*) Aeration
   (*b*) Deionization
   (*c*) Dehumidification
   (*d*) Drying
   (*e*) Sensible cooling

144. When moisture is removed from air at constant dry bulb temperature, the process is known as
   (*a*) Humidification
   (*b*) Sensible cooling
   (*c*) Wetting
   (*d*) chilling

145. In a refrigerator evaporator is located
   (*a*) On the bottom of refrigerator cabinet
   (*b*) Behind the refrigerator cabinet
   (*c*) Adjacent to compressor
   (*d*) Inside the refrigerator cabinet
   (*e*) None of the above

146. The effect of under cooling the refrigerant is to
(*a*) Reduce the COP of the cycle
(*b*) Increase the COP of the cycle
(*c*) Reduce the refrigerating effect
(*d*) Increase the super heat the vapour
(*e*) None of the above.

147. White ice is
(*a*) Sub cooled water
(*b*) Fast cooled water
(*c*) Due to dissolve air gases and impurities
(*d*) Produced by blowing air during freezing
(*e*) None of the above.

148. Clear ice
(*a*) Is pure ice
(*b*) Contains dissolved air
(*c*) Contains dissolved gases
(*d*) Is produced by blowing air during freezing
(*e*) None of the above.

149. Dry ice is
(*a*) Free from water
(*b*) Free from dissolved air gases
(*c*) Dose not contain impurities
(*d*) Solidified form of carbon dioxide
(*e*) None of the above.

150. A cooling tower in a refrigeration system
(*a*) Removes heat from refrigerant
(*b*) Removes heat from brine
(*c*) Removes heat from lubricating oil
(*d*) Removes heat from hot water coming out of condenser
(*e*) None of the above.

151. At 100% humidity
(*a*) Wet bulb temperature is same as atmospheric temperature
(*b*) Dry bulb temperature is same as atmospheric temperature
(*c*) Wet bulb temperature is same as dry bulb temperature
(*d*) Wet bulb temperature is exactly half the dry bulb temperature
(*e*) Dry bulb temperature is exactly half the wet bulb temperature.

152. The difference between wet and dry bulb temperature
(*a*) Increases at atmospheric temperature
(*b*) Increases as air gets wet
(*c*) Increases as air gets drier
(*d*) Remains constant always
(*e*) None of the above.

153. A low bulb temperature indicates
(*a*) Very cool air
(*b*) Very low humidity
(*c*) Very low atmospheric temperature
(*d*) (*a*) and (*b*)
(*e*) (*a*) and (*c*)

154. Capacity of a domestic refrigerator is in the range
(*a*) 0.1to 0.3 tonnes
(*b*) 1 to 3 tonnes
(*c*) 3 to 5 tonnes
(*d*) 5 to 7 tonnes
(*e*) 7 to 10 tonnes.

155. The COP of a domestic refrigerator as compared to that of an air conditioner is generally
(*a*) More
(*b*) Same
(*c*) Less
(*d*) Air conditioner depends on whether conditions
(*e*) Cannot be compared.

156. Presence of moisture in a refrigeration cycle will show its effect at
(*a*) Compressor section
(*b*) Compressor discharge
(*c*) Expansion valve
(*d*) Condenser
(*e*) All of the above.

157. If a refrigerator is directly exposed to sun rays
(*a*) It will stop working
(*b*) The refrigerant will leak
(*c*) There will be less cooling effect
(*d*) Frost will be formed on cabinet surface
(*e*) (*a*) and (*b*) above

158. The secondary refrigerant used in milk chilling plants is usually
(*a*) Brine
(*b*) Ammonia solution
(*c*) Glycol
(*d*) Sodium silicate
(*e*) Any of the above

159. Which type of compressor is commonly used in room air-conditioners?
(*a*) Reciprocating
(*b*) Rotary
(*c*) Screw
(*d*) Centrifugal
(*e*) None of the above

160. Refrigerant No. R-717 is
(*a*) Ammonia
(*b*) Freon 12
(*c*) Freon 22
(*d*) Methyl chloride
(*e*) Carbon dioxide.

161. In case, air is cooled by a process in which the humidity ratio of air does not change, the process is called
(*a*) Humidification
(*b*) Dehumidification
(*c*) Ventilation
(*d*) Chilling
(*e*) Sensible cooling

162. Brine is always used as a secondary refrigerant in
 (*a*) Cold storage plant milk chilling plant
 (*b*) Cinema hall air conditioning plant
 (*c*) Ice plant
 (*d*) All of the above.

163. One tonne of refrigeration is equivalent to SI units of
 (*a*) 1kW
 (*b*) 2.5 kW
 (*c*) 3.5 kW
 (*d*) 5 kW
 (*e*) 10kW

164. The freezing point of ammonia is
 (*a*) 0° C
 (*b*) – 20° C
 (*c*) – 77.8° C
 (*d*) – 132.5° C
 (*e*) – 202.5° C.

165. Which of the following refrigerant is the cheapest?
 (*a*) Ammonia
 (*b*) Freon 12
 (*c*) Freon 22
 (*d*) Methyl chloride
 (*e*) Air

ANSWER KEY (REFRIGERATION AND AIR CONDITIONING)

1.E	2. B	3.C	4.C	5.B	6.D	7.A	8.C	9.E	10.E
11.D	12.A	13.C	14.E	15.D	16.C	17.C	18.B	19.D	20.E
21.E	22.E	23.B	24.B	25.D	26.B	27.B	28.A	29.B	30.D
31.C	32.A	33.C	34.C	35.D	36.A	37.A	38.B	39.A	40.C
41.C	42.B	43.C	44.A	45.C	46.B	47.C	48.B	49.C	50.B
51.B	52.A	53.B	54.A	55.B	56.A	57.C	58.A	59.B	60.B
61.D	62.D	63.B	64.A	65.B	66.B	67.C	68.C	69.B	70.A
71.B	72.A	73.A	74.D	75.D	76.A	77.A	78.B	79.B	80.A
81.A	82.A	83.A	84.D	85.B	86.A	87.C	88.C	89.B	90.B
91.D	92.D	93.A	94.C	95.A	96.B	97.B	98.B	99.A	100.B
101.A	102.C	103.B	104.C	105.A	106.B	107.B	108.E	109.B	110.A
111.C	112.D	113.E	114.E	115. B	116. A	117.C	118.E	119.A	120.C
121.C	122.C	123.D	124.A	125.D	126.B	127.D	128.C	129.E	130.D
131.D	132.D	133.D	134.A	135.D	136.A	137.C	138.C	139.C	140.E
141.C	142.E	143.C	144.A	145.D	146.B	147.C	148.D	149.D	150.D
151.C	152.C	153.D	154.A	155.C	156.C	157.C	158.C	159.A	160.A
161.E	162.C	163.C	164.C	165.E					

Computer and Statistics

1. _________________ is the product of data processing.
   (*a*) Data
   (*b*) Information
   (*c*) Software
   (*d*) Computer
   (*e*) None Of These

2. The CPU (Central Processing Unit) consists of :
   (*a*) Input, Output and Processing
   (*b*) Control unit, Primary storage and Secondary storage
   (*c*) Primary storage, Arithmetic-Logical unit and Control unit
   (*d*) Input Processing and Storage
   (*e*) None of the above

3. Advances in computer hardware and software are generally classified into generations. We are currently in which generation.
   (*a*) First
   (*b*) Second
   (*c*) Third
   (*d*) Fourth
   (*e*) Fifth

4. Which of the following piece of hardware is used the most in the input phase of a computer based information system?
   (*a*) Printer
   (*b*) Diskette
   (*c*) Monitor
   (*d*) Keyboard
   (*e*) Main memory

5. Monitor is an _______________ device.
   (*a*) Input
   (*b*) Output
   (*c*) Input output (I/O)
   (*d*) Processing
   (*e*) None of these

6. Compared with secondary storage, primary storage is :
   (*a*) Slow and inexpensive
   (*b*) Fast and inexpensive
   (*c*) Fast and expensive
   (*d*) Slow and expensive
   (*e*) None of the above

7. Software intended to satisfy a user's specification processing needs is called ______________.
   (*a*) System software
   (*b*) Utility software
   (*c*) Operating software
   (*d*) Application software
   (*e*) All of the above

8. Which of the following is commonly used in academic testing?
   (*a*) MICR
   (*b*) POS
   (*c*) OCR
   (*d*) OMR
   (*e*) CRT

9. __________ is a non impact printer, which can produce high quality, letter-perfect printing.
   (*a*) Dot matrix printer
   (*b*) Daisy wheal printer
   (*c*) Line printer
   (*d*) Ink jet printer
   (*e*) Laser printer

10. ROM stands for __________.
   (*a*) Read only method
   (*b*) Read on memory
   (*c*) Read only memory
   (*d*) Remember only memory
   (*e*) None of these

11. ICON stands for __________.
   (*a*) Integrated circuit of networks
   (*b*) Image creation on network
   (*c*) Image on computer network
   (*d*) Images on CONsole
   (*e*) None of these

12. MICR stands for __________.
   (*a*) Many Inks Character Recognition
   (*b*) Multiple Inks Code Recognition
   (*c*) Magnetic Ink Code Recognition
   (*d*) Magnetic Ink Character Recognition
   (*e*) None of these

13. DVD stands for __________.
   (*a*) Direct Video Disk
   (*b*) Digital Video Disk
   (*c*) Digital Versatile Disk
   (*d*) Developed Video Disk
   (*e*) None of these

14. CASE stands for
   (*a*) Computer Aided Software Engineering
   (*b*) Computer Applications Software Engineering
   (*c*) Computer Architecture Software Engineering
   (*d*) Computer Applied Software Engineering
   (*e*) None of these

15. Nibble is group of __________ bytes.
    (*a*) 2 (*b*) 4
    (*c*) 6 (*d*) 8
    (*e*) None of these

16. One GB is equal to __________ KB.
    (*a*) 1000 (*b*) 1024
    (*c*) 1000000 (*d*) 1048576
    (*e*) None of these

17. What is the storage capacity of a CD-ROM
    (*a*) 500 MB (*b*) 500 GB
    (*c*) 700 MB (*d*) 700 GB
    (*e*) None of these

18. Which of the following is not GUI based software?
    (*a*) Windows (*b*) UNIX
    (*c*) MS Office (*d*) Corel Draw
    (*e*) PageMaker

19. The number of digits used in a system is known as its
    (*a*) Power (*b*) Weights
    (*c*) Radix (*d*) Base
    (*e*) Either (*c*) or (*d*)

20. __________ number system is used by the computer systems.
    (*a*) Decimal (*b*) Binary
    (*c*) Octal (*d*) Hexadecimal
    (*e*) None of these

21. Computer has two type of memory
    (*a*) Internal and external (*b*) RAM and ROM
    (*c*) Floppy and CD (*d*) CD and Hard disk

22. Cache memory is logically placed between the
    (*a*) CPU register and Primary memory (*b*) Primary memory and secondary memory
    (*c*) RAM and ROM (*d*) RAM and Hard disk

23. Four bits is equivalent to
    (*a*) 1 Byte (*b*) 1 Nibble
    (*c*) 1 character (*d*) None

24. One byte is equal to
    (*a*) 4 bits (*b*) 8 bits
    (*c*) 1024 bits (*d*) 16 bits

25. The capacity of compute memory is measured in
(*a*) KB (*b*) MB
(*c*) GB (*d*) All of the above

26. The storage capacity of the 3 ½" floppy disk is about
(*a*) 1.44 MB (*b*) 2.44 MB
(*c*) 700 MB (*d*) 80 MB

27. Secondary memory of computer is also known as
(*a*) RAM (*b*) ROM
(*c*) Internal memory (*d*) External Memory

28. A disk is made of circular platter coated with a magnetisable material
(*a*) Magnetic disk (*b*) RAM
(*c*) CD (*d*) DVD

29. The size of macro floppy is
(*a*) 8" (*b*) 5¼"
(*c*) 3½" (*d*) 8½"

30. A power point presentation is saved with a file name extension of
(*a*) .prt (*b*) .pst
(*c*) .ttp (*d*) .ppt

31. A collection of interrelated relevant data which is stored together to serve multiple applications is called
(*a*) Database (*b*) Memory
(*c*) Hyperlink (*d*) Hard disk

32. A collection of software packages for processing and managing a database is called
(*a*) Database management system (DBMS) (*b*) Memory
(*c*) Hyperlink (*d*) Hard disk

33. Main feature of the FOXPRO package is as
(*a*) It provides application wizard and builder
(*b*) It provides component gallery, OLE drag and drop
(*c*) It support web
(*d*) All of the above

34. An unauthorized use of computer and network resource is called
(*a*) Hacking (*b*) Hijacking
(*c*) Chatting (*d*) Cheating

35. The ratio of the sum of the observational values to its number of observations presented in a given series or distribution is defined as
(*a*) Arithmetic mean (*b*) Median
(*c*) Mode (*d*) Frequency

36. If the observation are arranged in ascending or descending order of magnitude array the value of centrally located observation is considered as
(*a*) Arithmetic mean (*b*) Median
(*c*) Mode (*d*) Frequency

37. The value of variate, which occurs most frequently, is called
(*a*) Arithmetic mean (*b*) Median
(*c*) Mode (*d*) Frequency

38. The positive square root of the variance is defined as
(*a*) Arithmetic mean (*b*) Median
(*c*) Mode (*d*) Standard deviation

39. The ratio of the sum of the square of deviation measured from the arithmetic mean of a given series to the number of observations minus one or degree of freedom is called
(*a*) Variance (*b*) Standard deviation
(*c*) Mean deviation (*d*) Geometric mean

40. Each observation of a given set of data is to be placed against the different classes according to magnitude by horizontal or vertical or cross line which is known as
(*a*) Mode (*b*) Range
(*c*) Frequency (*d*) Tally mark

41. The manner in which frequencies are distributed over the different classes is considered as
(*a*) Frequency distribution (*b*) Range
(*c*) Frequency table (*d*) Tally mark

42. A class of a distribution having a maximum frequency is defined as
(*a*) Mode class (*b*) Mode
(*c*) Median (*d*) Range

43. The ratio of number of observations to sum of the reciprocal of the value of different observations is called
(*a*) Arithmetic mean (*b*) Geometric mean
(*c*) Harmonic mean (*d*) Weighted mean

44. If all the items are not of equal importance in the calculation of mean at that time they are given proper weights according to their relative importance and the average (mean) which is calculated on the basis of these weights is called
(*a*) Arithmetic mean (*b*) Geometric mean
(*c*) Harmonic mean (*d*) Weighted mean

45. The difference between maximum value and the minimum value of the given distribution is called
(*a*) Mode class (*b*) Mode
(*c*) Median (*d*) Range

46. The ratio of the sum of the absolute deviation measured from the arithmetic mean of a given series to the number of observations present in a given mean deviation.
(*a*) Standard deviation (*b*) Mean deviation
(*c*) Absolute mean deviation (*d*) Variance

47. The value lying half way between upper and lower limit of a given class is called
(*a*) Mid value or class mid point (*b*) Mode
(*c*) Median (*d*) Range

48. The number of observations included in a given class or number of times an observation is repeated is called
(*a*) Mode (*b*) Range
(*c*) Frequency (*d*) Tally mark

49. When the principles of statistics are applied to living organisms or things like age of cow or height of plant, the science is called
(*a*) Biometry (*b*) Biostatistics
(*c*) Biology (*d*) Applied statistics

50. The difference between upper limit and lower limit of given class is called
(*a*) Range (*b*) Class interval
(*c*) Mid value (*d*) Mode class

51. The groups which are to be formulated on the basis of similarity or individual are termed as
(*a*) Class (*b*) Classification
(*c*) Class interval (*d*) Tabulation

52. The process by which the classified data are presented in an orderly manner by being placed in proper rows and columns of a table in order to bring out its chief characteristics and essential features is called
(*a*) Class (*b*) Classification
(*c*) Class interval (*d*) Tabulation

53. The process by which the data are arranged in groups or classes according to resembleness or similarities is termed as
(*a*) Class (*b*) Classification
(*c*) Class interval (*d*) Tabulation

54. A statistical method of process which is to be utilized for testing the difference under study is significant or not, is called as
(*a*) Co-relation coefficient (*b*) Test of significance
(*c*) Level of significance (*d*) Regression

55. The measure of degree of association or relation between variables is termed as
(*a*) Co-relation coefficient (*b*) Test of significance
(*c*) Level of significance (*d*) Regression

56. The ratio between the estimated values to the parameter value of the population to the standard error mean of the population is considered as
(*a*) Z-test (*b*) Test of significance
(*c*) Level of significance (*d*) Regression

57. It is committed by rejecting hypothesis on the basis of samples evidences it is true in reality and denoted by α %.
(*a*) Type-I error (*b*) Type-II error
(*c*) Standard error (*d*) Co-relation

58. It is committed by accepting null hypothesis on the basis of sample when it is false in reality and denoted by β.
(*a*) Type-I error (*b*) Type-II error
(*c*) Standard error (*d*) Co-relation

59. The maximum probability of rejection the hypothesis when it is true. Maximum probability of type-I error is known as
(*a*) Z-test (*b*) Test of significance
(*c*) Level of significance (*d*) Regression

60. An average change in the dependent variable without change in the independent one is defined as
(*a*) Z-test (*b*) Test of significance
(*c*) Level of significance (*d*) Regression coefficient

61. It is the measure of the relation between two or more variables in the original units of the data.
(*a*) Z-test (*b*) Test of significance
(*c*) Level of significance (*d*) Regression

62. The ratio of the difference between the sample mean and hypothetical or population mean to the standard error of which is to be worked out from the information.
(*a*) t-test (*b*) F-test
(*c*) Z-test (*d*) Chi square test

63. A ratio of bigger variance to smaller variance is called as
(*a*) t-test (*b*) F-test
(*c*) Z-test (*d*) Chi square test

64. The sum of the ratio of square of difference between observed frequency and expected frequency of the respective.
(*a*) t-test (*b*) F-test
(*c*) Z-test (*d*) Chi square test

65. The number of observation minus independent restriction which is to be imposed is defined as
(*a*) Degree of freedom (*b*) Limit
(*c*) Regression (*d*) Correlation

66. It is the standard deviation of sampling distribution of means.
(*a*) Degree of freedom
(*b*) Standard error of means
(*c*) Regression
(*d*) Correlation

67. It is the defined as the ration of standard deviation to the mean of a given series expressed into percentage.
(*a*) Degree of freedom
(*b*) Standard error of means
(*c*) Regression
(*d*) Coefficient of variance

68. It is the statement of equality ($H_o : A = B$) or it is hypothesis which is tested under assumption.
(*a*) Degree of freedom
(*b*) Standard error of means
(*c*) Null hypothesis
(*d*) Coefficient of variance

69. Any statement is to be made in respect to the parameter is termed as
(*a*) Degree of freedom
(*b*) Hypothesis
(*c*) Null hypothesis
(*d*) Hypothetical population

70. It is departure from symmetry.
(*a*) Kurtosis
(*b*) Skewness
(*c*) Moment
(*d*) Coefficient of skewness

71. The ratio between the numbers of favourable cases to the total number of equally likely cases is termed as
(*a*) Kurtosis
(*b*) Skewness
(*c*) Moment
(*d*) Probability

72. The measure of the degree of the departure from the symmetry (asymmetry) is known as
(*a*) Kurtosis
(*b*) Skewness
(*c*) Moment
(*d*) Coefficient of skewness

73. The number of individuals in a sample are called
(*a*) Sampling
(*b*) Population
(*c*) Sample size
(*d*) Random sample

74. Any quantity which is to be worked by using the information of entire population is called
(*a*) Sampling
(*b*) Population
(*c*) Sample size
(*d*) Random sample

75. In statistics, a difference between population parameter and sample values is referred as
(*a*) Sampling
(*b*) Population
(*c*) Standard error
(*d*) Error

76. A part of population under study is known as
(*a*) Sample
(*b*) Population
(*c*) Sample size
(*d*) Random sample

77. It is the process or method of selecting the individuals or sample from population is known as
(*a*) Sample size (*b*) Population
(*c*) Sampling (*d*) Random sample

78. Every individual of the population has an equal and independent chance of being selected or chosen for sample is called
(*a*) Sampling (*b*) Population
(*c*) Sample size (*d*) Random sample

79. The first step of summarizing the data is
(*a*) Tabulation (*b*) Graphical representation
(*c*) Classification (*d*) Collection

80. Which of the following statements for classification of data is correct?
(*a*) It is always necessary
(*b*) It is only necessary when data corresponds to discrete variable
(*c*) It is only necessary when, there is voluminous data
(*d*) None of the above

81. We study measures of central tendency
(*a*) To study central values (*b*) To study arithmetic mean
(*c*) To calculate standard deviation
(*d*) To represent the whole data by only a single value

82. The formula for G.M. on n values $X_1, X_2 \ldots X_n$ is
(*a*) $(X_1, X_2 \ldots X_n)^n$ (*b*) $(X_1, X_2 \ldots X_n)^{4/9}$
(*c*) $(X_1, X_2 \ldots X_n)^{1/n}$ (*d*) $(X_1, X_2 \ldots X_n)^N$

83. The measure of central tendency to be used to study average rate of change in the population is
(*a*) A.M. (*b*) Mode
(*c*) G.M. (*d*) Median

84. To find out the average of quantity prices, the measure of central tendency to be used is
(*a*) A.M. (*b*) Median
(*c*) G.M. (*d*) Mode

85. To find the average size of shoes sold in the market, we should use
(*a*) A.M. (*b*) Median
(*c*) Mode (*d*) H.M.

86. To find the average speed of a vehicle when same distance is covered with different speeds, the suitable measures is
(*a*) A.M. (*b*) H.M.
(*c*) G.M. (*d*) None of these

87. The suitable measure to find average speed when time for each speed is fixed would be
(*a*) A.M. (*b*) H.M.
(*c*) G.M. (*d*) None of these

88. The A.M. of age of 20 students in a class is 15 years. If we include the age of class teacher also, then the mean the increase by one year. The age of class teacher would be
(*a*) 16 years (*b*) 21 years
(*c*) 36 years (*d*) 25 years

89. The range of the value –6, –8, –2, –12, –20 will be
(*a*) +18 (*b*) –18
(*c*) –26 (*d*) –14

90. The standard deviation fro the values 4, 5, 6, 7, 8, will be
(*a*) $(5)^{1/2}$ (*b*) $(2)^{1/2}$
(*c*) (2) (*d*) $(2.5)^{1/2}$

91. The relation between A.M., G.M., and H.M. is
(*a*) AM < HM > GM (*b*) AM > HM > GM
(*c*) AM > HM < GM (*d*) AM < HM < GM

92. Coefficient of variation is calculated by the formula
(*a*) (Mean × 100)/S.D. (*b*) (S.D. × 100)/ Mean
(*c*) S.D. / Mean (*d*) Mean / S.D.

93. If all the variate values are negative, the S.D. will be
(*a*) Positive (*b*) Negative
(*c*) No definite sign (*d*) Impossible to calculate

94. In food industries for quality control, the most important measure of dispersion is
(*a*) Standard deviation (*b*) Mean deviation
(*c*) Range (*d*) Coefficient of variance

95. The student's '*t*' test was discovered by
(*a*) R.A. Fisher (*b*) F. Yates
(*c*) P.C. Mahalanabis (*d*) W.S. Gosset

96. For testing the hypothesis about mean of one population we use t-test when
(*a*) Sample size is large and population S.D. is known
(*b*) Sample size is small and population S.D. is unknown
(*c*) Sample size is small and population S.D. is known
(*d*) Sample size is large and population S.D. is unknown

97. For testing the significance of correlation coefficient, we use
(*a*) Chi square test
(*b*) F-test
(*c*) S.N.D. test
(*d*) 't' – test

98. For testing the agreement between observed and expected frequencies we use
(*a*) '*t*' – test
(*b*) S.N.D. test
(*c*) Chi square test
(*d*) F-test

99. The minimum sample size for using chi square test should be
(*a*) 100
(*b*) 30
(*c*) 50
(*d*) 5

100. The analysis of variance technique is used for
(*a*) Comparing the means of two population
(*b*) Comparing the means of more than two population
(*c*) Comparing the variance of more than two population
(*d*) Comparing the variance of two population

101. The degree of freedom for error in R.B. design with 10 treatments and 4 replications will be
(*a*) 30
(*b*) 27
(*c*) 36
(*d*) 40

102. Latin square design is suitable for comparing
(*a*) 5 to 12 treatments
(*b*) 3 treatments
(*c*) 15 treatments
(*d*) 2 to 4 treatments

103. In field experiments the commonly used design is
(*a*) CRD
(*b*) RBD
(*c*) LSD
(*d*) None

104. The relation between mean deviation and standard deviation is
(*a*) M.D. is greater than S.D.
(*b*) M.D. is less than S.D.
(*c*) M.D. is equal to 2/3 S.D.
(*d*) M.D. equal to S.D.

105. Under the assumptions for analysis of variance, the parent population should be
(*a*) Binomial
(*b*) Normal
(*c*) Poisson
(*d*) Any type

106. For comparing 4 treatments with RBD, the 15 d.f. error can be obtained, when the number of replications are
(*a*) 5
(*b*) 6
(*c*) 11
(*d*) 12

107. To reduce the experimental error with heterogeneous material, we need
(*a*) More replications
(*b*) Use of randomization
(*c*) Use of local control techniques
(*d*) More replications and use of local control techniques

108. For any two values a and b , the following relationship among arithmetic mean (A), Geometric mean (G) and Harmonic mean (H) exists
(*a*) $A = H.G^2$ (*b*) $G = A.H.$
(*c*) $G^2 = A.H.$ (*d*) $A = G.H.$

109. The following relationship among the correlation coefficient (r) and the two regression coefficient b_{yx} and b_{xy} exists
(*a*) $r^2 = b_{yx}.b_{xy}$ (*b*) $r = (b_{yx}.b_{xy})^{1/2}$
(*c*) $r = b_{yx} / b_{xy}$ (*d*) none of these

110. The degree of freedom to test, the significance of difference between two mean based on n_1 and n_2 observation is
(*a*) $n_1 + n_2$ (*b*) $n_1 = n_2 - 1$
(*c*) $n_1 + n_2 - 2$
(*d*) none of these

111. The most commonly used measure of central tendency is
(*a*) Harmonic mean (*b*) Mode
(*c*) Geometric mean (*d*) Arithmetic mean

112. Which of the following is considered as best measure of dispersion?
(*a*) Range (*b*) Standard deviation
(*c*) Mean deviation (*d*) Quartile deviation

113. The correlation coefficient lies between
(*a*) 0 to +1 (*b*) –1 to 0
(*c*) –1 to +1 (*d*) 0 to 1.5

114. The regression coefficient lies between
(*a*) –1 to +1 (*b*) –0 to 8
(*c*) 0 to –1 (*d*) None of above

115. For testing the independence of two attribute, the test used is
(*a*) '*t*' – test (*b*) S.D. test
(*c*) χ^2 - test (*d*) F-test

116. Correlation of continuity in 2 × 2 contingency table should be used when expected frequency of any cell is
(*a*) Greater than 5 (*b*) Less than 5
(*c*) Less than 10 (*d*) None of the above

117. The value of χ^2 always lies between
(*a*) 0 to 8 (*b*) 0.00 to 00
(*c*) 0 to +1 (*d*) –1 to 0

118. When individual in a same have two characters and a frequency distributes to be made by classifying them on the basis of two the characters so as to show the relation between the two characters, the resulted termed as
(*a*) Attributes (*b*) Bivariate population
(*c*) Contingency table (*d*) Tabulation

119. A characteristics of population which can not be measured quantitatively but described qualitatively *e.g.* honesty, morality, skill, sex is called
(*a*) Sample (*b*) Attributes
(*c*) Population (*d*) Class

120. FORTRAN is a problem oriented ______________ programming language.
(*a*) Computer (*b*) CPU
(*c*) UPS (*d*) Hard disk

ANSWER KEY (COMPUTER AND STATISTICS)

1.B	2. C	3.D	4.D	5.C	6.C	7.D	8.D	9.E	10.C
11. D	12.D	13.C	14.A	15.B	16.D	17.C	18.B	19.E	20.B
21.A	22.A	23.B	24.B	25.D	26.A	27.D	28.A	29.A	30.D
31.A	32.A	33.D	34.A	35.A	36.B	37.C	38.D	39.A	40.D
41.A	42.A	43.C	44.D	45.D	46.C	47.A	48.C	49.A	50.B
51.A	52.D	53.B	54.B	55.A	56.A	57.A	58.B	59.C	60.D
61.D	62.A	63.B	64.D	65.A	66.B	67.D	68.C	69.B	70.B
71.B	72.D	73.C	74.B	75.D	76.A	77.C	78.D	79.C	80.D
81.D	82.C	83.C	84.D	85.A	86.B	87.A	88.C	89.A	90.B
91.B	92.B	93.A	94.C	95.D	96.C	97.D	98.C	99.C	100.B
101.B	102.A	103.B	104.C	105.B	106.B	107.D	108.C	109.A	110.C
111.D	112.B	113.C	114.B	115.C	116.B	117.A	118.C	119.B	120.A

Short Notes/Terminology

- **Engineering:** The business of an engineer or the art of making and using the machinery in optimum way.
- **Technology:** The science of the industrial art.
- **Agricultural Engineering:** Application of principles, knowledge and techniques of one or all branches of engineering in finding out the solution to the problem in the field of agriculture for increasing agricultural productivity, increasing the standard of living of farmers and increasing the overall earnings per worker.
- **Agricultural Processing:** Agricultural processing may be defined as an activity which is performed to maintain or improve the quality or to change the form or characteristics of an agricultural product. Processing operations are undertaken to add value to the agricultural materials after their production. The main purpose of agricultural processing is to minimize the qualitative and quantitative deterioration of the material after harvesting. Some of the agricultural processing operations performed on various crops (cereals, pulses, and oilseeds) are cleaning, grading, sorting, drying, cooling, storage, milling, size reduction, expelling, mixing blending, packaging, waste utilization, seed treatments etc.

  Agricultural processing includes those operations, which maintain or raises the quality of raw material, change its form, or prepare it for market. Its importance is growing and will continue to increase as farming became more commercialized, specialized and mechanized in India. Some of the advantage are:

  * Produce, which is at present wasted because of inadequate transport facilities and excess supply can be preserved and used later.
  * The processing may be serving on a small-scale industry, providing additional employment facilities for idle labour.
  * Perishable products can be changed to a staple commodity, *e.g.* canning of fruits, drying of fruits and fodder making of silage, etc.
  * Product can be prevented from being affected by insects and pests during storage, rendering them completely unusable.
- **Food Processing:** Food processing covers general characteristics of raw material, principle of food preservation, processing factor, which affect quality, packaging, water and waste management and good manufacturing practices and sanitation procedure.
- **Food Engineering:** It involves study of engineering concept and unit operation used in food processing. Engineering principle should include materials, energy balance, thermodynamic,

fluid flow, heat and mass transfer, prerequisites should be one course in physics and two in calculus.

- **Abrasivity:** The property of particles of bulk materials to wear away the surface they are contact with when in motion.
- **Abrasive polisher:** The whitening machine used to remove the bran from the kernel using the principle of abrasive action between the kernel and the emery coated stone.
- **Aeration or Ventilation:** It mean blowing small quantity of air through the bulk of grain for cooling and equalizing the grain temperature and to prevent moisture migration and for preserving the grain from biological deterioration.
- **Aleurone layer:** It is the peripheral layer of the endosperm, one to several cells thick containing oil and protein but no starch.
- **Angle of Internal friction:** It is the angle whose tangent is the coefficient of the internal friction between particles and which reaches the limit a right angle. The angle is shown when part of the pile of non-free flowing materials, the angle of internal friction and the angle of repose are not synonymous because of the pressure of interlocking and cohesive forces.
- **Angle of repose:** The angle of repose is the angle between the base and the slope of the cone formed on a free vertical fall of the granular material to a horizontal plane.

  The size, shape, moisture content and orientation of the grain affect the angle of repose, there are two angles of repose

  (*i*) static angle of repose

  (*ii*) dynamic angle of repose
- **Static angle of repose:** It is the angle of friction taken up by granular material to just slide upon itself.
- **Dynamic angle of repose:** It comes in a picture when bulk of grain is in motion like discharge of grain from bins and hoppers. The dynamic angle of repose is more important than static angle.

  The angle of repose for a few important grains are :

Grain	Terminal velocity (m/s)	Angle of repose (degree)
Wheat	9–11.5	23-28°
Paddy	-	30-45°
Corn	34.9	30-40°
Barley	8.5–10.5	28-40°
Oats	8-9	31-44°
Millets	-	20-25°
Rye	8.5–10.5	23-28°

- **Engineering Properties of Agricultural Materials:** Engineering properties are the properties, which are useful and necessary in the design and operation of various equipments employed in the field of agricultural processing and also for design and development of other farm machinery. The properties which play an important role are physical, mechanical, frictional, rheological, aero and hydrodynamic, electrical and optical properties of the biomaterials.

❖ **Physical properties:** The physical properties such as size, shape, surface area, volume, density, porosity, colour and appearance are important in designing particular equipment or determining the behaviour of the product for its handling.

* Various type of cleaning, grading and separation equipment are designed on the basis of physical properties of seeds such as size, shape, specific gravity, surface roughness, colour, etc.
* For designing an air screen grain cleaner, the shape and size of the grain determine the shape and size of screen opening, angle of inclination, vibration amplitude and frequency of screens.
* The density of the grains decides the size of screening surface.
* The frontal area and related diameters are essential for determination of terminal velocity Reynolds's number and drag coefficient.
* The density and specific gravity are needed for calculating the thermal diffusivity in heat transfer operation, in determining Reynolds's number, in pneumatic and hydraulic handling of the agricultural materials.

❖ **Bin:** A structure made of cement concrete, metal or any other material in which agricultural product is stored in loose form. **Shallow bins** are those bins in which the plane of rupture meets the grain surface before it strikes the opposite wall. **Deep bin** are those in which the plane of rupture strike opposite wall before it emerges from the top of the filling.

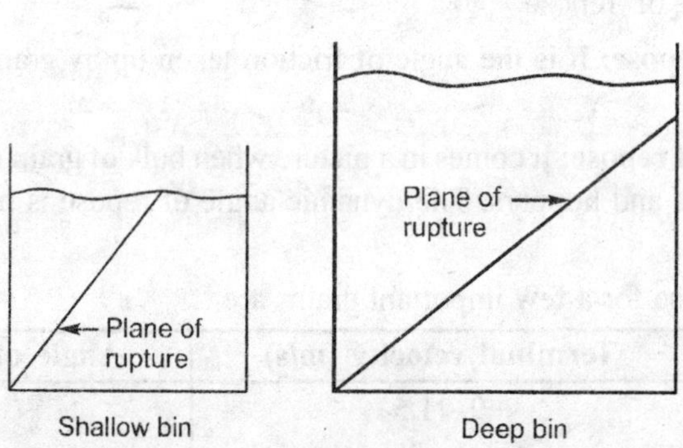

❖ **Bold Rice:** Rice with whole milled kernels having length breadth ratio between 2.0 and 3.0.

❖ **Bran:** The outer covering of the brown rice. It consists of the pericarp, seed coat and part or whole of aleurone layer depending upon the polish given. It is removed during milling process; sometime, small pieces of endosperm and germ may be also present.

❖ **Bran Oil:** The oil that is extracted form rice bran.

❖ **Brokens:** Pieces of kernel that is less than ¾th of the size of full kernels.

❖ **Brown Rice:** When paddy or rough rice is separated in to husk and kernel, the kernel is called brown rice or shelled rice, or dehusked rice or cargo rice.

- **Head Rice:** It refers to the kernels of milled rice, which is more than ¾th of the full kernel.
- **Head Yield:** It is the amount of head rice obtained by milling an unit amount of clean paddy, expressed in percentage.
- **Milled Rice:** Rice obtained from paddy after milling operation *i.e.* removal of husk and also removal of bran layers.
- **Milling Yield:** It is the outturn of rice after milling expressed in percentage with respect to clean paddy.
- **Polished Rice:** The rice obtained after removing the bran layer from brown rice. It is also called *white rice* or milled rice or polished rice.
- **Red Rice:** Rice kernel whole or brokens which have 25% or more their surface having red pericarp.
- **Cleaning:** Cleaning in agricultural processing generally means the removal of foreign and undesirable material from the desired grains/product. This may be done by washing, screening, hand picking, etc.
- **Grading:** Grading refers to the classification of cleaned product into various quality fractions depending upon the various commercial values and other usage.
- **Sorting:** Sorting refers to the separation of cleaned product in to various quality fractions that may be defined as on the basis of size, shape, density, texture and colour.
- **Scalping:** Scalping refers to the removal of few large particles in an initial process.
- **Screening:** Screening is a method of separating grain/seed in to two or more fractions according to size alone. The widely used device is called screen.
- **Aspirating:** It is the process of cleaning by air blast and separating the foreign material which is substantially lower in specific gravity than the product to be cleaned.
- **Milling:** Milling means reduction of food grain into various end products like meal, flour, splitted product.
- **Wheat:** Wheat grains are 5-8 mm long and 3.7-4.5 mm in equivalent diameter. The shape is ovoid and round at both ends. The pericarp and outer endosperm (or aleurone) layer consisting of 3-15% of the kernel weight are removed during milling operation. The removed portion is called "bran". The germ constitutes 2-3% of the kernel weight and is extracted during milling operation.
- **Paddy:** Paddy grains are 5-10 mm long and 4.2-4.6mm in equivalent diameter. A paddy grain consists of hull which is about 20-22% of weight of grain and brown rice kernel. The outer most six layers of cells of kernel form peri-carp which is about 5-7% of the brown rice weight. The seed coat together with aleurone layer form 4-6% of rice kernel. The entire mass of grain below the aleurone layer forms the endosperm. This is starchy portion of a rice grain. It consists of oblong rectangular cells and constituents about 89-94% of the grain mass. At one end of the vertical side of the grain is embryo or germ, responsible for germination. It constituents about 2-3% (by weight) of the rice grain.
- **Pulses:** Pulses contain 17-25% of proteins by weight. The protein content of pulses is about double of the cereals and more than meat and egg. Pulses contain lipids and fats in low

quantities and are good source of thiamin, niacin, calcium and iron. Though the ascorbic acid (vitamin C) content of pulses is low in dry grains, but during their germination stage ascorbic acid is available.

❖ **Application of Heat Energy to Food**

* Cooking
* Blanching
* Pasteurization
* Sterilization
* Drying
* Evaporation/Concentration

❖ **Cooking:** Cooking is a heat treatment process by which to produce a more palatable (digestible) food. Cooking is done in six forms *i.e.*

* Baking	dry heat
* Broiling	,,
* Roasting	,,
* Boiling	in boiling water
* Frying	in oil
* Stewing	in boiling water

Several preservative changes occur in food as a result of cooking:

* Destruction or reduction of microorganism
* Inactivation of undesirable enzymes
* Destruction of potentially hazardous toxins present naturally or through microorganism
* Changes in colour, odour, flavour and texture
* Improved digestibility of food

❖ **Blanching:** Blanching is a heat treatment process in which to partially cooking of product and to inactivates or destroy the enzymes like peroxidase and catalase present in food. Blanching is done at 100°C for 2 to 5 minute. *The purpose of blanching is* :

* Inactivate enzyme causing hardness, discolouration, off flavour, softening and loss of nutritional value. Discolouration is done by poly phenol oxidase and off flavour by peroxidase enzyme.
* The reduction of area of leafy vegetables and facilitate their easy handling.
* To remove the tissue gases.
* To kill the microorganism by about 99%.
* Enhances green colour of vegetable like peas, broccolis, spinach, etc.
* To remove undesirable acids and astringent or taste of peas and thus improve the flavour.

Blanching is done prior to freezing, drying or canning in hot water or exposed in steam. The time of blanching of product depends upon the characteristics of product.

❖ **Pasteurization:** Pasteurization of milk is a heat treatment process in which partially destroy the microorganism or harmful pathogen present in milk at predetermined time and temperature. The time and temperature treatment combination used in pasteurization depends upon.

* The heat resistance of the particular vegetative or pathogenic microorganism.
* The sensitivity of product quality to heat

Main Objectives of milk pasteurization:

* To make milk safe for human consumption
* To increase the keeping quality of milk

Operation	Temperature	Time
LTLT	62.8°C	30 min
HTST	71.7°C	15 sec
UHT	89°C	1sec
	90°C	0.5 sec
	94°C	0.1 sec
	95°C	0.05 sec
	100°C	0.001 sec

❖ **Sterilization:** Sterilization is a heat treatment process in which we can destroy the microorganisms and harmful pathogens completely from product at specified time and temperature combination.

Sterilization of product is done at 121.1°C for 2 to 5 minute. The thermal conditions needed to produce commercial sterility depend on many factors including;

* Nature of the food (*e.g.* pH)
* Storage condition of the food
* Heat resistance of the microorganisms or spores
* Heat transfer characteristics of the food, its container and the heating medium
* Initial load of microorganisms

❖ **D-Value:** The D-value is defined as the time in minutes at a specified temperature required to destroy 90% of the total organism in a population. Thus, the D-value or decimal reduction time, decrease the surviving population by one log cycle.

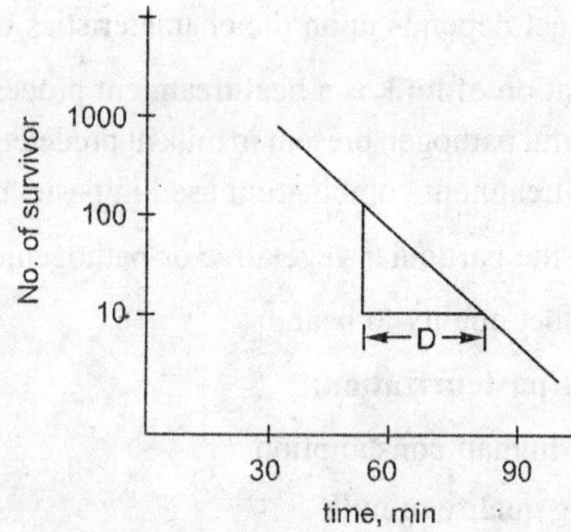

❖ ***Z*-Value:** The *Z*-value is the number of degree required a specific thermal death time curve to pass through one log cycle (change by a factor of 10). It is also negative slope index of the thermal death rate curve. **Different organisms in a given food have different *Z*-value.**

❖ ***F*-Value:** It is defined as the number in minute at a specific temperature required to destroy a specified number of organisms having a specific *Z*–value. The *F*-value is a measure of the capacity of heat treatment to sterilize.

❖ **Parboiling:** Parboiling is a hydrothermal treatment followed by drying before milling for the product of milled parboiled grain. The objective of parboiling is to provide extra strength to the grain kernel consequently improving its milling characteristics.

❖ **Hydrothermal treatment:** Hydrothermal treatment of grain refers to the addition of moisture and heat to the grain for improving the quality and quantity of their end product or to facilitate different milling operation for the desired product.

Hydrothermal treatment is commonly called conditioning or pre-milling treatment. It can be used for improving shelling efficiency, nutritional quality and milling qualities of paddy and for facilitating degermination and dehulling of corn and wheat.

❖ **Canning:** The process of hermetically sealing of fruits ad vegetables stuff in container and sterilizing by heat treatment for long period of storage is called caning or canned of the product. Ascorbic acid or acetic acid is used in canned fruit to create flavour in product. It is destruction of spoilage microorganism and biochemical's reactions in sealed container by means of heat.

❖ **Rheology:** Rheology can be defined as a science devoted to the study of deformation and flow. Therefore, the action of forces result in deformation and flow in the material. The mechanical properties will be referred to as Rheological properties. *e.g.* time dependent, stress-strain behaviour, creep-stress relaxation and viscosity.

- **Restandardization:** After homogenization, the evaporated milk is cooled to 45°F. The product is held in large storage tanks for restandardization of milk fat and / or total solids by the addition of water, skim milk, and/ or homogenized cream.
- **Culture:** A population of microorganisms cultivated in medium is called culture.
- **Medium:** A substrate used to provide nutrients to the organism for their growth and multiplication.
- **Gelatinization of starch:** It is a process of irreversible swelling of starch granules in water at suitable temperature, accompanied by a loss of birefringence and crystalinity. Galatinization is also known as the thickening of a liquid.
- **Gelatinization Temperature:** It is the temperature at which gelatinization of starch takes place and it varies depending on the variety of grain from 55 to 79°C.
- **Capacity of Machine:** It refers to the mass flow rate of clean product through the machine. In case of rice mill, tonnes of clean paddy passing through the mill perform.
- **Optimum Cooking Time:** It is the time required for a particular rice to cook to a desired degree of softness in boiling water. Rice is considered optimally cooked when its moisture content reaches to about 250% dry basis, depends on the consumers preference for soft or hard cooked rice.
- **Thawing:** It is the reverse process of freezing. It is done at the time of cooking or a part of preparation. The process by which a frozen product becomes less cold by using in to an unfrozen state is known as thawing. The rate of freezing is greater than rate of thawing at the same temperature differences between centre of product and outer surface. Ice has good conductivity than water. Since rate of thawing and freezing depends upon conductivity of water and ice.
- **Milk:** Milk can be defined as a complex consisting of fat, proteins, lactose, minerals in continuous phase of water as major constituents, and vitamins, enzymes, pigments, gases and several other organic and inorganic substances. The milk fat exists in the form of emulsion, which is stabilized by the presence of phospholipid membrane on the surface of the fat globules. Protein, particularly casein is found in the form of colloidal dispersion, lactose and certain minerals are present in true solution. Milk contains several substances, which could be classified under two major heads viz. liquid and solid. Water in milk forms the liquid base in which all the solids are dissolved. Milk solids such as fat, protein, lactose and minerals, all taken together are termed as total solids (TS) in milk. The solids can be regrouped as fat and solid-non-fat (SNF). The fat consists of true fat and substances associated with fat viz. lecithin, carotene, cholesterol, glycerides and vitamins like A, D, E and K. The SNF can be grouped as lactose (alpha and beta); protein like casein (alpha, beta, gamma and Kappa), lactalbumin and Lactoglobulin; minerals like phosphates, citrates and chlorides of Na, Ca and Mg with traces of Fe, Cu and I, and minor constituents of importance like vitamins, B_1, B_2, and B_6 and dissolved gases, enzymes and microorganism. The average chemical composition of milk of different species in Table 1 are given below:

Average composition of Milk

Species	Fat (%)	Protein (%)	Lactose (%)	Ash (%)	Total Solids (%)	Water (%)
Cow	4.9	3.2	4.6	0.75	12.80	87.20
Buffalo	7.2	3.8	4.8	0.76	15.85	84.12
Goat	4.8	4.2	4.1	0.89	13.00	87.00
Sheep	6.0	5.2	4.7	0.90	16.80	83.20
Camel	5.4	3.0	3.3	0.75	13.05	87.50
Human	3.5	2.0	6.8	0.25	12.55	87.45
Ass	2.5	2.0	6.1	0.41	-	89.00

Transmittance (T): It is the ratio of the radiant energy transmitted by sample (P) to the energy incident upon the sample (P_0). Both radiant energies should be obtained at same wavelength and with same spectral slit width.

$$T = P/P_0 \quad \text{or}$$

Transmittance or Transmission is the ratio of the transmitted light to that of the incident light. Transmittance (T_s) is the ratio of the transmittance of cell containing the coloured solution to that of an identical cell containing the solvent or a blank solution.

Optical Density or Browning Index: It also called the extinction (E) or the absorbency (A) of a medium is the logarithmic ratio of the intensity of the incident light to that of the emergent light. The per cent transmittancy may be converted into OD by using the expression 2-logT_s.

Colour in foods is usually due to the presence of natural pigment like anthocyanins, carotenoids or chlorophyll.

Colorimetry: Colorimetry is based on the determination of the concentration of a substance by measuring the relative absorption of light with respect to a known concentration of the substance.

A spectrophotometer on the other hand, consists of a spectrometer and a photometer combined. A spectrometer produces coloured light of any selected colour or wavelength, and it forms the part of a spectrophotometer, it is called a monochrometer and is generally calibrated in wavelengths (nm, hithero designated as mi.). A photometer measures the intensity of monochromatic beam produced by the monochromator.

Lambert's Law: Lambert law states that when a beam of monochromatic light is passed through a solution of constant concentration, the absorbency (A) is directly proportional to the thickness of solution (b). *i.e.*

$$A \propto b \tag{1}$$

or $$A = a_s b,$$

where a_s = constant absorbency index

or $$A = a_s b = \log_{10} I_0/I$$

where, I_0 = Intensity of light emerged

I = Intensity of transmitted light

Bear's Law: when a monochromatic light passed through a solution of constant length, the absorbance (A) is directly proportional to the concentration of solution (C).

i.e. $$A \propto C$$

or $$A = a_s C = \log_{10} (I_0/I)$$

or $$A = a_s bC = \log_{10} (I_0/I)$$

$$A = a_s C = \log_{10} (I_0/I)$$

Quality of Food: Quality of foods can be defined as the 'composite' that differentiate individual units of a product and have significance in determining the degree of acceptability of that unit by the buyer. In general, the quality control is the maintenance of quality at levels and tolerance acceptable to the buyer, minimizing the cost for the vender. But, from the scientific angle, the overall quality refers to technological, physical, chemical, microbiological, nutritional and sensory parameters to achieve the wholesome food. These quality factors depends on the specific attributes such as sensory properties based on flavour, colour, aroma, taste, texture; quantitative properties viz. percentage of sugar, protein, fibre etc. and hidden attributes like peroxides, free fatty acids , enzymes.

Enzyme: Enzymes are proteins produced by living cells and utilized by these cells to catalyze specific biochemical reactions. It is capable of catalyzing a reaction in which substrate (S) are converted to product (P) through the formation of an intermediate enzyme-substrate complex.

Substrate: The substrate upon which an enzyme acts is known as substrate. They are produced by the living protoplast of a cell.

Co-enzyme: When only organic molecules are attached to the enzyme is called Co-enzyme.

Prosthetic Group: It is the non-protein group of enzyme. The prosthetic group is firmly bound to protein component of enzyme by chemical bonds and is not removed by dialysis. Haem, biotin and pyridoxal phosphate like flavin, usually function as prosthetic group.

Apoenzyme: If, however, the prosthetic group is removed, the remaining protein part of the enzyme is called as Apoenzyme and it is inert.

Cofactor: Some enzymes don't require any chemical group other than amino acids for activity but some enzyme require additional chemical compound called Cofactor.

Post Harvest Technology: PHT is interdisciplinary "Science and Technique" applied to agricultural produce after harvest for its protection, conservation, processing, packaging, distribution, marketing and utilization to meet the food and nutritional requirements of the people in relation to their needs.

Importance of PHT: Importance of PHT lies in the fact that it has capability meet food requirement of growing population by elimination avoidable losses making more nutritive food items form low grade row commodity by proper processing and fortification, diverting portion of food material being fed to cattle by way of processing and fortifying low grade food and organic wastes and by products into nutritive animal feed. Post Harvest Technology has potential to create rural industries, the purpose of post harvest processing is to maintain or enhance quality of the products and make it marketable.

Homogenization: The homogenization can be defined as "The process designed to reduce the size of fat globules (droplets) to form many smaller ones, and making a permanent emulsion of milk fat, serum by the use of machine is called homogenizer. The size of homogenized fat globules should be around 2μm in diameter. Generally double stage homogenizer are used and the usually pressure for milk are 175-210 kg/cm^2 in first stage and 35 kg/cm^2 in second stage. The curd tension is reduced by 70-73% by homogenization. Homogenization slightly increases the viscosity due to clumping of fat globules. The effect of homogenization is more in pasteurized or sterilized milk as compared to raw milk. The surface tension is slightly increased when heat treated milk is homogenized than raw milk.

Standardization: Standardization refers to a process by which the components of milk is adjusted to a desired standards. According to PFA, Standardization of milk means, milk containing not less than 4.5% Fat and not less than 8.5% SNF. Standardized milk can be prepared form cow, buffalo or mixed or from dried milk solids with fat.

Toned Milk Toning: The main objective of preparing toned milk was to increase the quantity of whole milk by reducing its fat content and maintaining the SNF level in milk.

Hurdle Technology: Shelf stability of minimally processed food products can be enhanced by using a combination of treatments such as blanching, osmo-dehydration, addition of chemical preservatives, modification of pH and appropriate packaging. This technology is known as *Hurdle technology*. The different treatments called as hurdles are used in preventing the chemical as well as microbiological deterioration of the food products. A combination of these hurdles is very effective, while individually each hurdle is ineffective in protecting food from deterioration.

Equilibrium moisture content (EMC):

* If the vapour pressure of water present in food is more than the vapour pressure of water vapour in air, the water present in food vaporizes and diffuses in the atmosphere.
* If the vapour pressure of water present in food is less than the atmospheric vapour pressure, food will absorb moisture from the atmosphere.
* This property of gaining or loosing of moisture as per the atmospheric conditions is known as hygroscopicity.

Desorption Vs Adsorption: A product which attains its moisture equilibrium with the surroundings by losing moisture is said to have reached the desorption EMC. When a relatively dry material adsorbs moisture from a high humidity environment, it will eventually reach the adsorption EMC. There may be a significant difference at certain relative humidities and temperatures between desorption and adsorption EMC values for cereal grains, the desoption moisture contents are higher than the adsorption values. The difference between desorption and adsorption isotherms is called "hysteresis effect".

Mechanical hysteresis: The energy absorbed by a material in a cycle of loading and unloading evaluated as area between loading and unloading curve. Mechanical hysteresis is a measure of damping capacity or the ability of material to dissipate the strain energy as heat. (cm – kg/cycle).

Visco-elasticity: A combined solid like and liquid like behaviour in which the stress-strain relationship is time dependent.

Linear Visco-elasticity: A Visco-elasticity behaviour in which the ratio of stress to strain is a function of time alone and not of the stress magnitude.

Stress relaxation: Decay of stress with time when the material is suddenly deformed to a given deformation at constant strain.

Relaxation Time: The rate of stress decay in material subjected to a sudden strain. It is the time required for the stress in the Maxwell models, representing stress relaxation behaviour to decay to 1/e or approximate 37 % of its original value.

Creep: Deformation with time when the material is suddenly subjected to a dead load at constant stress.

Retardation Time: The rate at which the retarded elastic deformation takes place in a material creeping under dead load. It is the time required for Kelvin model, representing creep behaviour, to deform (1 – 1/e) or about 63 % of its deformation.

Viscosity: Resistance to flow indicated by coefficient of viscosity (μ). Unit is poise.

Viscosity coefficient: The ratio of shearing stress to shearing rate in Newtonian fluid.

Kinematic viscosity: Coefficient of viscosity (η) divided by the mass density (ρ).

Newtonian Liquid: An ideal fluid in which the relationship between shear stress and shear rate is a straight line passing through origin.

Non Newtonian liquid: An ideal fluid in which the relationship between shear stress and shear rate is nonlinear. If the flow curve is concave to the shear stress axis, the flow is called **Dilitant.** If the flow curve is convex to the shear stress, the flow is called **pseudo plastic.**

Apparent viscosity: As applied to Non Newtonian liquid is the viscosity of a Newtonian liquid exhibiting the same resistance to flow at the chosen shearing stress or shear rate. It is determined from slope of a straight line connecting the chosen point on the non-linear curve to the origin.

Thixotropic fluids: Those fluids which show a decrease in shear stress with time of shear at a given shear rate.

Rheopectic fluids: Those fluids which show an increase in shear stress with time of shear at a given shear rate.

Refrigeration: Refrigeration may be defined as the process of removing heat from a space at a temperature lower than the surrounding temperature. In other terms, refrigeration is the process of transferring heat energy from lower to higher temperature region. The refrigeration which is working on the reverse Carnot cycle is called Carnot refrigeration.

Ton Refrigeration: It is equivalent to the production of cold at rate at which heat is to be removed from 1 Ton of water at 0°C to freeze it to ice at 0°C in 24 hour or one day.

$$Q = 1000 \text{ kg of water}$$

$$Q = \frac{1000 \text{ kg} \times \text{Latent heat}}{24}\left(\frac{\text{kJ}}{\text{hr}}\right)$$

$$1 \text{ TR} = \frac{1000 \text{ kg} \times \text{Latent heat}}{24}\left(\frac{\text{kJ}}{\text{kg}}\right)$$

Refrigeration Effect: The rate of heat absorbed in the cycle from the body or space to be cooled is called refrigeration effect (N).

Mechanical Work done: The action of a force through a distance (or a torque on an angle) is called mechanical work done (W).

Coefficient of Performance (COP): It is the ratio of refrigeration effect (N) and mechanical work done (W). It is always more than 1.00.

Porosity: Porosity is defined as the percentage of volume of inter grain space to the total volume of grain bulk. The per cent void of different grains in bulk are often needed in drying, airflow and heat flow studies of grains. It depends on shape, dimension, and roughness of grain surface. Changes in porosity occurs depending on the bulk density, the degree of filling the amount of foreign material, the amount of fine material and the moisture content of the grain. Porosity can be determined by using the bulk density to true density values. The porosity of the crops is as wheat (50-55 %), paddy (46-60 %) and corn (38.5-47.6 %).

Post Harvest Management: Post Harvest Management concerns with ways and means to minimize post harvest losses and ensure utilization of agriculture with enhanced utilization efficiency. Every operation of management sequence is associated with an enterprise (big, small or cottage) and value addition to the produce, and increase in net availability to the consumer.

Food Additive: Food additives are non-nutritive substances which intentionally added to food during processing in small quantity to improve the shelf life or for improving colour, flavour and texture of food as processing aids. Food additives are generally classified based on their function in foods.

Function of food additives

* Enhancement of the attractiveness of foods by means of colouring and flavouring agents, emulsifier, stabilizer, thickeners, clarifier and bleaching agents.
* Maintenance of nutritional quality such as by the use of antioxidant.
* Facilitating food processing by means of acids, alkalies, buffers, sequestrants and various other chemicals, and
* Enhancement of keeping quality or stability by the use of antioxidant, antimicrobial agents, inert gases, meat curers etc.

Antioxidants: Antioxidant is a substance which when added to fat and fat containing foods prevents their oxidation and thus prolong their shelf-life, wholesomeness and palatability. Antioxidants are defined by the United States Food and Drug Administration Act (FDA) as "The substances used to preserve food by retarding deterioration, rancidity or discolouration caused due to oxidation" (Code of Federal Regulation, CFR). The function of antioxidants by interrupting the free radical reaction involved in lipid oxidation. An antioxidant should not have any harmful physiological effect and should not impart an objectionable flavour, odour or colour to the food in which it is present. It should be effective in low concentration (0.01 to 0.02%) and be fat soluble. Some antioxidants used in foods are butylated hydroxyanisole (BHA), butylated hydroxytoluene (BHT), tertiary-butylated hydroquinone (TBHQ) and propyl gallate (PG), thiodipropionic acid, dilauryl thiodipropionate, stannous chloride, tocopherols, more expensive naturally occurring vitamin E, sulphur dioxide and ascorbic acid. Ascorbic acid prevents browning caused by enzymatic oxidation of phenolic compounds. Acids such as citric and phosphoric increase the effectiveness of ascorbic acid. The prominent effects of oxidation include the development of rancid off-flavour by fat, oils and lipid containing foods, possibility of toxic oxidation products, discolouration of pigments, loss of product flavour or odour, changes in texture and the loss of nutritional value from the destruction of vitamin (A, D and E) and essential fatty acids (*e.g.* linoleic acid). Aldehydes and ketones which are generated from the chain reactions are responsible for the off flavour and odour. Addition of antioxidant compounds is widely used in food products to inhibit the progress of lipid oxidation.

Preservatives: Preservatives means a substance which is capable of inhibiting, retarding or arresting the growth of microorganisms in foods is known as preservatives. It may be chemical or natural substances (sugar, salt, acid, and oil). Sulphur dioxide (SO_2) and sodium benzoate, sodium and calcium propionate (as mould inhibitor in bread and cake), sorbic acid (as mould and yeast inhibitor in cheese and baked products, fruit juices, wines and pickles) and chlorine compound (as germicidal wash for fruits and vegetables) are also used as preservatives. The term preservatives includes fumigants *e.g.* ethylene oxide and ethyl formate used to control microorganisms from spices, nuts and dried fruits. Preservatives may be a single compound or mixture of compounds. Preservatives may be further classified in Class I and Class II preservatives.

Class I Preservatives: These preservatives are added in any amount in foods without any restriction and also used extensively in the household products like pickles, jam, jelly, curry, butter etc. Their addition is governed by their taste and flavour of the foods. These have been used since time immemorial. Class I preservatives include common salt, sugar, dextrose, glucose syrup, spices, honey, edible oils, and vinegar.

Salt: Salt is used as flavouring and taste enhancing compound in combination with other preservatives, it is also widely employed in foods for enhancing their shelf life. Common salt is known to decrease the water activity of the food product which restricts the growth of microorganisms.

Sugar: Sugar (sucrose) lowers down the water activity (a_w) of foods and thereby does not allow the microorganisms to grow and survive, and also lowers the solubility of oxygen in water.

Vegetable oils: Vegetable oils mainly consist of triglycerides and to a lesser extent of diglycerides mono glycerides and free fatty acids, while triglycerides have hardly any preservative action but mono diglycerides and free fatty acids have mild preservative action. The preservative action of vegetable oils in vegetable pickles is to avoid entry of microorganisms by forming and impervious layer and barrier properties against oxygen. Allyl isothicynate in mustard oil may have some preservative action.

Total soluble solid (TSS): The TSS value is defined as the amount of sugar and soluble minerals present in fruits, vegetables and their processed product (*i.e.* beverages). This is measured by Refractometer in percentage or degree Brix (°Brix) at 20°C temperature.

Jam: Jam is a product made by boiling fruit pulp with sufficient amount of sugar to a reasonably thick consistency, firm enough to hold the fruit tissue in position. Jam contains 0.5 -0.6 % acid, invert sugar should not more than 40 % and TSS about 68-70%.

Jelly: Jelly is a semi-solid product prepared by boiling a clear, strained solution of pectin containing fruit extract, free from pulp, after the addition of sugar and acid. A perfect jelly should be transparent, well set, but not too stiff, and should have the original flavour of the fruits. pH= 3.2 and TSS = 65 °Brix

Marmalade: This is a fruit jelly in which slices of the fruit or its peel are suspended. The product make from citrus fruits like oranges and lemons in which shredded peel is used in suspended material. TSS = 65° Brix.

Packaging: Packaging is essential for the safe and easy marketing of any food products and for retention of their natural characteristics till consumed or utilized otherwise. The reason for indiscriminate adoption of plastics as the main packaging material by the food industry is: they are cheaper, easy to fabricate, can be easily applied and the packaging process can be mechanized, loss of moisture from the food is easily nil, it protects the food from attack by microorganisms, insects and rodents; humidity control is not necessary etc. Packaging can be defined as "techno-economic function arrived at minimizing cost of delivery while maximizing sales. Food packaging is used to protect the food from physical, chemical and microbiological spoilage, environmental factors and to provide the food attractive.

Edible films Coating or edible packaging: An edible coating film or coating is simply defined as a thin layer of edible material formed on, placed on, or between the foods or food components. Edible packaging refers to the use of edible films, coatings, pouches, bags and other containers as a means of ensuring the safe delivery of food product to the consumer in a sound condition. Selection of material for use in edible packaging is based on its properties to act as a barrier to moisture and gases, mechanical strength, physical properties and resistance to microbial growth. Films with very low oxygen transmission rate (OTR) are necessary to protect rancidity in a product high in fat content whereas light barrier films protect pigments, flavours and nutrients in foods. The types of materials used for edible packaging include lipids, proteins and polysaccharides. Lipid components that are used as edible packaging materials include natural waxes such as carnauba wax, candelilla wax, rice bran wax, bee wax, vegetable oil (such as corn oil, soybean oil and palm oil) and acetylated monoglycerides. Lipid films and coatings are excellent barriers to moisture and gas, but their permeability is dependent on the packing of the lipid crystals and their orientation to the direction of the permeate flow. Permeability of lipid films to water increases with increasing polarity, insaturations and branching of lipids or as water is absorbed into the polar portion of the films. Edible coating materials such as waxes and lipid can be applied as stable emulsion or micro-emulsions with water or by direct applications.

Aseptic Packaging: Aseptic processing and packaging is a continuous sterilization technology whereby the food and the container are sterilized separately and then food is packaged in a sterile environment. The word 'aseptic' can be defined as the complete absence of any living microorganisms. But in the food industry, it only refers to the condition in which the food is packaged in an environment free from viable microorganisms of public health significance, capable of growing in the food under normal handling and storage condition. The sterilization is done by exposing the food products to elevated temperature over a short period of time. The aseptic product is shelf stable and does not require refrigeration. During aseptic processing, the food material is rapidly heated to an elevated temperature and held for certain time to inactivate microorganisms. Sterilization of food products can be done by (*i*) direct heating, (*ii*) indirect heating, and (*iii*) ohmic heating.

Food Irradiation: Irradiation preservation of foods is based on the inactivation of spoilage and pathogenic microorganisms by ionizing radiations *i.e.* high energy electrons, X-rays or gamma rays, Food irradiation is a physical treatment in which food is exposed to ionizing radiation, *i.e.* radiation of high energy to expulse electrons from atoms and to ionize molecules. It is used to destroy the bacteria and parasites that cause human illness. It can also make food last longer by destroying or inactivating insects, molds and yeasts that spoil food, delaying the ripening of fruits and vegetables, and limiting unwanted spouting of potatoes and other similar vegetables. Food irradiation is safe and keeps the food in fresh condition. The process is quite simple and technical. The irradiation process involves passing of food through a radiation field allowing the food to absorb desired radiation energy. The food itself never comes in contact with the radioactive materials.

Application of ionizing radiation in food preservation

Levels	Doses (kGy)	Utilization
Low dose	0.03-0.15	Sprout inhibition in bulbs and tubers
	0.25-0.75	Delay in fruit ripening
	0.25-1.00	Insect disinfection and elimination of food borne parasites
Medium dose	1.5-3.0	Reduction of spoilage microbes to improve shelf-life of meat, poultry and sea foods under refrigeration
	3.0-7.0	Elimination of pathogenic microbes in fresh and frozen meat, poultry and sea foods
	10.0	Reduction of microbes in spices to improve hygiene
High dose	25-70	Sterilization of packaged meat, poultry and their products which are self stable without refrigeration
	25-70	Sterilization of hospital deals

Radiation processing: Radiation processing of food strengthens food preservation, improves food hygiene, and helps food exports overcome quarantine barriers. It facilitates packing, storage, transport and distribution of food.

Use of radiation processing

- Extension of shelf life of food products by killing microorganisms causing spoilage.
- Enhancement of safety of food by killing food borne pathogens and parasites
- Inhibiting spouting in tubers, bulbs and rhizomes
- Delaying ripening of fruits.
- Destroying insect pests in stored products, fruits and vegetables.
- Inhibition of greening of potatoes
- Reduction of microbial load.
- Control of post harvest diseases
- Arresting of post harvest growth of vegetables

Effect of ionizing radiation of fresh fruits and vegetables

Dose, kGy	Observed effects
0.05-0.15	Sprouting inhibition in tubers, bulbs and root vegetables, inhibition of growth in asparagus and mushroom.
0.05-0.75	Insect disinfection
0.25-0.75	Delayed ripening of some tropical fruits *e.g.* banana, mango and papaya.
>1.75	Control of post harvest diseases.
1.00-3.00	Accelerating softening, development of off flavour in some commodities.
>3.00	Excessive softening, abnormal ripening, incidence of some physiological disorders, improve flavour

Minimal Processing: Minimal processing is used for the preservation of short shelf life of fruits and vegetables, and meat products, with minimum damage to the freshness of the products. Minimal processed foods retain their freshness and they are more acceptable to consumer than conventionally processed products. Minimal processing can be achieved by using of a combination of mild processing methods, which control the growth of spoilage and pathogenic microorganisms (hurdle technology), without serious damage to the product quality. Hurdles (obstacles) used in food preservation include temperature, water activity, pH, preservatives and competing microorganisms (*e.g.* lactic acid bacteria). Recent non thermal preservation methods such as irradiation, high pressure and pulse electric field may also be used. Minimal processed fruits can be produced using the following hurdles: steam blanching for 1-3 min, reduction of water activity (a_w) to 0.98 -0.93 by sugar addition, lowering the pH to 4.1-3.0 by addition of citric acid or phosphoric acid, and addition of antimicrobials *e.g.* 1000 ppm of sorbate and sulfite. The treated fruits can be packaged and stored at ambient (room) temperature (upto 35°C) for about 4 months. Minimal processed raw vegetables can be produced by modified atmosphere packaging (MAP) and storage at 1-8°C for 5-7 days.

Vegetables and potato dishes (salads), vacuum packaged (sous vide) and heated mostly at 65-95°C, can be stored under refrigeration (1-8°C) for about 42 days. The growth of anaerobic microorganisms such as toxin producing *Clostridium botulinum* should be prevented by some additional hurdles. Development of minimally processed foods by hurdle technology should be combined with GMPs (good manufacturing practices) and HACCP (hazard analysis) for better process control.

Pulse electric fields processing: Pulse electric fields can inactivate vegetative cells by dielectric breakdown of bacterial membrane *e.g.* electric field of 30-60 kV/cm and pulse duration of about 1µs.

Good Manufacturing Practices (GMP): Hygiene (sanitation) is a fundamental requirement of all food plant operations *i.e.* processing, packaging, storage, building and personnel. Microbial and non microbial contamination should be prevented by proper design and operation of all processing equipment and the entire plant. The principles and practices of Good Manufacturing Practices (GMP) should be taken into serious consideration in food plant and equipment design. GMPs are a combination of manufacturing and management practices aimed at ensuring that food products are consistently produced to meet specifications and customer expectations. They are practical rules and recommendations, based on experience, which, when followed in the various food processing operations will results in safe and high quality responsibilities on food processing plants and processing equipments. Current Good Manufacturing Practices (CGMP) cover wide spectrum of manufacturing practices, but the main emphasis is on food plant hygiene (sanitation); while food quality receives the proper consideration.

Browning reaction: Browning reaction takes place mostly in processed food products. The coloures developed range pale yellow to dark brown or black, depending on the type of product and the extent of the reaction. Browning reaction may be two types *i.e.* enzymatic and non-enzymatic. The presences of reactive reducing sugar are responsible for browning in foods and food products.

Enzymatic browning: In this, the formation of brown colour is due to the action of poly phenol oxidase enzymes or phenolic substances. When foods containing such substances are cut and exposed to air rapid browning of the cut surface takes place. *e.g.* apples, banana, potatoes.

$$\text{Polyphenols} + O_2 \xrightarrow{\text{Polyphenole oxidase}} \text{Brown colour}$$

The enzymatic browning is due to the oxidation of phenols to orthoquinones, which is turn rapidly polymerize to form melanin (the brown pigment). Tyrosine is the major phenolic substrate for phenolase action in foods. When the substrate the phenol, it is converted by hydroxylation into orthodiphenol and then oxidized to orthoquinones.

Pigment formation (melanin) can be eliminated by inhibiting enzyme action, thermal treatment, addition of Sulphur dioxide (KMS) and ascorbic acid are the commonly used methods of inactivating these enzymes. Both KMS and ascorbic acid are strong reducing agents. They consume oxygen and browning activity of enzymes. Browning can be also prevented by excluding oxygen from the reaction site or protecting the phenolic substrates. The exclusion of oxygen achieved by immersing the tissues in brine or syrup, or by processing under vacuum and the phenolic substances are protect by the use of certain enzymes which modify orthophenolic substrates.

Non-enzymatic browning: The non- enzymatic browning reaction is responsible for the colour and flavour of foods and food products. The distinctive flavour *i.e.* coffee beans, groundnuts and breakfast cereal develop after roasting is due to browning reactions. Roasting of potatoes, like toasting of biscuit and baking of bread and cakes, produces a golden brawn colour. The presence of reactive reducing sugars are responsible for browning of foods. The sugars under goes ring opening, endisation dehydration and fragmentation under heating. The unsaturated carbonyl compounds that are formed react to produce brown polymer and flavour compounds. The heat- induced browning reaction is subjected in two ways: Caramelization and Maillard reaction.

Maillard reaction: The maillard reaction is actually a complex group of many reactions between (*i*) nitrogenous compounds and sugars (*ii*) nitrogenous compounds and organic acids, (*iii*) sugars and organic acids, (*iv*) among organic acids themselves. The Carbonyl group of acyclic sugars readily combines with basic amino group of proteins, peptides and amino acids, resulting in sugar amines. The set of various reactions that sugar amines undergo resulting in browning is known as the maillard reaction. The sugar amines have a brown colour at a low temperature.

Inhibition of the maillard reaction can be accomplished by keeping the pH of the amino acids, peptides and proteins and by keeping the temperature as low as possible during processing and storage. Use of non reducing sugar (sucrose) under condition not flavouring inversion also helps bring down maillard browning. Sulphur dioxide (KMS) and Sulphite used in extending the storage life of dehydrated and processed foods, fruit juices and wines also inhibit the browning reaction.

Zero Energy Evaporative Cooling/Storage: Zero-energy cool chamber, is based on the principle of evaporative cooling could be a good substitute over the other storage methods for short-term storage of fruits and vegetables. The evaporative cooled (EC) storage structures can be utilized in places where the fresh perishable commodities are readily spoiled by high temperature and low relative humidity. The use of evaporative cooled storage can considerably reduce dry bulb temperature and raise the humidity and thus aid in the preservation of fruits and vegetables after harvesting. It requires low initial investment and low cost of repair and maintenance. This type of storage structures can give sufficient time to the farmers to hold back their produce during peak season to avoid distress sale. As water evaporates it has a considerable cooling effect, and faster the rate of evaporation, greater the cooling. Evaporative cooling occurs when air that is not already saturated with water passes over a wet surface. The efficiency of an evaporative cooler depends on the humidity of the surrounding air. Very dry, low humidity air can absorb a lot of moisture as considerable cooling occurs. Theoretically, the lowest temperature that can be achieved by the evaporation of water is the wet bulb temperature.

Controlled Atmosphere Storage (CAS): The storage of fruits and vegetables in control atmosphere (CA) is one of the most advanced methods of storage. It reduces nutrient and quality losses, increase the shelf life and best preserves the taste and aroma of fruits. Extension of storage life under refrigeration storage system is widely in practice despite high losses in quality due to physiological disorders resulting from ill effect of extremely low temperature of storage environment. Low temperature storage is delay ageing by retarding metabolic activity of commodities and suppresses the activity of spoilage agents (biochemical and phytophathogenic). The method of storing commodities (fruits and vegetables) at slightly higher temperature and retarding metabolic process by controlling the gaseous composition of storage atmosphere is so called controlled atmosphere storage. Air contains about 20.9 % O_2, 78.1% N_2, 0.03% CO_2 and trace amounts of other gases including Ne, He, CH_4 and water vapour. In regular CA storage, oxygen is reduced and CO_2 is increased and ripening is slow down. Respiration rates serves as indicator of metabolic activity. Delayed ageing of fruits and vegetable leads to better maintenance of nature resistance against infestation resulting in better storage life.

Benefits of controlled atmosphere: The controlled atmosphere storage offers several advantages as:

* Retardation of senescence and associated biochemical and physiological changes (slowing down respiration and ethylene production).
* Reduction of fruit sensitivity to ethylene action at oxygen (O_2) levels below 8 % and/or carbon dioxide (CO_2) levels above 1 %.
* Useful tool for insect control in some commodities.

Harmful effects of controlled atmosphere:

* Initiation of certain physiological disorders such as blackheart in potatoes, brown stain on lettuce.
* Irregular ripening of fruits such as banana, pear, tomato etc.
* Development of off-flavour and off odours at very low O_2 concentration

Effect of Concentration of Constituent Gases on Fruits and Vegetables

Effect of Oxygen (O_2): Lower level of oxygen in plant cells decreases the rate of biochemical reaction and reduce the metabolic activity. This reduction is however or not significant upto 14 % concentration (air has O_2 concentration of 21 %). But if O_2 level in cells is too low, there may be undesirable changes in the chemicals which contributes to the off flavour and aroma. The level of O_2 at which anaerobic respiration may occur in plant cells may be as low as 0.2%. Fruits such as pears, stone fruit, blueberries and straw berries stored between 9 to 5°C upto 10 days would tolerate oxygen levels of 0.25-1.0 % while apples could tolerate these levels for longer periods. The effect of oxygen on post harvest responses of fruits, vegetables and flower may be summarizes as follows

* *Reduced respiration rate*
* *Reduced substrate (glucose) oxidation*
* *Delayed ripening of climacteric fruits*
* *Prolonged storage life*
* *Delayed breakdown of chlorophyll*

* *Reduced rate of production of ethylene*
* *Changed fatty acid synthesis*
* *Reduced degradation rate of soluble pectin*
* *Formation of undesirable flavour and odour*
* *Altered texture*

Effect of Carbon dioxide: CO_2 accumulation in the intercellular spaces of fruits acts as ethylene antagonist. High level of CO_2 (10-15%) control grey mold in grapes and has shown less chilling injuries in fruits. Prolonged exposure to these level can damage the fruit but upto 2 weeks exposure did not have a detrimental effect. Effect of high CO_2 level in stored fruits may be summarized as follows:

* *Decreased synthetic reaction in climacteric fruits*
* *Delaying the initiation of ripening*
* *Inhibition of some enzymatic reactions*
* *Decreased production of some organic volatiles*
* *Modified metabolism of some organic acids*
* *Reducing the rate of breakdown of some pectin substances*
* *Inhibition of chlorophyll breakdown*
* *Production of off flavour*
* *Induction of physiological disorders*
* *Retarded fungal growth on the crop*
* *Inhibition of the effect of ethylene*
* *Change in sugar content (potatoes)*
* *Effect of sprouting*
* *Inhibition of post harvest development*
* *Retention of tenderness*
* *Decreased discolouration levels*

Effect of Nitrogen : Treatment of fruits in atmosphere of total nitrogen prior to storage was shown to retard the ripening of tomatoes and avocados. Pretreatment of mangoes in 97 % N_2 and 3 % O_2 have been reported to be remaining firmer and ripening more slowly than those untreated.

Effect of Carbon mono oxide (CO): This is colourless, odourless and flammable gas which is highly toxic to animals having hemoglobin's even at very low concentration. If added to CA storage with level of oxygen between 2 and 5 %, it can inhibit discolouration of lettuce on cuts and damages, but effect was lost when lettuce was removed to normal air at 10°C. Respiration rate in lettuce reduced during 10 days storage period at 2.5°C when CO was added to the storage. A combination of 4 % O_2, 2 % CO_2 and 5%CO shown to be optimum in delaying ripening and maintaining good quality in mature tomatoes at 12.8°C after they subsequently ripened at 20°C.

effect of Ozone : It has been found that fungal growth and rotting can be controlled by ripening fruit to 1-2 ppm of ozone. Its effectiveness is increased in high humidity conditions.

Essential Features of CA Storages:

- *Mechanical refrigeration is used to maintain temperature of 30-32°F or 36-38°F.*
- *The CA storage room is constructed gas tight.*
- *Nitrogen gas is introduced into the storage by portable equipment to hasten reduction of oxygen after storage room is filled and sealed. Carbon dioxide is added into storage from CO_2 gas cylinder.*
- *In CA storage, excess CO_2 is removed by dry hydrated lime.*
- *Oxygen level is held at 3% and CO_2 at 2-5 % depending on the variety of apple stored.*
- *To load or unload fruit, the atmosphere must be flushed to raise oxygen level upto at least 18 % and reduce CO_2 to a low level.*
- *Atmosphere samples are taken daily for CO_2 and O_2.*

Table: Recommended CA or MA conditions during transport and/or storage of selected Fruits and Vegetables.

Commodity	Temperature range, °C	CA+	
		% O_2	% CO_2
Fruits			
Apples	0-5	2-3	1-2
Apricot	0-5	2-3	2-3
Grape	0-5	None	None
Kiwifruit	0-5	2	5
Peach	0-5	1-2	5
Persimmon	0-5	3-5	5-8
Plum and prune	0-5	1-2	0-5
Straw berry	0-5	10	15-20
Nuts and dried fruits	0-25	0-1	0-100
Avocado	5-13	2-5	3-10
Banana	12-15	2-5	2-5
Lemon	10-15	5	0-5
Olive	8-12	2-5	5-10
Orange	5-10	10	5
Mango	10-15	5	5
Papaya	10-15	5	10
Pineapple	10-15	5	10

Commodity	Temperature range, °C	CA+	
		% O_2	% CO_2
Vegetables			
Asparagus	0-5	Air	5-10
Broccoli	0-5	1-2	5-10
Cabbage	0-5	3-5	5-7
Cauliflower	0-5	2-5	2-5
Corn, sweet	0-5	2-4	10-20
Cucumbers	8-12	3-5	0
Leeks	0-5	1-2	3-5
Lettuce	0-5	2-5	0
Mushroom	0-5	Air	10-15
Okra	8-12	3-5	0
Onion green	0-5	1-2	10-20
Peppers, bell	8-12	3-5	0
Potato	4-12	None	None
Spinach	0-5	Air	10-20
Tomatoes partially ripe	8-12	3-5	0

Modified Atmosphere Packaging (MAP): Modified atmosphere packaging (MAP) is indented to create an appropriate gaseous atmosphere around a commonly packaged in film packages to enhance shelf life and to conserve the quality of the product. One of the most important factors to be considered during design of MAP is selection of packaging film whose permeability to different gases would help in regulating the gaseous exchange during storage. However, MAP is only a supplement to good temperature and humidity management as any change in storage environment can push upto the respiration rate of packaged produce to dangerous levels resulting in spoilage. MAP is the replacement of the inpack air with a gaseous mixture. During the entire storage period no control is exerted over the gaseous composition and it changes with owing to the diffusion of gases into and out of the package and also to the effects of product and microbial metabolism. Storage under CAP/MAP is known to extend the shelf life of food by retarding the physiological metabolism leading to senescence by the increased CO_2 and decreased O_2 concentration in the storage atmosphere, slowing down the rate of ethylene synthesis subsequently helping in reducing microbial contamination and by creating high humidity resulting in less moisture loss and better quality retention.

Designing a MAP system for a fruit and vegetable at a selected temperature and relative humidity conditions necessities to know the optimum O_2 and CO_2 concentrations for that commodity in order to design an ideal package capable of creating the desired gaseous atmosphere inside the film packages. The choice of desired gas mixture is also influenced by the microflora capable of growing on the product, product's tolerance to O_2 and CO_2 and colour stabilizing requirement.

Table : Recommended gas conditions for CAP and MAP for selected Fruit and Vegetable

Fruit / Vegetable	Temperature, °C	Gas Concentration, %	
		Oxygen (O_2)	Carbon dioxide (CO_2)
Apple	0-5	2-3	1-2
Straw berry	5-13	10	15-20
Avocado	0-5	2-5	3-10
Broccoli	0-5	1-2	5-7
Lettuce	0-5	2-5	0-2
Tomato	8-12	3-5	0

Techniques for creating MAP: MAP can either be commonly generated /passive MA or active MA.

***Commodity-generated or passive MA*:** If the rate of respiration of the stored commodity and film permeability characteristics of the package is suitably matched by appropriate design of MAP, it can create an atmosphere within the sealed package through consumption of O_2 and production of CO_2 by the stored fruits or vegetables. The film package must allow O_2 to enter inside the package at a rate offset by consumption of O_2 by the packaged produce. Similarly, CO_2 must be vented from the package to offset the production of CO_2 to prevent any physiological damage to stored produce.

Active MA: The passive established atmosphere has a draw back that it can not be regulated effectively and is largely dependent upon film permeability characteristics and stored produce respiration rates. Therefore, it necessities that MAP should be actively established and adjusted. This can be done by pulling a slight vacuum replacing the package atmosphere with the desired gas mixture. This mixture can be further adjusted through the use of absorbing and adsorbing substances in the package to scavenge O_2, CO_2 or C_2H_4. This method ensures the rapid established of the desired atmosphere.

Requirement of packaging films: One of the most important parameters to be considered for MAP studies is the need for proper packaging films. An atmosphere of high CO_2 and low O_2 concentration in package decreases product respiration and quality deterioration. The ideal film must posses following characteristics:

* *Ability to change the gas permeability properties in case of rise in temperature*
* *Controllable Moisture Vapour Transmission Rate (MVTR) in order to prevent supersaturated vapour accumulation and condensation problems*
* *Ability to warm the consumer that the quality is not at its best*
* *Good thermal and ozone resistance*
* *Commercial suitability and ease of handling and application*

* *Non reactant with produce and nontoxic*
* *Ease of printing for labeling purpose*

The developed/selected packaging materials must be suitable in terms of optimal gaseous exchange which would help in creation of an appropriate atmosphere around the produce, thus increasing the shelf life of the packaged produce. The development of edible films with selective gas permeability could be very promising for controlling respiratory exchange and improving the conservation of fresh or minimally processed fruits and vegetable, *e.g.* hydrocolloid- based films such as gluten films appear to have suitable O_2 barrier properties while remaining relatively permeable to CO_2.

Low Temperature Storage: Fruits and vegetables are mostly seasonal crops but they are demanded by consumers throughout the year. They being very high their moisture content is difficult to preserve for the longer period of time under normal conditions. This is necessitated special storage condition and to be maintained for such products. The lower temperature conditions retard the activity of various biological and biochemical agents of spoilage and provided longer shelf life to the products. These spoilage agents are microorganism such as (bacteria, yeast and molds) and larvae's. Optimum storage temperature for most products is one slightly above the freezing point of the products. Duration for which any product could be stored depends on many factors such as its maturity level, moisture content and its hygienic conditions. Although the effect of in correct storage temperature generally is to lower the product quality and shortens storage life, some fruits and vegetables are particularly sensitive to storage temperature and are susceptible to called storage disease when stored at temperature below or above the critical temperature. For example, citrus fruits develop rind pitting when stored at high temperature.

On the other hand they are subjected to cold (browning of rind) and watery breakdown, when stored at temperature below their critical temperature. Banana suffers peel injury when stored below 13°C. Although, onions tend to sprout at temperature above 0°C, Irish potatoes tend to become sweet at storage temperature below 4°C. Squash, green bean and peppers develops pits on their surface when stored at or near to 0°C, whereas the best storage temperature for most of the apples varieties is -1°C to 0°C. Some varieties are subjected to soft scold and soggy break down when stored below 2°C. Other develop brown core at temperature below 2°C, and still others develop internal browning when stored below 4°C. The storage of perishables in their natural state (unpacked) requires close control not only of the space temperature but also the space humidity and air movement. One of the main causes of deterioration of unpacked fruits and vegetables is loss of moisture from the surface of the product by evaporation into surrounding air. This process is called dissociation drying or dehydration. In fruits and vegetables, dissociation is accomplished by shriveling and wilting and product undergoes a considerable loss in both weight and vitamin contents.

Design Consideration of Cold Storage

Cooling load calculation: To determine the capacity of the refrigeration system of cold storage. The knowledge of cooling load is must. The cooling load on a refrigeration system is the summation of heat sources which usually evolves from several different sources. Some of the common sources of heat supply are :

* Heat transfer by conduction
* Heat load gain through infiltration/ventilation
* Product heat load/field load
* Heat gain due to respiration
* Load due to air changes
* Effect of solar radiation

1. **Heat through conduction/wall grain load**: The quantity of heat transmitted through the walls of the refrigerated space per unit time is the function of three factors and determined by expression as

$$Q = UA\,(\Delta T)$$

Where, Q = rate of heat transfer (walls) unit
U = Overall heat transfer coefficient unit
A = outside surface area of walls (sq. m)
ΔT = Temperature differences across the wall unit

Thermal conductance of any materials is given as

$$C = K/x$$

Where, C = Thermal conductance unit
K = Thermal conductivity unit
x = thickness of materials in meter unit

Determination of overall heat transfer coefficient (U) by

$$1/U = 1/f_i + x_1/K_1 + x_2/K_2 + \ \dots +1/f_o$$

Where, f_i & f_o = convection coefficient at inside and out side air or wall
K_1 & K_2 = thermal conductivity unit
x_1 & x_2 = thickness of material of wall unit

2. **Heat gain through infiltration/ventilation:** Design table gives infiltration air quality for various types of doors and windows for observable cracks and for infiltration due to people opening the door. This load is generally added to the load due to fresh air intake for ventilation.

3. **Product heat load/Field load:** The heat gain due to product being brought at higher temperature than the storage temperature depends upon the temperature of product, the recommended storage temperature for that product. The amount of heat that to be removed depends upon specific heat of product, quality of product and recommended storage temperature.

$$Q = MS\Delta T$$

Where, M = mass of product to be stored unit
S = specific heat of product unit
ΔT = temperature difference unit

4. **Heat gain due to respiration:** Fruit and vegetable given after harvesting have living cell which respires during the storage and produce heat as a result of respiration. However, rate of respiration depends upon storage temperature high respiration rate at high temperature takes place.

5. **Load due to air changes:** The temperature and moisture content of outside air has to be brought down to the storage conditions, these results in a food on the system. The frequency of opening doors is dependent upon the volume of storage *e.g.* a cold storage of volume 150 m^3 will have seven air changes in 24 hours while for 3000 m^3 storage it is only 1.5.

6. **Effect of solar radiation:** Solar transmission gain through walls and roofs during day time. Due to solar radiation the temperature of walls and roofs rises above the ambient temperature. This results in larger temperature difference with respect to air conditioned space. Therefore, it is required to make necessary correction in the temperature difference in outside and inside temperature. Allowable of solar radiation, the temperature of outside wall is found to be more than that of outside air during summer, because of direct incident of solar radiations on the walls or the actual temperature of outside wall depends upon several factor such as (1) colour and smoothness of wall, and (2) its orientation.

Design Consideration for Bag Storage Structure/Godowns: In India, food grains are stored in conventional godowns designed for begged storage. These godowns have side walls of brick or stone, masonry and sloped roofing in asbestos or corrugated galvanized iron sheets over steel trusses. Bagged food grains are arranged in stacks with a base of 6m × 9m with stack height varying from 4 to 5 m, leaving 27 % of the floor area for alley ways.

Even when all precautions are taken, the period of storage of bagged grains is severely restricted through the grain can be held in bulk without any damage for extensive period of time, while wheat can be stored in bags under dry climatic conditions for a period upto 2 years (8 to 12 months in humid conditions), it can be kept in bulk safely even upto 5 years. Moreover, bulk storage is preferable both from qualitative and quantitative aspects. While, in case of bulk storage damage by birds, insects etc. can be almost eliminated, these losses can not be fully avoided in bagged storage. The only attractive feature of bagged storage system is the higher storage capacity and less initial cost. Countries like India where economics is main consideration and storage of food grains is not for very long period of time, bagged storage system is extensively adopted.

(A) Location: While selecting the location for the construction of godowns following points should be given consideration.

(*i*) Structure shall be located on a raised and well drained site not liable to flooding, inundation and seepage.

(*ii*) It should be located near on road side or should have railway siding.

(*iii*) At the side of construction, there should be sufficient parking and maneuvering space for vehicles.

(B) Storage capacity and dimension of structure: The net storage capacity of the structure is computed by deleting space for alley ways and top clear head left from total available space. It may be determined by using following empirical formula-

$$\text{Net storage capacity} = \frac{80\% \text{ of total floor area (plinth area) } (m^2) \times \text{stack height (m)}}{0.065}$$

In the term of number of bags that could be kept in the structure the space may be computed using following thump rule.

Space requirement (floor area) is 0.6 m^2 per 10 bags, if stack is 12 bags high and floor area required in 0.7 m^2 per 10 bags if stack is 10bag high.

At any point a clear space of 2 m height should be left above the stack from inside of the roof. The maximum height to which the bags are placed in a stack shall depend on the load bearing capacity of the floor, nature of commodity and strength of bags.

(C) Stack spacing (arrangement): The floor can be divided into rectangular or square section. There should be a minimum gap of 74 cm between the two stacks and also there should be a minimum gap of 150 cm between stacks and side walls. The width of main alleys should also be a minimum of 150 cm. The length and breath of stack should not exceed 9 m and 6 m, respectively.

Dunnage are made of timber matting with a larger of polyethylene sandwiched between two layers of matting should be layed on each stack space. Pallets or square shall be made of hard and well seasoned timber and shall be free from any cracks which may, otherwise, harbours insects. They shall be layed parallel to each other and shall not be more than 15 cm apart. Before laying dunnage, it should be brushed and cleaned.

Dimension of godowns

Type of godowns	Approx. No. of bags	Capacity (tonnes)	Internal dimension (L × B) sq. m
Small Godowns	1100	100	12 × 7.5
	2700	250	20 × 9
	5400	500	34 × 12
	10500	1000	35.5 × 18
	28500	2500	97.19 × 14.48
Large Godowns	57,020	5000	129.74 × 21.34 (4 compartment)
			120.31 × 21.34 (3 compartment)

For storage capacity beyond 2500 tonnes, the godown may be divided in suitable number of compartments depending on the availability of land. The stack dimension for large capacity godowns are maximum 10.915 m × 6.10 m.

(D) Foundation of structure: Suitable foundation depending upon size and condition may be provided. The type of foundation will depend upon the property of subsoil and may provide as per BIS code. The foundation in no case may be less than 1 meter deep unless, it meets a rock. If at all place rock is met within the foundation depth it shall be properly dressed, leveled and if necessary cut in horizontal strips so as to receive the footing of the foundation. The foundation consists of lean cement concrete 1:5:10 for the walls and RCC 1:2:4 for column.

(E) Footing: Leveling cause of lean cement concrete (1:5:10) at least 75mm thick shall be provided under reinforced cement concrete footings.

(F) Plinth: It is generally kept about 80 cm above the finished level both for road and rail fed structure. The minimum width of platform should be 305 cm for rail siding and 245 cm for road fed godowns.

To prevent rain water entry into the godowns through the doors and window, the platform shall be provided with a slope of 1 in 4 from it wall to outer edge. In case no railway siding is provided the platform should be at least 500 mm above the height more flood level. The plinth should be constructed of either stone or brick masonry in lime mortar 1:2 or cement mortar 1:6.

(G) Floor: The flooring in the godowns should be damp proof, rigid, durable and free from any cracks and gabbies. The following type of flooring may be provided in a godown.

Selected earth filling-well consolidated and stabilized to avoid possibilities of settlement and cracks.

- A layer of sand filling 230 mm thick thoroughly watered and well consolidated.
- A layer of cement concrete (1:5:10), 75 mm thick.
- A layer of bitumen and maximum phalt 80-100 mm thick or equivalent spread uniformly @ 1.7 kg/sq.m.
- A further layer of cement concrete (1:5:10), 75 mm thick over a water proofing layer of bitumen.
- A top coat of 50 mm thick cement concrete 1:2:4 finished with a floating coat of neat cement is provided. The cement concrete is layed in pands should not exceeds area of 3.5 sq.m.

(H) Walls: The design of walls shall be in accordance with general constructional practices and care should be taken that tensile strength do not exceed the cracking limit. The longitudinal walls shall be brick or stone masonry in cement mortar in ratio of 1:6 and shall be at least 5600mm high for road fed and 6350 m for rail fed godown form floor level. Minimum thickness of 340mm with plaster provided at truss supported. For span 10 m and above RCC column may be provided at regular intervals. Gabbles wall are made of either bricks masonry or stone masonry with cement mortar 1:4. It should be provided with water proofing compound to a minimum thickness of 40 mm for brick and 50 mm thickness of stone, masonry upto a height of atleast 3 m walls should be rendered smooth on both outer and inner surface, there should be no offset and projections in walls.

(I) Doors and Ventilators: A door shall be provided preferably opposite to each alley way. The door should be of rolling of shutter and fixed into a suitable frame. The door should not be less than 1.83 m × 2.45 m dimension. Regular and periodic maintenance of rolling shutter should be provided.

In longitudinal walls two steel ventilators of opening not less than 1.5m × 0.5 m shall be provided in each bay between RCC columns spaced at 4.5 m from centre to centre. The ventilator should be fixed at 150 mm below the top edge of wall measured from inside the godowns. They shall be provided with glazed centre hung with fixed wire mesh shutter. The frame of ventilator should be provided with suitable beading to avoid any chances of moisture and air leakage. In between above ventilators, air inlets of steel ventilator 620 mm x 620 mm glazed openable outside double huge in each bag shall also be provided at 600 mm from floor level of the godown except those which have rolling shutter. These shall be protected by expanded metal frame out side, when close shutter should fit tight on the frame.

(J) **Roofing:** Roof should be single span structural steel. These trusses are fixed on RCC column at a height of 5.6 m for the road fed and 6.35 m for rail fed, from floor level to tie level. RCC sheet type roofing is well suited for grain storage structure. Experties is needed in structural, centering and shuttering design as well as the execution for such type of roof.

Roofing of platform should be cantilever structure of angle iron or tubular trusses fixed to RCC column. These platform cantilever trusses should be fixed at a height of 3.35 m for road fed and 4.35 m for rail fed godowns (for broad gauge line). All the height should be measured from floor level of the godown. The width of the platform for rail fed and road fed godown shall be 3.05m and 2.45m, respectively.

(K) **Roofing materials:** Roofing materials may be corrugated asbestos sheet or galvanized corrugated sheet, Corrugated Aluminium sheet or black corrugated sheet, not-thinner than 0.6 mm. The sheet should be projected at least 46 cm from the outer space from longitudinal walls is both the side. The sheet should be well anchored and secured on purlins by means of galvanized Iron- J or L-shaped hooks sufficiently long to have good grip on sheets and purlins and accommodate nuts and washer. In area liable to excessive heat use of heat reflecting point may be considered.

Aeration and Fumigation: Aeration is the process of moving air through stored grain at low flow rates to maintain or improve its quality.

Benefits of Aeration

* It cools the grain and slows down insect activity.
* By cooling the grain, aeration prolongs the effectiveness of pesticides.
* It can be provide an appreciable drying function
* It prevents storage odours. Grains stored for longer periods often develop objectionable odours. Aeration provides the grains a new fresh smell.
* It reduces moisture accumulation. Moisture condensation by changes in temperature and relative humidity can be reduced by aeration.
* Application of fumigants through an aeration system is an easy and practical method for controlling insects in stored grains.

Purpose of Aeration

* To removal of heat generated in the bulk due to biological activity
* Removal of excess moisture
* Modification of gaseous environment in the bulk
* To facilitate the fumigation
* To removal the of bad odour

Physiological Development in Fruits and vegetables: The lives of fruit and vegetables can be conventionally divided into three physiological stages following initiations or germination. These are

Growth: Growth involves cell division and subsequent cell enlargement, which accounts for the final size of the produce.

Maturation: Maturation usually commences before growth ceases and includes different activities in different commodities. Growth and maturation are often collectively referred to as the development phase.

Senescence: Senescence is defined as the period when anabolic (synthetic) biochemical processes give away to catabolic (degradable) processes, leading to ageing and finally death of the tissue.

Fruit Ripening: The ripening of fruit undergoes many physiological changes after harvest. Among the important substitutes of the fruits are enzymes. These bring about changes, such as softening of the edible tissues, change in colour, flavour, odour, pectic substances, protein, carbohydrate, organic acids, ethylene production and also in respiration rate during the ripening of fruits. With ripening, changes occur in colour, flavour and texture which render the fruit acceptable to the consumer. Pectic substances are cementing substances in fruit cell which decompose to pectin and pectic acid. These decomposes are partially responsible for the softening of fruit tissues during ripening. Compositional changes in fruits during ripening are ideal if the fruits are harvested at proper maturity. Immature fruit will have unsatisfactory quality. Colour changes in fruit and vegetable is due to degradation or break down of chlorophyll and consequent unmasking of carotenoid pigments in case of orange. During ripening, there is a decrement in starch and increment in fructose, glucose and sucrose resulting in sweetness of the fruit. Ripening also decreases the organic acids and phenolics to reduce the acidity and astringency and makes it more acceptable.

Chilling of fruits & vegetables: Chilling is the unit operation in which the temperature of a food is reduced to between –1°C and 8°C. It helps to reduce the rate of bio-chemical and microbiological changes, and hence extend the shelf life of fresh and processed foods. It cause minimal changes to sensory characteristics and nutritional properties of foods and, as a result, consumers as being convenient, easy to prepare, high quality and "healthy", perceive chilled foods natural and fresh.

Chilling is often used in combination with other unit operations *i.e.*, fermentation or pasteurization to extent the shelf life of mildly processed foods. There is a greater preservative effect when chilling is combined with control of the composition of storage atmosphere than that found using either unit operation alone. However, not all foods can be chilled and tropical subtropical and some temperature fruits. *e.g.* suffer from chilling injury at 3-10°C above their freezing point.

Chilling foods are grouped into three categories according to their storage temperature range as follows

1. –1°C to +1°C (fresh fish, meats, sausages and ground meats, smoked meats and breaded fish.)

2. 0°C to +5°C (pasteurized canned meat, milk, cream, yogurt, prepared salad, sandwiches, baked goods, fresh pasts, fresh soups and sauces, pizza, pastries and unbaked dough).
3. 0° to + 8°C (fully cooked meats and fish pies, cooked or uncooked cured meat, butter, margarine, hard cheese, cooked rice, fruit juice and soft fruit).

Cryogenic Chilling: A cryogen is a refrigerant that changes phase by absorbing latent heat to cool the food. Cryogenic chillers use solid carbon dioxide (CO_2) or liquid N_2. Solid CO_2 removes latent heat of vaporization (352 (kJ)/kg at –78°C), and liquid cryogenics remove latent heat of vaporization (358KJ/kg at – 196°C) for liquid N_2; liquid CO_2 has a similar latent heat to the solids.

The gas absorbs sensible heat as it warms from –78°C (CO_2) or from –196°C (liquid N_2) to give a total refrigerant effect pf 565 KJ/Kg and 690 KJ/Kg, respectively. The advantage of CO_2 includes:

(*i*) A higher boiling and sublimation point than N_2, and therefore a loss severe effect on the food.

(*ii*) Most of ethnology (heat capacity) arises from the conversion of solid or liquid to gas.

Other application of cryogenic cooling include sausage manufacture, where CO_2 snow remove the heat generated during size reduction and mixing and cryogenic grinding where the cryogen reduces dust levels, prevents dust explosive and improves the throughout of mills. In spice milling, cryogen also prevents the loss of aromatic compounds. In the production of multiplayer chilled foods (*e.g.* triffes and other desserts) the first layer of product is filled and the surface is hardened with CO_2. The next layer can them be added immediately, without waiting for the first layer to set, and thus permit continuous cooling the case hardening of hot bakery products and chilling flour to obtain accurate and consistent flour temperature for dough preparation.

Effect of chilling on foods: The process of chilling foods to their correct storage temperature causes little or no reduction in thė eating quality or nutritional properties of food. The most significant effect of chilling on the sensory characteristics of processed foods is hardening due to solidification of fats and oils. Chemical, bio-chemical and physical changes during refrigerated storage may lead to loss of quality, and in many instances it is these changes rather than microbiological growth that limit the shelf life of chilled foods. These changes include enzyme browning, lipolysis, colour and flavor deterioration in some products and retro gradation of starch to cause staling of baked foods products (which occurs more rapidly at refrigeration temperature then at room temperature). Lipid oxidation is one of the main causes of quality loss in cook chilled products, and cooked meats in particular rapidly develop and oxidized chemical changes including colwsaw, syneresis in sauces and gravie due to changes in starch thickness, evaporation of moisture from unpacked chilled meats and cheeses, more rapid staling of sandwich bread at reduced temperature and moisture migration from sandwich fillings may each result in quality deterioration.

In cook-chill systems,, nutritional losses are reported by Bognar (1980) as insignificant for thiamine, riboflavin and retinal, but vitamin C losses are 3.3-16% per day at 20°C. The large variation is due to difference in the chilling time, storage temperature oxidation (the amount of food surface exposed to air) and reheating conditions. Vitamin C losses in cook-pasteurize-chill procedures are lower than cooked –chilled foods (*e.g.* spinach lost 66% within 3 days at

2-3°C after cook –chilling composed with 26% loss with in 7 days at 24°C after cool-pasteurizing - chilling).

Dehydro-freezing: In which, the product is first dehydrated to about 50 % moisture content using by drying method and after dehydration, the product frozen by appropriate method of freezing. In case of some fruits and vegetables is about 50 % moisture content is removed by dehydration prior to freezing. This has been found to improve the quality of the food. Dehydration does not cause deterioration and dehydration foods are relatively more stable. The advantages of this method:

i) Saving the packaging, storage, handling and transportation cost.

ii) Reduction in thawing and dripping losses

iii) Increase freezing capacity of freezer.

Freeze drying: In this method, food is first frozen at –18°C on trays in the lower chamber of a freeze drier and then frozen material dried (initially at 30°C for 24 hr and then at 20°C) under high vacuum (0.1 mm Hg) in the upper chamber. Direct sublimation of ice takes place without passing through intermediate liquid stage. The product is highly hygroscopic, excellent in taste and flavour and can be reconstitute readily. Mango pulp, orange juice concentrate, passion fruit and guava pulp are dehydrated by this method,.

Quick Freezing: In this process the food attains the temperatures of maximum ice crystal formation (0 to -4°C) in 30 min or less. Such a speed results in formation of very small ice crystal and hence minimum disturbance of cell structure. The yield of a good quality product after thawing, to achieve quick freezing. The temperature to which product is exposed is much lower. This may be achieved by following methods

* By direct immersion
* By indirect contact with refrigerant
* By air blast

Field Heat Reduction techniques (Pre-cooling): The main role of pre-cooling is to remove the field heat from the horticultural produce. This process not only removed field heat but also reduces the respiration rate, rate of ripening, ethylene production, and loss of moisture and spray of decay. There are various techniques to reduce the field heat and increase the storability of fresh fruit and vegetables are given below:

Air Cooling: When the temperature of the produce is above the desired level, and if the temperature of the outside air is lower, air is circulated through the store by conventional or mechanical means through bottom inlet vents and tops outlets fitted with dampers. Fans may be insulated and are controlled manually or automatically with differential thermostat. The air may be humidified, a process that can also be automated. Air cooled stores are cheap to construct and to operate and are still widely used for the storage of potato and sweet potato, both of which need relatively high storage temperature to avoid accumulation of sugar and chilling injury respectively. The efficiency of room cooling can be improved by the use of ceiling jets or fans to increase air velocities. Cabbage and artichokes are primarily cooled with this method, although all vegetables can be room cooled.

Hydro Cooling: Hydro cooling, in which water acts as the heat transfer medium, is a rapid method for cooling produce become of the high heat capacity and thermal conductivity of water, which makes it better cooling medium than air. Hydro cooling may also clean the produce but contamination of the produce with spoilage microorganism occurs if soil is not removed from the system regularly and the water must be chlorinated. A further advantage is that the commodities losses little weight during hydro cooling. Its use is limited for packed commodities because the containers must withstand water soaking. The produce must be tolerant of wetting.

Vacuum Cooling: Vacuum cooling is a rapid and uniform process, where produce are subjected to very low pressure and water evaporates from the commodities giving off the latent heat of vaporization, thus cooling itself. Cooling is achieved by reducing the atmospheric pressure inside a large, steel chamber containing the product. This reduction in atmospheric pressure causes loss of water vapour from the commodities resulting in cooling. It is used to cool cauliflower, celery, mushrooms, sweet corn, carrots and bell peppers. The rate of cooling by this process is largely dependent on the surface to volume ratio of the produce and the ease with which the produce loses water. Fruits, which have a low surface to volume ratio and a waxy cuticle and therefore, lose water slowly do not benefit by vacuum cooling.

Ice Refrigeration: An advance on air-cooling storage is the use of natural ice as a refrigerant. The lower temperature obtained enable longer storage of perishable foods *i.e.*, horticultural produce. The melting of one kg of ice absorbs 325 kJ heat, but the considerable bulk of ice needed and disposal of the melt water are demerits. The introduction of small 'ice box' or ice chest is a great advance in the domestic and commercial preservation of foodstuffs.

Package Icing: Package icing is used to cool some produce that is field packed into shipping containers. The ice may be finely crushed, flake ice, or a slurry of ice and water called liquid-ice. Liquid-ice is injected in the container and has better contact with the produce than the other forms. More expensive water-tolerant containers are required, and the added weight of the ice decreases the weight of actual produce that can be shipped.

Forced-air cooling: Forced-air cooling is similar to room cooling in that produce is placed in a cold storage room. Forced-air cooling is designed to force cold air through produce containers instead of around them. Converting existing cooling facilities to forced-air cooling is practical and feasible. There are variations of forced-air cooling that fit specific container needs. The key to forced-air cooling is moving the cold air through the container and its contents. Important factors in container ventilation are location of container vents, stacking of containers, and size of the vents. Container vents should be aligned whether the containers are straight-stacked or cross-stacked, to maximize airflow through the containers. If vents are too small or too few, air flow is slowed. If there are too many, the container may collapse. In this method, containers are stacked close together (tight). Five per cent vent-hole space per side and/or end is best. Liners, bags, wrappers, or dividers can slow the flow of air through the container, so pre-cooling produce is usually recommended prior to additional packing.

***Anti-nutritional factors in foods*:** Substance which, from nutritional point of view, interferes with normal growth, reproduction or health when consumed regularly in amount existing in a normal component of diet should be considered as harmful and toxic. A significant part of human population rely on legumes as staple food for subsistence, particularly in combination with cereals. They are unique foods because their rich nutrient content including starch, protein, dietary fibre, oligosaccharides, phytochemicals (especially the isoflavones in soybean) and minerals. Their nutritional contents contribute too many health benefits to human. The anti-nutritional factors present in food, including their contents of enzyme inhibitors, lectins, flatulence factors, tannins, phytic acid and saponins. Anti-nutritional factors can cause detrimental effects to humans and animals growth and performance by impairing intake, uptake, or utilization of other food and feed components, or by causing discomfort and stress to humans and animals. The anti-nutritional factors mainly occur in pulses, grain legumes, foods and feed material prepared from grain legumes and pulses. Anti-nutritional factors in grain legumes can be divided into several groups based on their chemical and physical properties such as non protein amino acids, quinolizidine alkaloids, cyanogenic glycosides, pyrimidine glycosides, isoflavones, tannins, oligosaccharides, saponins, phytates, lectins or protease inhibitors. Their elimination can be achieved either by selection of plant genotypes with low levels of such factors, or through post-harvest processing (germination, boiling, leaching, fermentation, extraction etc.).

Eliminating methods of Anti-nutritional factor from foods: There are various methods to reduce the activity of antinutritional factor from the foods are described below

Thermal treatment: Legume seeds are hardly been consumed raw; they are usually cooked and by this procedure lectins and protease inhibitors are inactivated. Low molecular weight compounds are leached out into the cooking water, to be discarded afterwards. All these simple techniques have been "invented" by man (even without a profound knowledge of the underlying toxicology) to make legume seeds more palatable and digestible. Today, a deeper knowledge of the chemical structure of the anti-nutrients involved can help to devise technological strategies to process legume seeds in order to obtain toxin free products. Since a diversification will increase the economic value of the overall crop plant, food technology and rational processing are an alternative to breeding of ANF free plants which can be more susceptible to pests and pathogens. Followed by modern food technology (separation, filtration etc), pure nutritionally valuable dietary products, such as protein, dietary fibres, oil and other fine chemicals can be generated. The remaining fractions containing the anti-nutrients don't need to be discarded: some of them are useful for the pharmaceutical industry (see below), others might be used in agriculture as biorational pesticides. Because of higher costs these considerations are related to the use of legume seeds which will be utilized for human consumption (an example is soybean for which a specialized industry has been developed). The situation is different in the case of feed industry (which is a major outlet of legume seeds in Europe) where pulses are simply ground or pelleted. For this purpose we need to develop varieties which are low in heat stable nutritional factors (*e.g.*, alkaloids, saponins, phytates, isoflavones, and non-protein amino acids). If heat labile compounds (protease inhibitors, lectins) can be denatured by heat treatment during grinding and pelleting these compounds might be maintained since they confer resistance to the plants. Thus many fields remain to be explored even in a widely known group like the temperate grain legumes.

Table : Heat-Labile Antinutritional Factors Inactivated by Thermal Processing

Antinutritional factor	Common food sources	Effects of anti-nutritional factor
Avidin	Egg whites	Binds biotin, making it biologically unavailable
Hemagglutinins	Red kidney beans, yellow wax beans	Induces red blood cell clumping
Lathyrogens	Chick peas	Disrupts collagen structure
Goitrogens	Sweet potatoes, beans cabbage, turnips	Causes goiters by limiting iodine absorption
α-Amylase inhibitors	Cereal grains, peas, beans	Slows starch digestion
Trypsin inhibitor	Legumes, egg whites, potatoes	Inhibits activity of trypsin
Thiaminases	Fish, shellfish, brussel sprouts, red cabbage	Destroys thiamin

Chemical detoxification: Deaminocanavanine is a well known untoxic deamination product of canavanine. The degradation of canavanine to deaminocanavanine under alkaline conditions provides therefore a chemical strategy for the detoxification of this compound and has already been successfully employed for the processing of the canavanine containing seeds of *Canavalia ensiformis.* The major antinutrient in the seeds of *V. sativa* (besides vicine) can be rendered inactive by mild acid hydrolysis. It is reasonable to propose that in principle post-harvest detoxification procedures can be developed for these anti-nutritional factors.

Fermentation: The use of fermentation as an integral part of food detoxification processes is widely practiced. A wide variety of fermented foods are produced and eaten around the world. The further development of fermented foods has been advocated amongst nutritionists because of the intrinsic nutritional benefits associated with such products. Fermentation is also an effective means for food preservation. Fermented foods can be prepared at both, an industrial and the household scale. Indeed, many fermented foods are prepared by very simple techniques and represent grass roots technology which is already widespread; a fact which facilitates their further refinement, transfer and adoption in underdeveloped countries for the detoxification of alternative food sources. The various Indian household food preparation techniques for their effectiveness in detoxifying *L. sativus*, and their methods which included a fermentation step were the most effective in reducing ODAP levels, eliminating 95% of this toxin. Further improvements in detoxification are likely to be made with selection for better ODAP degradation. Such methods can, in principle, also be used for the post-harvest detoxification of *Vicia* seeds, thus providing an alternative approach to the wider utilization of these grains without the need for genetic removal of their low molecular weight antinutritive and unpalatability factors. The incorporation of fermentation processes into other simple food technologies also offers good prospects for a detoxification of food sources while simultaneously giving flexibility in the manipulation of flavour, texture and colour of the raw material.

Germination: Pea and lentil sprouts (even lupin sprouts in Germany) have gained popularity in recent years. Traditionally, mediterranean grain legumes have not been used as sprouts. The potential toxicity of beta-isoxazolin-5-one-alanine (BIA), the biosynthetic precursor for the lathyrism toxin beta-ODAP may be a risk factor if consumption of such sprouts is excessive. The concentration of this compound increases during the germination of lentils and peas. This kind of processing, however, which reduces the contents of oligosaccharides and of other N-containing ANFs, has a long history in Asia, where it has served to improve the palatability of soybeans. Pea sprouts are a very recent addition to Chinese cuisine.

Hazard Analysis and Critical Control Points (HACCP): HACCP system is designed to identify systematically specific hazards and preventative measures for their control to ensure the safety and quality of food. The hazards that have been identified and assessed and the control systems that have been developed and installed, focus on preventative measures rather than relying mainly on end-product testing. It should be versatile and capable of accommodating change, such as advances in equipment design, processing procedures, or technological developments. They should also be applied from the production of the raw material to final consumer to attain any real assurance of food safety, optimal use of resources, and more timely. Before proceeding further it is important to knowledge about the certain terms related to HACCP system.

Hazard: A hazard is defined as the potential to cause harm to the consumer and/or the product and can be biological, chemical or physical, which has the potential to classify as severe, moderate and low depending on the type and extent of likely harm and the rapidity with which they may occur and spread.

Hazard Analysis: The identification of potentially hazardous ingredients, storage conditions, packaging, critical control points and relevant human factor which may affect product safety and quality.

Critical Control Point (CCP): A Critical Control Point is the location identifying a particular processing step in a procedure or an operational practice which is required to be controlled in order to totally eliminate the possibility of a hazard occurring or to prevent its occurring or to minimize it to acceptable level.

Risk: A risk is the probability of hazard occurring.

Critical limit: The critical limit is a value which defines the difference between acceptability and unacceptability.

Monitoring: A planned sequence of observations or measurements of critical limits designed to assess whether a CCP is under control and to produce an accurate record.

The Hazard Analysis and Critical Control Points (HACCP) is a quality assurance system for ensuring the safety of the manufacturing products. HACCP system which identifies, evaluates and controls hazards that are significant for food safety can be applies throughout the food chain from the primary production to final consumption. It enhances food safety besides better use of resources and timely response to problems. This is the reason why HACCP system is now widely embraced by the food industries and by the government regulatory agencies around the world as a most cost effective means of minimizing the occurrence of identifiable food borne biological, chemical and physical hazards and maximizing product safety. The primary objective of a HACCP programme is to produce reliably a safe food. This means a product which is free of microbiological, chemical or physical hazards. Industry is fairly familiar with various microbiological, chemical and physical hazards and trying to avoid various common hazards. They result in serious consumer safety exposures and expensive product recall and retrieval. It is, therefore, necessary to have technical orientation on common microbiological, chemical and physical hazards that may cause serious problems in foods. It is a system which targets critical areas of processing and in doing so, reduces the risk of manufacturing and selling unsafe products. The HACCP system consists of seven principles as follows:

* Hazards analysis.
* Identify the critical control points (CCPs) in food preparation.
* Establish critical limits (thresholds) which must be met at each identified critical control point.
* Establish procedures to monitor CCPs.
* Establish the corrective action to be taken when monitoring indicates that a critical limit has been exceeded.
* Establish procedures to verify that the HACCP system is working.
* Establish effective record keeping that will document the HACCP system.

The system is based on monitoring are outside the pre-set limit. HACCP is used throughout each stage of a process, and includes raw materials, processing, storage and distribution. It can be used for all potential hazard, including inadequate quality as well as safety and can identify areas of control where failure has not yet been experienced, making it particularly useful for new operations. Implementation of HACCP systems involves the following stages:

1. Produce a detailed flow diagram of the process, including production methods and schedules, preparation and transport of raw materials, in process stock etc., confirmed by a site visit.
2. Identify the essentials characteristics of the product and its use and define the hazards or potential that could threaten the consumer or the product. Included in this is the product formulation, handling procedures and storage conditions, packaging, expected customer handling and the target consumer groups.
3. Consider all stages in the process, including realistic process deviations, and identify critical stages (critical control points, CCPs) which must be controlled to assure safety. The judgment of risk is made using one of three methods: probabilistic, comparative or pragmatic, and should be made by people who have a high degree of expertise and experience.
4. Devise target levels and critical limits for each CCP and produce effective procedures to monitor them, to verify them and to implement corrective action. Monitoring may be by physical, microbiological or chemical tests or by visual or sensory observations. All monitoring procedures should be recorded and they should also include the location of the CCP, the frequency of monitoring and satisfactory compliance criteria. Examples include

cleaning procedures (what is cleaned, how and when it is cleaned, who cleans it and what with), temperature of foods, processing conditions, recipe formulation, hygienic practices, opportunities for cross contamination and workers illness or infections.

5. Establish documentation to record the HACCP system and the procedure needed to review it.

There are a many advantages of implementation of HACCP in food processing industry:

1. It is a proven system for providing safety.
2. It gives confidence that food safety is being effectively managed and nothing adverse is going to happen.
3. It ensures prevention of anything going wrong without waiting for problems to appear and which need to be controlled.
4. It also avoids lot of botheration of analyzing products in the end before sending products out in the market.
5. It has been found to be cost effective system which encompasses everything between the source of materials and the critical processing steps to final handling so that risk in manufacturing unsafe products is eliminated.
6. HACCP system not only ensures safety but also other attributes of foods.
7. HACCP is well proven and very flexible system, if followed correctly and by trained personnel it creates confidence of consumers in the producing facility and product itself.

High pressure processing: High Pressure Processing (HPP) is a method of food processing where food is subjected to elevated pressures (up to 87,000 pounds per square inch or approximately 6,000 atmospheres), with or without the addition of heat, to achieve microbial inactivation or to alter the food attributes in order to achieve consumer-desired qualities. Pressure inactivates most vegetative bacteria at pressure above 60,000 pounds per square inch. HPP retains food quality, maintains natural freshness, and extends microbiological shelf life. High (hydrostatic) pressure in the range of 1-8 kbar (100-800 MPa) can inactivate vegetative cells of pathogenic and spoilage microorganisms without heat damage to the quality of sensitive foods. High pressure processing (HPP) can be applied to pasteurization and minimal processing. Destruction of microbial spores any require combined heat and HPP treatment.

The process is also known as high hydrostatic pressure processing (HHP) and ultra high-pressure processing (UHP). Most processed foods today are heat treated to kill bacteria, which often diminishes product quality. High pressure processing provides an alternative means of killing bacteria that can cause spoilage or food-borne disease without a loss of sensory quality or nutrients. In a typical HPP process, the product is packaged in a flexible container (usually a pouch or plastic bottle) and is loaded into a high pressure chamber filled with a pressure-transmitting (hydraulic) fluid. The hydraulic fluid (normally water) in the chamber is pressurized with a pump, and this pressure is transmitted through the package into the food itself. Pressure is applied for a specific time, usually 3 to 5 minutes. The processed product is then removed and stored/distributed in the conventional manner. Because the pressure is transmitted uniformly (in all directions simultaneously), food retains its shape, even at extreme pressures. And because no heat is needed, the sensory characteristics of the food are retained without compromising microbial safety. High pressure processing causes minimal changes in the 'fresh' characteristics of foods by eliminating thermal degradation. Compared to thermal processing, HPP results in foods with fresher taste, and better appearance, texture and nutrition. High pressure processing can be conducted at ambient or refrigerated temperatures, thereby eliminating thermally induced cooked off-flavors. The technology is especially beneficial for heat-sensitive products. Like any other processing method, HPP cannot be universally applied to all types of foods. HPP can be used to process both liquid and solid foods. Foods with a high acid content are particularly good candidates for HPP technology. At the moment, HPP is being used in the United States, Europe, and Japan on a select variety of high-value foods either to extend shelf life or to improve food safety. Some products that are commercially produced using HPP are cooked ready-to-eat meats, avocado products (guacamole), tomato salsa, applesauce, orange juice, and oysters. HPP cannot yet be used to make shelf-stable versions of low-acid products such as vegetables, milk, or soups because of the inability of this process to destroy spores without added heat. However, it can be used to extend the refrigerated shelf life of these products and to eliminate the risk of various food-borne pathogens such as *Escherichia coli*, *Salmonella* and *Listeria*. Another limitation is that the food must contain water and not have internal air pockets. Food materials containing entrapped air such as strawberries or marshmallows would be crushed under high pressure treatment, and dry solids do not have sufficient moisture to make HPP effective for microbial destruction.

Canning and Bottling: Canning involves cooking fruits or vegetables, sealing them in sterile cans or jars, and boiling the containers to kill or weaken any remaining bacteria as a form of pasteurization. Various foods have varying degrees of natural protection against spoilage and may require that the final step occur in a pressure cooker. High-acid fruits like strawberries require no preservatives to can and only a short boiling cycle, whereas marginal fruits such as tomatoes require longer boiling and addition of other acidic elements. Many vegetables require pressure canning. Food preserved by canning or bottling is at immediate risk of spoilage once the can or bottle has been opened. Lack of quality control in the canning process may allow ingress of water or micro-organisms. Most such failures are rapidly detected as decomposition within the can causes gas production and the can will swell or burst. However, there have been examples of poor manufacture and poor hygiene allowing contamination of canned food by the obligate, *Clostridium botulinum* which produces an acute toxin within the food leading to severe illness or death. This organism produces no gas or obvious taste and remains undetected by taste or smell. Food contaminated in this way has included Corned beef and Tuna.

Food Preservation: The process of treating and handling food in a way to stop or cut down spoilage to prevent the food borne illness without hampering the texture, nutritional value and flavour is called as food preservation. In other words food preservation is method of preparing food to be stored for future use. Food preservation is practiced from the early ages to make food edible for a long time. The food preserved in early times was cheese, butter, raisins, pemmican, sausage, bacon and grains. Food preservation involves preventing the growth of bacteria, fungi and other micro-organisms and retarding the oxidation of fats causing rancidity. Food preservation to a great extent means preventing the contamination of food substances.

Preservation processes include:

* *Heating to kill or denature organisms (e.g. boiling)*
* *Oxidation (e.g use of sulphur dioxide)*
* *Toxic inhibition (e.g. smoking, use of carbon dioxide, vinegar, alcohol etc)*
* *Dehydration (drying)*
* *Osmotic inhibition (e.g use of syrups)*
* *Low temperature inactivation (e.g. freezing)*
* *Ultra high water pressure (e.g. fresherized, a kind of "cold" pasteurization, the pressure kills naturally occurring pathogens, which cause food deterioration and affect food safety.)*
* *Many combinations of these methods*
* *Chelation*

Membrane processing: This is a technique that permits concentration and separation without the use of heat. Particles are separated on the basis of their molecular size and shape with the use of pressure and specially designed semi-permeable membranes. Membrane separation processes are based on the ability of semi permeable membranes of appropriate physical and chemical nature to discriminate between molecules primarily on the basis of size, and to lesser extent, on shape and chemical composition. Depending on the chemical/physical properties of the membrane (primarily pore size distribution) concentration or dewatering (reverse osmosis, RO), fractionation of components in solution (ultrafiltration, UF) and clarification of slurries or removal of suspended matter (microfiltration, MF) are accompanied. Membrane filtration is a pressure driven technology with pore sizes ranging from 100 molecular weight to 5 microns.

Reverse Osmosis/Hyperfiltration: RO is a process where the fluid is passed through a semi-permeable membrane with very small pores that permit only the transfer of water from other solutes. Most systems consists a membrane cast on a solid porous backing-usually in the form of a tube. High pressure is applied to force the water (called permeate) through the membrane the concentrated fluid (called retentate) is retained in the tubing. The rate of water removal decreases, as the fluid is concentrated, until it is no longer economically feasible to remove more water. Reverse osmosis (RO) designates a membrane separation process, driven by a pressure gradient, in which the membrane separates the solvent (generally water) from other components of a solution. The membrane configuration is usually cross-flow. With reverse osmosis, the membrane pore size is very small allowing only small amounts of very low molecular weight solutes to pass through the membranes.

Ultrafiltration: Ultrafiltration is a fractionation process separating molecular species on the basis of size using a membrane, which retains the larger molecules and allows water and smaller molecules to pass. Ultrafiltration, UF has membranes with larger pores and will retain protein, lipids and colloidal salts, while allowing smaller molecules to pass through to the permeate phase. It designates a membrane separation process, driven by a pressure gradient, in which the membrane fractionate components of a liquid as a function of their solvated size and structure. The membrane configuration is usually cross-flow. In UF, the membrane pore size is larger allowing some components to pass through the pores with the water. It is a separation/ fractionation process using a 10,000 MW cutoff, 40 psi, and temperatures of 50-60°C with polysulfone membranes. In UF milk, lactose and minerals pass in a 50% separation ratio; for example, in the retentate would be 100% of fat, 100% of protein, 50% of lactose, and 50% of free minerals.

Diafiltration: It is a specialized type of ultrafiltration process in which the retentate is diluted with water and re-ultrafiltered, to reduce the concentration of soluble permeate components and increase further the concentration of retained components. It is more or less a purification process in producing milk enriched in protein and fat, and very low in lactose and salt content. This most efficient diafiltration, is carried out by adding water at the same rate as the permeate is removed.

Microfiltration: Microfiltration (MF) designates a membrane separation process similar to UF but with even larger membrane pore size-allowing particles in the range of 0.2 to 2 micrometers to pass through. The pressure used is generally lower than that of UF process. The membrane configuration is usually cross-flow. MF is used in the dairy industry for making low-heat sterile milk as proteins may pass through but bacteria can not. Microfiltration with pores > 0.1 micron is used to separate fat from protein and to reduce microorganisms from fluid food system.

Additional Short Notes

Nitrite and nitrate: Cured meat products should demonstrate a strong and stable red curing colour as the customer initially selects a product by sight. In order to obtain a stable red curing colour in meat products, sodium nitrite ($NaNO_2$) is generally the material of choice. $NaNO_2$ dissociates into Na^+ as well as NO_2^- and nitrite is the water-soluble, highly toxic salt of nitrous acid (HNO_2). For nitrite, and here specifically sodium nitrite, the lethal dose for humans is by around 1.1 g. For that reason, a mixture of salt and nitrite is commonly sold to manufacturers of cured meat products. The level of nitrite within such salt–nitrite mixtures varies generally between 0.5% and 20%, depending on the country's regulation.

Nitrite is a strong oxidizing agent and can cause pinking in non-cured meat products at levels as low as 3–5 ppm of nitrite per kilogram of meat product. Occasionally, potassium nitrite is utilized in applications such as low-sodium products. Based on the difference in molecular weight, around 30% more potassium nitrite has to be applied if the same impact on colour development is to be achieved compared with sodium nitrite.

Nitrate (NO_3^-) is the salt of nitric acid (HNO_3) and is also readily soluble in water. In meat products, potassium nitrate (KNO_3) is the most commonly applied source of nitrate but, having said that, in a large percentage of all cured meat products, nitrite is the material of choice. Potassium nitrate dissociates into K^+ and NO_3^- ions and is still sometimes introduced in products such as fermented salami dried for a long time as well as cured air-dried products. Nitrate does not contribute directly to the formation of the red curing colour and nitrite is the active component regarding the formation of the desired red curing colour. If nitrate is applied, it has to be reduced to nitrite, which then forms the curing colour in conjunction with myoglobin, the colour of meat. Nitrite is utilized in meat products for the following reasons.

- It promotes the red curing colour.
- It provides the curing flavour.
- It has a bacteriostatic impact.
- It also acts as an antioxidant.

In order to obtain a strong and stable curing colour in cooked cured meat products, around 30–50 ppm of nitrite are required per kilogram of meat product. Nitrite also acts as a preservative against bacteria such as *Salmonella* spp. and *Staphylococcus aureus* and especially against *Clostridium botulinum*. A concentration of 80–140 ppm of nitrite per kilogram of meat product is an effective hurdle against the growth of those bacteria especially in canned and retorted products. However, nitrite has relatively little impact against *Micrococcus* spp., *Lactobacillus* spp. and *Enterococcus* spp.

Alveograph: The alveograph test provides results that are common specifications used by flour millers and processors to ensure a more consistent process and product. The alveograph is well suited for measuring the dough characteristics of weak gluten wheats. Weak gluten flour with low P value (strength of gluten) and long L value (extensibility) is preferred for cakes and other confectionery products. Strong gluten flour will have high P values and is preferred for breads.

The alveograph test measures and records the force required to blow and break a bubble of dough.

- ***P Value*** is the force required to blow the bubble of dough. It is indicated by the maximum height of the curve and is expressed in millimeters (mm).
- ***L Value*** is the extensibility of the dough before the bubble breaks. It is indicated by the length of the curve and is expressed in millimeters (mm).
- ***P/L Ratio*** is the balance between dough strength and extensibility.
- ***W Value*** is the area under the curve. It is a combination of dough strength (P value) and extensibility (L value) and is expressed in joules.

Weak gluten flour has lower P values than strong gluten flour.

Extensigraph: Extensigraph test are useful in determining the gluten strength and bread-making characteristics of flour. The effect of fermentation time and additives on dough performance can also be evaluated.

Falling Number: The level of enzyme activity measured by the falling number test affects product quality. Yeast in bread dough, for example, requires sugars to develop properly and therefore needs some level of enzyme activity in the dough. Too much enzyme activity, however, means that too much sugar and too little starch are present. Since starch provides the supporting structure of bread, too much activity results in sticky dough during processing and poor texture in the finished product. If the falling number is too high, enzymes can be added to the flour in various ways to compensate. If the falling number is too low, enzymes cannot be removed from the flour or wheat, which results in a serious problem that makes the flour unusable.

- The falling number instrument analyzes viscosity by measuring the resistance of a flour-and-water paste to a falling stirrer.
- Falling number results are recorded as an index of enzyme activity in a wheat or flour sample and the results are expressed in time as seconds.
- A high falling number (for example, above 300 seconds) indicates minimal enzyme activity and sound quality wheat or flour.

- A low falling number (for example, below 250 seconds) indicates substantial enzyme activity and sprout damaged wheat or flour.

Farinograph: The farinograph test is one of the most commonly used flour quality tests in the world. The results are used as parameters in formulation to estimate the amount of water required to make a dough, to evaluate the effects of ingredients on mixing properties, to evaluate flour blending requirements, and to check flour uniformity. The results are also used to predict processing effects, including mixing requirements for dough development, tolerance to over-mixing, and dough consistency during production. Farinograph results are also useful for predicting finished product texture characteristics. Fór example, strong dough mixing properties are related to firm product texture.

- The farinograph determines dough and gluten properties of a flour sample by measuring the resistance of dough against the mixing action of paddles (blades).
- Farinograph results include absorption, arrival time, stability time, peak time, departure time, and mixing tolerance index.

Mixograph: The mixograph determines dough and gluten properties of flour by measuring the resistance of dough against the mixing action. Mixograph results include water absorption, peak time, and mixing tolerance. The mixograph curve indicates gluten strength, optimum dough development time, mixing tolerance (tolerance to over-mixing), and other dough characteristics. The amount of water added (absorption) affects the position of the curve on the graph paper. Less water increases dough consistency and moves the curve upward. The mixograph test quickly analyzes small quantities of flour for dough gluten strength. Wheat breeders use mixograph results to screen early generation lines for dough gluten strength. Flour water absorption measured by the mixograph often serves as bake absorption in bread baking tests.

Dough Conditioners: While many brands of dough conditioners are available, most contain a mixture of the following ingredients:

- Emulsifiers, such as DATEM and calcium stearoyl-2-lactylate, for increasing water absorption and gluten strength. DATEM stands for diacetyl tartaric acid esters of mono- and diglycerides.
- Salts and acids, such as calcium carbonate or monocalcium phosphate, for optimizing gluten development by adjusting water hardness and pH. Calcium carbonate increases both water hardness and pH; monocalcium phosphate increases water hardness while it decreases pH. Monocalcium phosphate, an acid salt, is also present in many baking powders.
- Maturing agents that strengthen, such as potassium bromate, ascorbic acid, potassium iodate, and azodicarbonamide (ADA), for increasing gluten strength.
- Yeast foods, such as ammonium salts, to improve yeast fermentation.
- Enzymes, such as amylase, to improve yeast fermentation and browning, and to delay staling.

Patent flour: Patent flour is the highest quality of all commercial grades of white flour. Bakers often use the term patent flour to mean patent bread flour, but most flours sold today—whether bread, pastry, or cake—are patent flours. Patent flour is made by combining the first few streams of flour from the milling process. It consists of the innermost part of the endosperm and is essentially free of bran and germ. This makes patent flour lowest in ash and whitest in color. Different grades of patent flours are available, depending on which streams of flour from the milling process are blended. The highest-quality patent flour is called extra short or fancy patent.

Clear flour: Clear flour is the lowest quality of all commercial grades of flour. It is milled from the outer part of the endosperm, made from flour streams that remain after patent flour is produced. While different grades are available, all clear flours are relatively high in bran, high in protein and ash, and slightly gray in color. A high grade of clear flour, called first clear, remains after the production of hard wheat first patent flour. Clear flour is less expensive than patent flour. While it is higher in total protein than patent flour, the gluten formed from clear flour is typically of lesser quality than that from patent flour. First clear is commonly added to rye and whole grain breads. Its protein provides needed strength to low-gluten grains, while its gray cast is hidden by the dark color of the rye or whole grain.

Straight flour: Straight flour is milled from the entire endosperm. It is made by combining all usable streams of flour from the milling process and contains bran and germ particles that are not easily separated from the endosperm. Straight flour is not commonly used by the baking industry in North America. French bakers, however, use a type of straight flour in breads.

Wheat Flour Terms: The Food and Drug Administration inspects and approves the use of flour treatments and additives that are used to improve the storage, appearance and baking performance of flour. The treatment additives are in no way harmful.

- **Enriched"** flour supplemented with iron and four B-vitamins (thiamine, niacin, riboflavin and folic acid) and may be with calcium. Reconstituting the nutritional status of a processed food ingredient to match that of the original raw materials.
- **"Fortified"** implies that something is added to a product that makes its nutritional status higher than the product made from "unprocessed" raw materials. i.e. Cereals.
- **"Pre-sifted"** flour is sifted at the mill, making it unnecessary to sift before measuring.
- **"Bromated"** flour is largely discontinued in the United States. Ascorbic acid is now being added to strengthen the flour for bread dough's.
- **"Bleached"** refers to flour that has been bleached chemically to whiten or improve the baking qualities. No change occurs in the nutritional value of the flour and no harmful chemical residues remain. It is a process which speeds up the natural lightening and maturing of flour.
- **"Unbleached"** flour is aged and bleached naturally by oxygen in the air. It is more golden in color, generally more expensive and may not have the consistency in baking qualities that bleached flour does. Unbleached is preferred for yeast breads because bleaching affects gluten strength.

- **"Patent"** flour, bleached or unbleached, is the highest grade of flour. It is lower in ash and protein with good color. Market-wise, it is considered the highest in value and mostly used by bakers.
- **"Organic"** or chemical-free flour is not standardized, so its definition varies from state to state. It may be grown and stored without the use of synthetic herbicides or insecticides. It may also mean no toxic fumigants were used to kill pests in the grain and no preservatives were added to the flour, packaging, or food product.
- **"Gluten"** is a protein formed when water and wheat flour is mixed. Gluten gives bread dough elasticity, strength and gas-retaining properties. Wheat is the only grain with sufficient gluten content to make raised or leavened loaf of bread.

Accelerated freeze drying: This is a process of food dehydration for certain foods such as peas, prawns and coffee. The process involves placing prepared pieces of quick-frozen food under vacuum and gently heating. As a result, 'sublimation' takes place where crystals of ice turn to water vapour without melting. On completion of the process, the food has a brittle and sponge-like texture. Foods treated in this way have a much longer shelf life providing the packaging is kept intact. When water is added to the product, it enters much faster than with conventional dried products. Furthermore, when reconstituted (or re-hydrated), the food is similar in nature to that of the original food. Re-hydrated foods should be treated in the same way as fresh foods.

'Best before' date: This is a system of marking food products with a specific date before which the food will retain its original quality or 'shelf life'. The system is generally applied to low risk foods, some of which have long shelf-lives, such as canned goods, drinks, biscuits and other dry products.

Blast chilling and freezing: Blast chilling is the process of using a fan-assisted refrigeration cabinet or room which lowers the temperature of cooked food very quickly to around 5°C prior to going into refrigerated storage. In order to prevent bacterial multiplication, the process should be completed within 90 minutes. Attention must be paid to the specific size of the food commodities being chilled, particularly in the case of meat joints. Ideally these joints should be no more than 2.5 kg in weight and regularly shaped. Blast freezing takes place in similar manner to blast chilling. Here hot cooked food is quickly reduced to between –18°C and –22°C before being put into refrigerated storage.

Botulism: This is the disease condition caused by the toxin of *Clostridium botulinum* which, in many cases, attacks the central nervous system. This micro-organism is associated with a range of foods including canned fish, certain pates and bottled vegetables. The disease has an incubation period of 18 to 36 hours and initial symptoms are characterized by diarrhoea, headache and an overall tiredness. In serious cases, these symptoms may be followed by double vision, loss of balance, constipation and speech difficulties which, in some cases, can be fatal. Patients take a long time to recover and this recovery may be incomplete.

Cleaning-in-Place (CIP) systems: CIP systems are commonly installed in those premises manufacturing and/or bottling liquid products, such as beer, milk and mineral waters. The system involves extensive pipework connected to all items of equipment which come into contact with the product at various stages of the process. At the completion of the food manufacturing process, or following the completion of a particular batch, the whole of the plant is flushed with a series of detergents and/or detergent/sanitisers followed by hot and cold water rinses. Detergents may be held within the system at a particular temperature for a specified length of time with a view to clearing soiling, food residues and other deposits. Most CIP systems tend to be fully automatic and controlled from a central location.

Clostridium botulinum (Clostridium welchii): This micro-organism is one of the most deadly food poisoning bacteria. It is a Gram-positive anaerobic bacillus which forms spores and can survive high temperatures. Consumption of food contaminated by *Clostridium botulinum* may result in botulism due to the release of a neurotoxin. There are seven types of botulism (A to G), each of which is recognised by the toxin produced in each strain, types A, B, E and F causing botulism in humans. Death may result due to disturbances of various brain centres, particularly that regulating the control of breathing. Many foods are associated with the micro-organism, principally those with a pH above 4.5. Those which are common are poorly processed or contaminated bottled or canned products, which provide the anaerobic conditions necessary. Contamination of meat and fish may also arise since spores of the organism are widely spread.

Cold chain: A term used to describe the continuing and uninterrupted storage requirements of certain high risk chilled foods from the final production stage to actual consumption. The cold chain incorporates a number of stages of storage, including that after production, during transport and at the premises of wholesalers, retailers and ultimate consumers.

Coliforms: Coliforms or, more correctly, faecal coliforms are a group of bacteria commonly found in the colon (large intestine) and include *Eischerichia coli* (*E. coli*), *Vibrio cholerae* and *Shigella dysenteriae*. Evidence of their presence in food may be an indication of sewage contamination. Coliforms are also responsible for much food spoilage, causing the production of 'off' flavours.

Colony count: A microbiological technique which, fundamentally, involves counting the number of colonies of micro-organisms present in or on a particular food sample after a specified period of time.

Cook-chill: This is a method of mass food preparation used by hospitals, airlines and supermarkets, in particular, where there is a high demand for ready-prepared meals on a continuing basis. In this process food is prepared, cooked to a core temperature of 70°C and then chilled rapidly to 3°C. It is then stored at that temperature and, on demand, regenerated (reheated) to 70°C. The success of this process is dependent on effective temperature control at all stages. This entails regular temperature checks using probe thermometers and should the food, during its storage period, reach 5°C, it must then be consumed within 12 hours. Where the food commodity temperature reaches 10°C, it must be rejected. This is, fundamentally, a five day process although some operators of the system may reduce this time. The storage period lasts three days with a further day for regeneration and eventual consumption on the fifth day.

Corrective action: Corrective action implies any form of action directed at preventing contamination of food and is an important feature of the HACCP process. Other food safety monitoring techniques, such as food safety risk assessments, food hygiene inspections and audits, may identify deficiencies in cleaning procedures, inadequate storage temperatures and other breaches of food hygiene legislation. Inspection reports should specify corrective action required and the time scale for such action.

Deep freeze: This is the process of storing a range of foods at –18°C and below at which the growth of many micro-organisms is virtually reduced and the shelf life greatly increased. However, storage at these temperatures does not necessarily kill pathogenic bacteria and, on thawing, bacteria will commence to multiply. Spoilage enzymes are also inhibited at these temperatures, but long term storage may result in loss of quality.

Endotoxin: These are hazardous toxic substances which are released from the cells of bacteria and other micro-organisms, including various fungi. They remain within the cell during its normal life and it is only after the death of the cell, damage to the cell wall or during the formation of spores, that they may be released. Endotoxins are associated with *Escherichia coli*, *Salmonellae* and *Clostridium perfringens* and the symptoms of infection include diarrhoea, fever and local tissue shock.

Enterotoxin: These are poisonous waste products which attack the stomach and intestine linings causing diarrhoea and vomiting. Both endotoxins and exotoxins fall within this group. Enterotoxins are particularly poisonous and resistant to heat, being capable of surviving at a temperature of 100°C for 30 minutes. However, the bacteria which produce enterotoxins, such as *Staphylococcus aureus*, are less heat-resistant and the presence of bacteria in a food product is not vital for illness to arise.

E-numbers: The term 'E-number' applies to the incorporation in food of a range of additives, such as antioxidants, colours, emulsifiers, stabilisers, flavour enhancers, flavourings, preservatives and sweeteners. Where a food additive is identified by an E-number, this implies it has been tested for food safety and is approved for use throughout the European Union. This approval is monitored, reviewed and amended in the light of new scientific data.

Escherichia coli (E.coli): This micro-organism takes a number of strains and is found in the intestines of man, farm animals, dogs and cats. It is commonly found in raw foods, such as meat, poultry and fish and their by-products. It is easily destroyed by heat, through pasteurization and effective cooking. The presence of *E. coli* is generally an indicator of poor standards of hygiene and/or inadequate processing of food. Whilst the majority of strains are harmless, some strains cause a range of ill health conditions due to the release of an enterotoxin, (which causes an illness similar to cholera in symptoms) and the invasion of the lining of the intestine (which causes an illness with symptoms similar to dysentery). Infection from E. coli food poisoning has an incubation period of 8 to 72 hours and the illness may last for 1 to 7 days. Maximum growth of the organism takes place between 10°C and 40°C. Symptoms include watery or blood-stained diarrhea, abdominal pain, vomiting and fever.

Extrinsic contamination: This is the contamination arising from sources other than the raw food itself, namely external sources of contamination, such as foreign bodies. During manufacturing and catering processes, food can be exposed to extrinsic contamination from items such as glass particles, wood splinters, hairs, small personal items, such as pens, rings and hair grips, rodent droppings, dead insects, cigarette ends, plastic items and metal objects falling from badly-maintained plant and equipment. Contamination may also be associated with poor standards of building maintenance resulting in flaking paintwork, rust from overhead pipes and contaminated water leaking from pipes, all of which could enter the product at some stage of preparation. Food safety management systems should explore the potential for extrinsic contamination at the various stages of processing.

Food poisoning (food-borne illnesses): Food-borne illnesses are defined as diseases; usually either infectious or toxic in nature, caused by agents that enter the body through the ingestion of food (WHO).Consumption of infectious or toxic food most commonly results in food poisoning. Food poisoning is an acute illness of sudden onset caused by the consumption of contaminated food or food which may contain some form of poisonous agent. The symptoms of food poisoning typically begin several hours after ingestion and, depending on the agent involved, can include one or more of the following body effects: *a*) nausea; *b*) abdominal pain; *c*) vomiting, *d*) diarrhea; *e*) fever; *f*) headache; or *g*) tiredness.

Food storage: Most foods will deteriorate if not stored correctly and, in many cases, at the correct temperature. Deterioration may be associated with the food losing its moisture content, as with bread and confectionery products, i.e. the process of staling; food absorbing odours from chemicals, such as disinfectants; or bacterial spoilage resulting in gradual decomposition, particularly in the case of meat, fruit and vegetables.

Fumigation: The process of filling a building or part of a building with an aerosol fumigating agent, such as methyl bromide or hydrogen cyanide, with a view to eliminating cockroaches and stored product pests, such as flour beetle and biscuit beetle. Much will depend upon the structure and layout of the building as to the effectiveness of this form of infestation control as re-infestation frequently occurs later. Several fumigation exercises, spaced out at monthly intervals, may, therefore, be necessary to be successful. Whilst such processes may be partially successful, fumigation must be used in conjunction with other infestation control techniques, such as efficient cleaning and housekeeping techniques. Clean-as-you-go procedures on the part of operators, and regular stock rotation, which creates a 'disturbance' effect and slows down reproduction.

High risk foods: Those foods, typically with a high moisture and protein content, that are at particular risk of contamination by bacterial spoilage or the action of pathogens. High risk foods will, under favourable conditions, support the growth of pathogenic bacteria. High risk foods include:

- meat, fish, eggs and prepared products incorporating these foods, together with smoked or cured meat which has been sliced and cured fish
- soft cheeses ripened with moulds or micro-organisms
- smoked and cured fish
- prepared vegetable salads

- fresh cream and artificial cream cakes
- ice cream and frozen desserts
- milk-based desserts
- cooked cereals, such as rice, beans and vegetables
- sandwiches

Strict temperature control is required with respect to these foods. Such foods should be stored at below 8°C or above 63°C.

Hydrogen swell: The buildup of hydrogen gas in canned goods arising from internal corrosion of the can and resulting in 'blowing' of one or both ends of a can. Hydrogen swell is particularly common with acidic canned fruits and tomatoes.

Layout of food premises: The term 'layout' refers to the space available for people working in a particular room or area and the situation of plant, machinery and equipment, etc in relation to the operator and the tasks performed. Well-designed layouts in food manufacturing and catering premises contribute significantly to the avoidance of various forms of contamination. Ideally, raw materials are introduced at one end of the premises and finished products leave at the other end, with the intermediate stages of storage, preparation, production/cooking and assembly taking place in a sequential flow between these two points. An efficient layout should make a material contribution to preventing or reducing overcrowding, minimising the effort required to perform operations, expediting the process in an orderly and sequential flow, and ensuring maximum hygiene and safety standards throughout, together with facilitating the cleaning process. Above all, the working areas should be large enough for the processes carried out, a problem encountered with many restaurant kitchens, for example.

Mycotoxin: A mycotoxin is a form of poison produced by some moulds and yeasts and which accumulate in food animals. Mycotoxins are capable of surviving conditions in which bacteria would not normally survive. They are associated with products with a low pH, such as fruit juices, and relatively dry products such as cereal products, dried beans, nuts, coffee and cocoa beans, soya beans and various cured meats, such as ham. They can tolerate low water activity down to 0.80 A_w, and are resistant to heat, such as that produced during heat treatment processes. They are generally present in all areas of the food chain. Mycotoxins thrive within the temperature range – 7°C to 35°C and their evidence at the lower end of this scale is seen in cold stores where black patches form on damp surfaces and in areas where cleaning has not been effective. Mycotoxins are highly resistant to preservation techniques and any measures to control same must be concentrated at source by preventing or controlling their formation on farms, supported by quick-drying after the harvesting process. Their control is further assisted by storage of products at temperatures as low as –20°C and maintaining pH at above 6.5 or below 3.0. Exposure to mycotoxins can cause mycotoxicosis, an illness caused by the toxins produced by some fungi.

Neurotoxin: A toxic substance which attacks the central nervous system and the most dangerous biotoxin known. They are produced by several organisms. Common neurotoxins are exotoxins produced by *Staphylococci, Clostridium botulinum* and ciguatoxin (*Ciguatera)* from dinoflagellates.

pH value: The logarithm to the base 10 of the hydrogen ion concentration of a substance and, fundamentally, the degree of acidity or alkalinity of a substance. The pH values below 7 are acidic, above 7 alkaline, with a pH of 7 representing neutrality. Most foods support bacterial multiplication and the relative acidity or alkalinity of a food is an intrinsic factor in the growth of bacteria. Most bacteria grow best at or near a neutral pH and, to this extent, high or low pH can have a serious effect on such growth. Bacterial spores, on the other hand, can survive at any level, but germination will only take place at a pH of 4.5 and above. The way a food is manufactured, processed and stored can have the effect of lowering the pH, thereby inhibiting bacterial growth except in the case of certain moulds which thrive at low pH values.

Rancidity: Rancidity is a cause of unfitness in fatty foods, such as dairy products and oils. It is associated with the formation of substances causing odours and off-flavours in the products concerned, such as butter.

Risk assessment: The identification of hazards, the measurement and evaluation of the risks arising from hazards, and the implementation of preventive and protective measures to prevent or control exposure to those hazards.

Rotaviruses: This type of virus is responsible for several forms of gastro-enteritis, particularly in young children, resulting in severe diarrhea, vomiting, fever and dehydration. Man is the source, together with sewage-contaminated water. Onset is within one to three days and the duration, four to eight days. Foods commonly associated with Rotavirus are salad, vegetables and raw shellfish.

Salmonellae: *Salmonella* is a Gram-negative, non-spore forming, short rod-shaped bacterium which is widely distributed in humans and animals. Their effects can vary widely and there are nearly 3,000 known serotypes. *Salmonellae* can be classified into three principal groups:

a. those that infect only people and which can be transmitted through food, *e.g.Salmonella Typhi* and *Salmonella Paratyphi*;

b. host-adapted strains, some of which are human pathogens and which can be transmitted by food, *e.g. Salmonella Dublin*; and

c. certain strains which do not have a particular host, but which may have pathogenic effects on people and animals.

The most common causes of food poisoning are those associated with *Salmonella Typhimurium* and *Salmonella Enteriditis* which result in the condition known as 'salmonellosis' The symptoms include sickness, fever, diarrhoea and severe abdominal pain lasting as long as seven days in some cases. A person suffering from salmonellosis excretes the bacteria in the faeces and may transmit the disease to others.

Salmonellae multiply quickly at a temperature between 7°C and 45°C and the route of infection is through raw animal products, such as milk, poultry, eggs and egg-based products and raw meat. They can also be carried on insects, rodents, birds and animal pets, together with water and sewage.

Prevention of infection is based on the maintenance of high standards of personal hygiene, thorough cleaning and disinfection of food processing areas, prevention of cross contamination, strict control of the time-temperature relationship in the preparation of high risk foods and effective pest control procedures.

Shelf life: The period within which food is both at its best quality and safe for human consumption. Shelf life means:

a. in relation to food with respect to which an indication of maximum durability is required in accordance with regulation 20 of the Food Labelling Regulations 1996 (form of indication of minimum durability), the period up to and including the date required to be included in that indication;

b. in relation to food with respect to which a 'use by' date is assigned in the form required in accordance with regulation 21 of the Food Labelling Regulations 1996 (form of indication of 'use by' date), the period up to and including that date; and

c. in relation to food which is not required to bear an indication of minimum durability or a 'use by' date, the period for which the food can be expected to remain fit for sale if it is kept in a manner which is consistent with food safety.

Shigella: A group of Gram-negative non-sporing bacteria found in the intestine of man and other animals, responsible for bacillary dysentery or shigellosis. Shigellosis is common in many tropical countries, being associated with sewage-polluted water supplies, and is commonly known as 'Delhi belly' and 'gyppy tummy'. The disease has an incubation period of 12 to 48 hours and symptoms include diarrhoea, vomiting, fever, abdominal pain and cramps. Severe cases may manifest pus and blood in the faeces.

Spoilage of food: Food spoilage is associated with the gradual process of decomposition which takes place in all foods over a period of time, the rate of decomposition varying substantially according to the type of food and the conditions, particularly temperature, under which it is stored. Decomposition takes place in meat following the slaughtering of animals, in vegetables and fruit after picking and in other food products following manufacture or processing. All foods reach a stage where the degree of decomposition makes them unpalatable and eventually unfit for human consumption. Spoilage may also arise through contamination caused by chemicals, insects and rodents, oxidation of the product and through accidental physical damage. Microbiological spoilage is, however, more important and is brought about by a complex degradation process in foods. Spoilage organisms will multiply particularly in high risk food products, such as dairy products, fruit and vegetables, meat, fish and poultry. Spoilage can result in the development of 'off' flavours and tastes in the food, together with changes in the appearance of the food, including consistency, nature and colour. Fermentation within food can result in the production of gases. Whilst refrigeration will delay the onset of spoilage, it will not stop these changes taking place.

Ultra-high-temperature (UHT) processing: Ultra-high-temperature (UHT) processing involves heating milk in a continuous-flow system to a high temperature (~135–145°C) and holding it at that temperature for a short time (1–10 seconds) followed by rapid cooling. This produces a "commercially sterile" product, that is, a product in which bacterial growth is highly unlikely to occur under normal storage conditions. Since heating and cooling take place through relatively small distances in the flowing liquid in the UHT process, the heat penetration problems of in-container sterilization are avoided. This rapid heat transfer rate minimizes undesirable changes in the taste and nutritional quality of the resulting product. Since UHT is a continuous process, it produces uniform product quality that does not depend on the size of a container, unlike in-container sterilization. This attribute is especially important for products containing heat-sensitive ingredients and highly viscous products with poor heat transfer properties.

Aseptic processing involves UHT processing followed by filling the product into sterile containers in a sterile environment and sealing the containers in a sterile manner in a continuous process. In practice, UHT-processed product is usually transferred to an aseptic tank before it is aseptically packed into containers. This provides greater operational flexibility and allows the use of processing and aseptic packaging equipment with different capacities. The most popular aseptic packages are the tetrahedral-shaped paperboard cartons exemplified by Tetra Pak and Combibloc products, although multilayered plastic bottles are also now popular. The introduction of UHT treatment of milk, coupled with aseptic packaging, made a fundamental improvement in the bacteriological safety of milk and extended its shelf life from the typical 2–3 weeks for refrigerated pasteurized milk to 6–9 months without refrigeration. No preservative is added to UHT milk for its extended shelf life at room temperature, an important point for today's additive-conscious consumer.

Aseptic packaging is the process that allows UHT-processed milk to have a non-refrigerated shelf life of several months. While UHT treatment destroys all microorganisms that are likely to grow under the normal conditions of storage, aseptic packaging prevents the ingress of contaminating organisms during packaging and subsequent non-refrigerated storage of the unopened package. Therefore, the full benefit of commercial UHT treatment cannot be realized without the use of aseptic packaging. In fact, aseptic packaging was revolutionary in taking UHT processing from the research laboratory to the grocery store.

Microwave Heating: In microwave and radio-frequency processing, the electromagnetic waves of certain frequency are used to heat food products. Generally two frequencies (915 and 2.450 MHz) for microwave food processing and three frequencies (13.56, 27.12, and 40.68 MHz) for radio-frequency processing are used.

Microwaves are nonionizing, time varying electromagnetic waves of radiant energy with frequencies ranging from 300 MHz to 300 Ghz; that is, wavelengths ranging from 1 mm to 1 m in free space. Due to possible interferences with TV and radio waves, microwave ovens operate at 915 and 2,450 MHz.

allocated by the International Telecommunications Union. Domestic ovens operate at 2,450 MHz. Microwaves travel similarly to light waves; they are reflected by large metallic objects, absorbed by some dielectric materials and small strips of metal, and transmitted through other dielectric materials. For example, water and carbon absorb microwaves well; on the other hand, glass, ceramics, and most thermoplastics allow microwaves to pass through with little or no absorption. They can also be refracted when travelling from one dielectric material to the next, analogous to the way light waves bend when passing from air into water.

There are two main mechanisms by which microwaves product heat in dielectric materials: ionic polarization and dipole rotation. Ionic polarization occurs when ions in solution move in response to an electric field. Kinetic energy is given up to the ions by the electric field. These ions collide with other ions, converting kinetic energy into heat. When the electric field is rotating at 2.45×10^9 Hz, numerous collisions occur, generating a great deal of heat. However, the dipole rotation mechanism is more important. It is dependent on the existence of polar molecules. The most common polar material found in foods is water. Water molecules are randomly oriented under normal conditions. In the presence of an electric field, the polar molecules line up with the field. Heat is generated by dielectric and ionic mechanisms. Dielectric heating is due to the oscillations of the water molecules in the food. Oscillatory migration of ions in the food also generates heat under the influence of the oscillating electric field. These provide rapid heating and thus require less time to bring the product to the desired temperature. These also provide relatively more uniform heating compared to conventional hot air heating. Equilibration of the product following heating can help to level the temperature distribution and improve uniformity.

Radio-Frequency Heating: Radio frequency (RF) heating involves the transfer of electromagnetic energy directly into the product, therefore inducing volumetric heating due to frictional interaction between molecules . In RF heating the food is placed between two capacitor plates, where it plays the role of a *dielectric* to which a high-frequency alternating electric field is applied. Such field will force polar molecules (*e.g.*, water) to constantly realign themselves with the electric field. This molecular movement is very fast due to the high frequency of the field and will provoke generation of heat within the food by energy dissipation caused by molecular friction. RF waves (as well as microwaves) are both within the radar range and this strongly limits the frequencies that can be used for applications other than communications. The allowed frequencies for RF applications are 13.56, 27.12, and 40.68 Mhz.

As compared with conventional heat-processing technologies, RF heating presents similar advantages to ohmic and microwave heating that are essentially due to the generation of heat throughout the volume of the material to be processed. However, there are some specific advantages of RF over those alternative volumetric technologies, namely:

- there is no need for electrodes contacting the food (in contrast with, *e.g.*, ohmic heating), allowing RF to be easily applied to both solid and liquid foods;
- due to the longer wavelength of RF, its power will penetrate more deeply in the foods as compared to microwave power;

- the construction of large RF heating systems is simpler than their microwave counterparts, and their application to continuous processes is straightforward; it is a technology particularly suited to large industrial applications.

RF's main disadvantages are:

- the higher equipment and operating costs for an equivalent power output when compared to conventional heating systems and also to ohmic heating;
- the reduced power density when compared to microwave heating, meaning that larger RF heating systems are needed for the same power rating and also that slower heating rates are achieved with RF as compared to microwaves;
- the so far limited research efforts regarding, for example, the determination of food RF dielectric properties.

Active Packaging: It is defined as

- Active packaging refers to those technologies intended to interact with the internal gas environment and/or directly with the product, with a beneficial outcome.
- Active packaging is intended to sense internal or external environmental change and to respond by changing its own properties or attributes and hence the internal package environment.
- Active packaging changes the condition of the packaged food to extend shelf life or to improve safety or sensory properties, while maintaining the quality of the packaged food.
- Active packaging is an intelligent or smart system that involves interactions between package or package components and food or internal gas atmosphere and complies with consumer demands for high-quality, fresh-like, and safe products.

Active packaging is a packaging system that possesses extra processing functions that do not exist in conventional barrier packaging. Due to these extra functions, active packaging performs its special task during storage, distribution, and consumption. Therefore, active packaging can be a very special non-thermal processing application. More effective technology to enhance the performance of the packaging system has been developed and will continue to be advanced in the future. Active packaging has many advantages and applications for the food industry. The biggest market of active packaging applications is in the area of antimicrobial packaging and oxygen-scavenging packaging. For producers and processors, better protection by extended shelf life and more convenient systems by reduction in process time and skilled labor would be preferred for the successful utilization of active packaging development and applications. For consumers, fresher quality of products and more convenient use of the products will be attractive. New active packaging materials or systems may have a large market in the food industry if the cost-to-benefit ratio is favorable. Since current packaging is not always optimum for many different food products, there are still many opportunities for packaging improvement and consequently the quality and safety management of packaged foods.

Intelligent Packaging: The term "intelligent" implies that the package has an indicator that monitors the conditions (*e.g.*, quality) of food products, while the "active" in active packaging indicates the procedures that consistently accommodate active interaction of the package with internal atmosphere or the package food. While active packaging changes the condition of the packaged food to give extended shelf life, improved safety, and desirably maintained quality, intelligent packaging monitors the condition of packaged foods to provide information about the quality of the foods during transportation and storage. Intelligent packaging may sense the condition of packaged foods specifically or respond to the changes of the food condition.

Lard: Lard refers to pig fat in both its rendered and unrendered forms. Lard was commonly used as a cooking fat or shortening, or as a spread similar to butter. Its use in contemporary cuisine has diminished because of health concerns posed by its saturated fat content and its often negative image. The culinary qualities of lard vary somewhat depending on the part of the pig from which the fat is taken and how the lard is processed. Lard can be obtained from any part of the pig as long as there is a high concentration of fatty tissue. The highest grade of lard, known as leaf lard, is obtained from the "flare" fat deposit surrounding the kidneys and inside the loin. The next highest grade of lard is obtained from fatback, the hard fat between the back skin and flesh of the pig. The lowest grade (for purposes of rendering into lard) is obtained from the soft caul fat surrounding digestive organs, such as small intestines, though caul fat is often used directly as a wrapping for roasting lean meats or in the manufacture of pates. Industrially produced lard is rendered from a mixture of high and low quality fat sources from throughout the pig. It is typically hydrogenated, and often treated with bleaching and deodorizing agents, emulsifiers, and antioxidants. Such treatments make the lard shelf stable. The fat contains palmitic (20%–32%), stearic (5%–24%), oleic (35%–62%), and linoleic (3%–16%) acids as major components. It is unusual in that 70% of the palmitic acid is in the *sn*-2 position. It also contains cholesterol (0.37%–0.42%).

Tallow: Animal edible tallow is normally obtained from beef but also from sheep and goats, processed from suet. Unlike suet, tallow can be stored for extended periods without the need for refrigeration to prevent decomposition, provided it is kept in an airtight container to prevent oxidation. It is used in animal feed, to make soap, for cooking, as bird feed, and was used for making candles. It can be used as a raw material for the production of biodiesel and other oleochemicals. Tallow contains mainly saturated acids (50%): palmitic acid (26%), stearic acid (14%), and myristic acid (3%). It also contains monounsaturated acids (42%), especially oleic acid (47%) and palmitoleic acid (3%), and polyunsaturated acids (4%), especially linoleic acid (3%) and linolenic acid (1%). Also present are odd-chain, branched-chain, and trans-fatty acids, and cholesterol (0.08%–0.14% in beef tallow and 0.23%–0.31% in mutton tallow).

Butter: Butter, a water-in-oil emulsion, comprises of >80% milk fat, but also contains water in the form of tiny droplets, perhaps some nonmilk fat, with or without salt (sweet butter); texture is a result of working/kneading during processing at appropriate temperatures, to establish a fat crystalline network that results in desired smoothness (compare butter with melted and recrystallized butter). It is used as a spread, a cooking fat, or a baking ingredient. Salt, flavorings, or preservatives are sometimes added to butter. When refrigerated, butter remains a solid, but softens to a spreadable consistency at room temperature, and melts to a thin liquid consistency at 32°C–35°C. Butter generally has a pale yellow color, but varies from deep yellow to nearly white. As a water-in-oil emulsion it contains 80%–82% milk fat and 18%–20% of an aqueous phase. Butterfat is very complicated in its fatty acid and triacylglycerol composition. In addition to the usual C16 and C18 acids, it contains short-chain and medium-chain acids (C4–C14), a range of trans monoene acids, mainly 18:1, and oxygenated and branched-chain acids. The trans acids represent 4%–8% of the total acids.65 Butter contains some cholesterol (0.2%–0.4%). Spreads with lower levels of fat are also available. Butter that spreads directly from the refrigerator is made by removing some of its higher melting glycerol esters or by blending with a vegetable oil.

Ghee: Ghee is a solid fat-based product made in India from cow- or buffalo-ripened milk. It is less perishable than butter and, therefore, more suitable for a tropical climate. Milk is curdled. The curd is then manually churned until it precipitates butter and leaves behind some whey. The butter is then heated on a low flame until a layer of white froth covers the surface. This state indicates the end of the process, and the liquid obtained on filtering the suspension is pure ghee. Ghee is made by simmering unsalted butter in a large pot until all water has boiled off and protein has settled to the bottom. The cooked and clarified butter is then spooned off to avoid disturbing the milk solids on the bottom of the pan. Unlike butter, ghee can be stored for extended periods without refrigeration, provided it is kept in an airtight container to prevent oxidation and remains moisture-free.

Fortification: Fortification usually refers to the addition of nutrients to foods from which they were either absent, or present in unimportant amounts. This is carried out because deficiencies or potential deficiencies have been shown in the population. It also includes additions to fulfil the role of another food in the diet. Thus margarine is fortified with vitamin A in many countries to replace that 'lost' when margarine is substituted for butter. Vitamin D is added at higher levels than are found in butter as a public health measure, since extra is considered necessary for the population as a whole.

Restoration: Restoration is self-evident, involving the replacement, in full or in part, of losses incurred in processing (*e.g.* loss of B vitamins and iron in milling of cereals to low extraction rates; loss of vitamin C in the preparation of instant potato). It should be noted that, up to now, restoration has been limited; for example, white bread has thiamin, niacin and iron added back, although other vitamins such as pantothenate, folate, pyridoxine and tocopherol are also partly lost in the milling process. Similarly, fruit juices might have vitamin C added back, but juices also supply folate and thiamin. In some countries, calcium and riboflavin are also added to bread: this is fortification, since wheat contains only small amounts of these two nutrients.

Enrichment: Enrichment is increasing the level of nutrients present to make the food a 'richer' source. Enrichment has frequently been interchanged with restoration and fortification as a general term. However, for labelling purposes the use of the term 'enriched' is controlled.

remains to be seen if at last some of the prejudices which have prevented NMR being widely taken up will at last be overcome.

Ultrasonics in Food Processing: Ideally, a food manufacturer would like to take a combination of raw materials and convert them into a high quality product at the lowest possible cost. This conversion is achieved by subjecting the raw materials to a number of processing conditions, e.g., heating, cooling, pressure, shearing or mixing. Inherent variations in the raw materials and of the processing conditions mean that the properties of the final product vary in an unpredictable manner. To control and minimize these variations food manufacturers need to characterize the properties of the raw materials, and to monitor the food at each stage of processing. Food processing operations are becoming increasingly sophisticated and are often computer controlled. Traditional chemical and gravimetric techniques are time consuming and laborious to carry out, and so there has been considerable motivation for the development of rapid analytical sensors for monitoring the properties of foods. Ultrasound utilizes interactions between high-frequency sound waves and matter to obtain information about the composition, structure and dimensions of materials through which it propagates. The power levels used in ultrasonic testing are so low that the properties of the material are not altered, thus the technique is *non-destructive*. This is in contrast to high energy ultrasonic applications which are sometimes used in the food industry for cleaning, cell disruption, heating, sterilization or emulsification purposes. A wide variety of different applications of ultrasound to foods have been developed over the past 50 years or so, reflecting the diversity and complexity of food materials, and the versatility of the ultrasonic technique. Even so, ultrasound has still not found wide spread use in

monitoring food processing operations. This situation will almost certainly change in the near future. Advances in microelectronics have made available sophisticated electronic instrumentation capable of making accurate ultrasonic measurements at relatively low-cost. The interaction between ultrasound and many micro-heterogeneous materials is fairly well understood, and there are mathematical formulae available for interpreting ultrasonic measurements in a number of systems relevant to the food industry. Finally, ultrasound offers a number of advantages over alternative techniques used to monitor food processing operations: it is capable of rapid and precise measurements, it is non-intrusive and noninvasive, it can be applied to systems which are concentrated and optically opaque, it is relatively inexpensive and it can easily be adapted for on-line measurements.

Staling: The two main components of staling (firming of the bread crumb) are loss of moisture, mainly due to migration of moisture from the crumb to the crust (which becomes soft and leathery), and the retrogradation of starch. The major chemical change that occurs during staling is starch retrogradation, but a redistribution of water between the starch and the gluten also has been proposed. The gelatinization of the starch that occurs during cooking or baking gradually reverses, and the starch molecules form intermolecular bonds and crystallize, expelling water molecules and resulting in the firming of crumb texture. Starch retrogradation is a time- and temperature-dependent process and proceeds fastest at low temperatures, just above freezing point. Since the rearrangement of the starch molecules is facilitated by a high water activity, staling is retarded by the addition of ingredients that lower water activity (*e.g.*, salt and sugar) or bind water (*e.g.*, hydrocolloids and proteins). The staling